Y0-AAG-915

Alcohol

Alcohol

THE DEVELOPMENT OF SOCIOLOGICAL PERSPECTIVES ON USE AND ABUSE

EDITED BY

Paul M. Roman

Publications Division
Rutgers Center of Alcohol Studies
New Brunswick, New Jersey USA

This book is the second in the Alcohol, Culture, and Social Control Monograph Series, funded in part by the Wine Institute. Editors of the series are David J. Pittman and Helene Raskin White.

ISBN 911290–23–0
Library of Congress Catalog Card Number: 91–061677

Printed in the United States of America

To Selden D. Bacon, Scholar

Contents

Preface

In many respects, the salience of a sociology of alcohol use and abuse in the United States is as high in the 1990s as it has ever been. The contrasts between the problems associated with the use of alcohol, the legal psychoactive substance, and those associated with the illegal substances targeted by the "War on Drugs" become increasingly sharp. Despite the vigor of this "war," virtually no significant constituency supports a return to a societywide prohibition of alcohol manufacture and distribution. Thus, for a world that must live with a multiplicity of psychoactive substances that will always carry some degree of desirability, there are lessons to be learned from drinking behaviors. A great deal remains to be specified about the dynamics of sociocultural integration that accompany the predominant pattern whereby alcohol is used in moderation without adverse consequences. At the same time, numerous strategies have shown varying degrees of effectiveness in dealing with alcohol problems. Further refinement of these offers a strong potential for greater societal cost savings in dealing both with alcohol and other substance use problems.

It is clear that sociological research on alcohol is needed. A commonsense prediction would be that the developments leading to such cultural readiness for social research would consolidate a body of scholarly interest and activity among sociologists and other social science scholars. Yet there is little if any evidence of such consolidation.

This sets the stage for an assessment of the sociological approach to the study of alcohol use and abuse. Nearly 50 years ago Selden Bacon set forth a programmatic statement—a grand plan and outline of what could be a scholarly interest area. This proposed focus would combine the theories and methods of sociology with the varied phenomena associated with the consumption of alcohol. This book contributes a range of perspectives on where we are today in this quest and offers some explanation of why we are where we are.

This book has had a very long gestation. Such is not unusual for academic endeavors that come to occupy a small corner pen in an otherwise disorderly barnyard of clucking and mooing commitments, all of which are being fed and groomed (however haphazardly) for eventual exhibition at the county fair of scholarship. It is difficult for me to recall exactly how this project began. I do remember the cheerleaders who said that such a project *had* to be done. On occasions, such as repeated missed deadlines, I have looked back with hardening resentment over the several years of its progress at those few persistent turkeys who kept egging me on.

To the best of my recollection, the project began in the midst of a quarter of teaching the sociology of alcohol and drugs when it occurred to me that nowhere could one find a collection of the work of Selden Bacon, despite its breadth and influence on alcohol studies. The first prospectus for this volume was exceedingly grand, encompassing a reprinting of many of Selden's major manuscripts. These reprints would be accompanied by Bacon's students' and other contemporary scholars' contributing commentaries on the influence of each of the works. Experience with prospective publishers led to a cutting back of the reprinted material in favor of the single previously published piece included in the present volume. The focus of the volume turned toward original essays.

The contributors represent individuals whose careers and work have been influenced in different ways and to different degrees by Bacon's work. An effort was made to extend invitations to as many sociologists as possible whose published work indicates a substantial career commitment to alcohol studies. The contributors to this volume comprise a solid cross-section of senior sociological scholars in alcohol studies. Each was asked to assess the influence of Bacon's work on their own research and to assess his influence on the broader endeavor of sociological studies of alcohol-related phenomena. Personal recollections of work with Selden were encouraged, as were projections of the future sociological study of alcohol-related issues. Few of the chapters follow this suggested format exactly, lending strength to the volume through its diversity.

My own personal recollections have relevance here. An accurate statement of where this project really began would name the Edgewater Beach Hotel in Chicago. Here, in 1965, as a graduate student I attended my first meetings of the Society for the Study of Social Problems (SSSP) and the American Sociological Association (ASA). I had just come on board as a graduate assistant at the School of Industrial and Labor Relations at Cornell, and had begun work on a monograph on schizophrenia and social class that I would eventually coauthor with Harry Trice. At that time Harry was trying to broaden his scholarly identity beyond alcohol issues, so he took on the schizophrenia project. At that point I had not had much exposure to the sociology of alcohol.

I was a passenger in a Cornell fleet car from Ithaca to Chicago where Harry had several slots on the program of the Committee on Drinking Behavior at the SSSP meetings; he may have even been chair of the committee at the time. He generously allowed me to follow him (most places) during the meetings and was even more generous in introducing me to his buddies who were active on the SSSP committee. I remember meeting, dining, and drinking with Dave Pittman, Chuck Snyder, Harwin Voss, Earl Rubington, Si Dinitz, Keith Lovald, George Maddox, Herman Lantz, Bob Straus, Ralph Connor, Hal Mulford, and Joe Gusfield.

Those encounters collectively became a kind of epiphany for me that has held its effects to this day. From my perceptions, these guys were really enjoying what they were doing. They were enthusiastic, and the enthusiasm was infectious. Furthermore, they were inclusive and projected a sense that any serious sociologist would want to be involved in alcohol studies. The message to a newcomer was "come on board, there's lots of room." Such an attitude is certainly rare in most circles of established scholars, especially when they encounter a lowly graduate student.

From those few days in Chicago I acquired the perspective that alcohol issues constitute a critical research site. Studies of drinking behavior can elucidate many facets of social organization and social interaction. Despite the typical range of disappointments in one's work that accompany any academic career, I have never wavered from this enthusiasm about the nearly unlimited value of alcohol studies in broadening "basic" knowledge about social phenomena.

Somehow this came to me during those Chicago meetings and was subsequently reinforced as I, a fledgling graduate student, attended the ASA/SSSP meetings in Miami Beach in 1966, San Francisco in 1967, and Boston in 1968, and had more of the gusto of the members of the Committee on Drinking Behavior "rub off" on my scholarly interests. While these sociologists were involved in alcohol studies, they were also doing exciting sociology. Furthermore, in contrast to some of the teachers I had had, they were field researchers, with a different sense of data than I had been accustomed to in lecture and seminar courses.

Behind this vibrant interest area it seemed there must be a "mother lode" from whence it sprung. I thought I had initially discovered this when (again with initiation through Harry Trice) I began getting into the field myself and interacting with some of the leading practitioners in the alcohol field. Among my earliest acquaintances was Dr. Vernelle Fox, a highly innovative and supercharged physician whose roles in both treatment and research afforded the genuine projection of charisma. Harry introduced me to Bill W., with whom I started a brief but intense correspondence around his promulgation of megavitamin therapy for recovering alcoholics, again exposing me to genuine charisma. There were many other encounters with alcoholism practitioners, almost all of whom were recovering. These individuals tended to project some degree of charisma. They were remarkably inclusive and enthusiastically supportive of the entry of a new "research type" into the alcoholism arena.

I learned through these encounters that contact and interaction with the world of practice was vital for the scholarly world. However, the charisma it carried was not the mother lode I was seeking. Alcoholism intervention is centered on ideology and beliefs that in the late 1960s could be challenged only gingerly (if at all) if one desired to maintain rapport and a sense of being an insider. These new friends were especially dear to a young sociologist trying to figure out his field. However, I felt then and now that one could not confine oneself to the narrow paradigm and undisputable assumptions of the practitioners. To sustain the excitement that I had seen in sociological work in alcohol studies, a strong dose of healthy skepticism, coupled with deep respect, was essential.

In 1967 or thereabouts I had my first encounter, albeit at a distance, with one who combined the scholarly and the charismatic: Selden Bacon. Again in the company of Harry Trice, I attended a conference for federal officials and alcoholism practitioners in Washington that was intended to set the stage for the adoption of an employee alcoholism policy for federal employees. Bacon was the keynote speaker. I could not believe the uninhibited manner in which this man addressed this group, obviously poking holes in cherished beliefs and peppering his remarks with fairly highbrow cynical humor. At the same time it was clear that he was keeping the respect of the audience by being generally supportive of the whole idea of a federal employee alcoholism program. This perhaps was the mother lode for the "spirit of Chicago" that I had observed two years earlier.

It was not until quite a few years later that I really read Bacon's 1944 plan for sociology and alcohol that is printed as the first chapter of this book. It is clear that my unarticulated vision about the prospects for sociological excitement in alcohol studies had already been well stated by Bacon 25 years before my epiphany. It is also clear that many of those in that inclusive circle in Chicago were familiar with Bacon's blueprint, although there is no certainty that all of them had read it or even agreed with it. Thus, while Bacon's work may not have been the mother lode, his scholarship and teaching have certainly been significant influences upon this specialty interest.

It may be of interest that despite the identification with Bacon's work that seems evident in my editing this volume, my professional and personal relationships with him have actually been minimal. I think Harry Trice probably introduced us at that Washington meeting, but no recollection sticks with me. Similarly, I was introduced again when Harry and I were at Rutgers University about a year later, but there was no interaction, although watching the intellectual parrying between Selden and Harry during their brief conversation was worth the trip.

I next recall a short bus ride to the National Council on Alcoholism (NCA) meeting banquet in Milwaukee in the mid-1970s during which I sat next to Selden and did some listening. In 1975 I had what should have been an important encounter when I was again at Rutgers, being interviewed as a candidate to succeed Selden in the director's chair. Naturally, I wanted to learn more about his job, but when my time for the interview with Selden arrived, his lecture to me took other directions. Most recently I worked with Selden through the mails in my role as editor of two different numbers of the *Journal of Drug Issues* to which he made contributions. We have also had numerous phone conversations about this book and its painfully slow emergence. But our time together in the same room probably totals less than three hours.

While perhaps a somewhat shopworn designation, there is no doubt that as an intellectual and a social scientist, Selden Bacon is *sui generis.* He is regarded as a kind and generous man who has made his presence known. Beyond what is reviewed in the chapters constituting this volume, Selden has contributed numerous seminal ideas that have been incorporated into the fabric of both alcohol research studies and intervention directed at alcohol problems. He has been a most significant godparent of the sociological study of alcohol-related phenomena, relating to his offspring and extended family in unique but important ways. For these contributions, we who have worked to make this volume a reality salute him and dedicate this effort to him.

As editor I owe many acknowledgments, first to the authors' efforts in preparing and revising their contributions to this volume. Special recognition is due the few conscientious "early birds" who heeded the first set of deadlines. These contributors then had the opportunity to live out the biblical parable where those who arrive to work at the vineyard at the end of the day received the same penny as those who arrived at dawn. Gratitude is also due to David Pittman and Helene Raskin White for their confidence in including this book in the special series of volumes, to Alex Fundock III for his patience in dealing with the manuscript, and to Selden Bacon for his own contribution and for his patience in awaiting the volume's publication. Finally, it should be noted that this volume was supported in part by a grant from the Wine Institute to the Rutgers Center of Alcohol Studies. The opinions expressed by the editor as well as the authors of the various chapters are their own and not necessarily those of either the Wine Institute or the Rutgers Center of Alcohol Studies.

During part of the time that I have worked on the preparation of this volume, I have received partial support from the National Institute on Alcohol Abuse and Alcoholism through Research Grant RO1-AA-07218 and Training Grant T32-AA-07473, for which gratitude is due. I owe thanks to my office staff at the Institute for Behavioral Research for retyping manuscripts and keeping the chapters in general order. My collaborator and spouse, Terry Blum, has been incredibly patient in allowing me time away from our joint research and writing to work on "the Bacon book." Finally, our son, Luke Paul, was conceived and born during the later stages of this project. He really has nothing to do with this book, except to carry the hope that someday he will be as intelligent and creative as Selden Bacon and the other godfathers and godmothers in alcohol studies who are the scholarly lights his Dad has tried to follow.

Paul M. Roman

Introduction

Paul M. Roman

We offer in this book a set of original essays by social scientists who have spent much or all of their careers studying alcohol-related phenomena. The book is an overview of the contributions made by the social sciences, particularly sociology, to the scientific study and understanding of phenomena related to beverage alcohol. While not an exhaustive coverage of those contributions, it provides a reasonable sampling. Much of the contents is directed toward a critical appreciation of the direction and influence brought to this area of study by the work of Selden Bacon, the sociologist who is generally acknowledged as the initiator of American sociological studies of alcohol-related phenomena in the post-Prohibition era.

Across the chapters, several important themes are developed. First, the volume is an assessment, albeit a partial one, of the status of sociological knowledge about alcohol-related phenomena. Particularly notable by its absence is any significant consideration of research on the relationships between alcohol availability and drinking-related social consequences, a topical area in which social science activity has escalated in the recent past. The coterie of authors included here share interests that are centered in and around the post-Prohibition medicalized conception of alcohol problems. For the most part, these researchers and their professional significant others have not been specifically concerned with the public health-oriented strategies centered on drinking opportunity.

Each chapter more or less pivots from the essay by Selden Bacon, "Sociology and the Problems of Alcohol," which appears as the first chapter. This essay was first published in 1943, with a somewhat explicit intention of providing a blueprint for the sociological investigation of phenomena related to alcohol. The contributors to this book were asked to comment on what they see as the fit between what Bacon suggested in 1943, what has come to be in 1990 as a sociology of alcohol, and what may explain the discrepancies.

This focus on Bacon's essay has provided a second theme, namely a consideration of the influence and impact of Bacon's contributions on sociological and cultural analyses of alcohol. This focus does not make the volume a *Festschrift* in any way near the traditional sense of that concept. The set of contributors is not limited to Bacon's students or even to persons who have worked closely with him over his career. Further, there is no systematic attempt to review or consider the impact of the array of Bacon's writings. Instead, we present a partial assessment of what has happened to the sociological approach over the past 48 years, the influences upon it, and the fate of certain ideas. Such an approach is more appropriate for a social science volume in which ideas and data are the primary concerns, leaving the historian to assess the impact of individuals.

Third, throughout the book there is a theme of understanding how and why knowledge is constructed. Many of the chapters are written from the perspective of the sociology of science and knowledge, although not necessarily deliberately or consciously. The authors' recounting of significant career experiences has generated observations on the sociocultural context within which data and knowledge about alcohol-related phenomena have come to be. Within this theme, it is very evident that particular cultural limitations and political forces have shaped knowledge about alcohol and its use and abuse.

Fourth, several of the chapters demonstrate the substantive content of the sociology of alcohol. These contributions not only offer valuable reference points but also demonstrate the distinctiveness of social science thinking about alcohol-related phenomena, showing in particular the importance of combined sensitivity to social structure and culture in understanding patterns of group and organizational behavior.

Fifth, many of the chapters indicate a distinctive concern with the application of sociological knowledge. As an historical note, an important point of entry of sociology into alcohol studies was Bacon's "jail" study, described below. This empirical analysis demonstrated the successful application of sociological method and substance and was followed by rapid research utilization in the context of manpower pressures of World War II. Many social scientists, including the present writer, have found their careers in alcohol studies to be deeply involved in the all-too-real world, easily losing sight of the distinction between applied and basic scholarship. The chapter authors include those who have been both hugged and battered by practitioners, and to whom the halls of academia have sometimes seemed more foreign than the schoolrooms, jails, treatment centers, and factories where we believe we have gotten "close to the phenomena." This pattern of involvement is changing, however, as will be discussed below.

The ordering of the chapters is to some degree arbitrary, as they do not all neatly fit into distinctive sections or parts. The text begins with a reprinting of Bacon's 1943 essay. Despite its influence and repeated citation, we believe that this is the essay's first reprinting. The next five chapters offer different yet complementary perspectives on sociology and alcohol generally and the influence of Bacon's work more specifically. Kaye Fillmore provides an historical overview of emphases in alcohol studies and the relative influence of social science research over time. Harry Levine deals specifically with Bacon's 1943 agenda, its influence, and barriers to its implementation. Following is a combined set of reminiscences and general observations by Robert Straus, one of Bacon's earliest students and a subsequent major collaborator. A discussion of the overlap and mutual influences of anthropological and sociological work is found in the chapter by Dwight Heath, another scholar directly influenced by Bacon. The last chapter in this section is a brilliant integration of basic sociological concepts in the understanding of drinking behavior, an essay in which James Orcutt demonstrates a distinctive continuity with Bacon's 1943 essay.

Specific applications of sociological research to alcohol-related social policy at different phases of the life cycle comprise the next part of the book. Florence Andrews discusses the influence of Bacon's work on a research topic of great interest to policy makers and the public, the epidemiology of alcohol problems among youth. An extremely engaging and somewhat cynical account of the relationship between research and policy is found in the chapter by Armand Mauss, who writes on organizational and political dimensions of the utilization of sociological research centered on the efficacy of alcohol-problem prevention programming. Looking at youthful alcohol problems from a somewhat different perspective, Earl Rubington examines the nature and control of drinking behaviors in residential dormitory settings.

The focus then shifts to adulthood, where David Pittman discusses the sociological dynamics of the decriminalization efforts of the 1960s directed toward the disengagement of the alcoholic from the criminal justice system. Paul Roman then considers definitional problems that may be eroding support for the medicalized conception of alcohol problems, together with the medicalization advocates' loss of a monopoly over scientific approaches to alcohol issues. He then undertakes a conceptual analysis of the continuing relevance of the "hidden alcoholic," a theme in Bacon's work, as a mechanism to attract societal resources to deal with alcohol issues. This is followed by a chapter by William Sonnenstuhl and Harrison Trice that extensively reviews research on the interrelations between drinking behavior and work roles. Complementing the Sonnenstuhl and Trice chapter from a very different perspective is Kaye Fillmore's critical consideration of three empirical issues that would not ordinarily be linked to workplace policies and interventions: the progression of antisocial behavior in youth to alcohol problems in adulthood, the role of genetics in the etiology of alcoholism, and the effects of alcohol availability on alcohol problems.

Finally, three chapters focus on the general application of sociological research in the design of social policy. Robin Room offers a general essay that was originally drafted more than a decade ago, but ironically remains as timely as if it were written today. Melvin Tremper, working from the unique position of a state official in alcohol and drug policy implementation who is also employed in academia, writes specifically about the application of Bacon's work. The final chapter in this section is a new work by Selden Bacon. Posing himself as an "old warrior" in the battle with alcohol problems for at least 50 years, Bacon offers an insightful essay as to what those involved in the "War on Drugs" might learn from experiences in combating alcohol problems. The book closes with a complete bibliography of the published work of Selden Bacon, compiled by Catherine Weglarz.

The contributions cover a wide range of issues and topics. Common across them is the issue of the importance of sociological theory and methods in alcohol studies. As a context for considering that theme, I briefly review the status of this specialty area of study over several points in time.

The Beginnings

Most historians and other scholars mark the beginning of a distinctively new and different sociocultural orientation to alcohol problems in the United States within the decade of the 1930s (Beauchamp, 1980; Roman, 1988; Schneider, 1978). Three tangible events of importance to this new orientation occurred during the decade: the repeal of Prohibition, the initiation of what was to become Alcoholics Anonymous, and the aggressive initiation of the study of alcohol-related phenomena with the Laboratory of Applied Physiology on the prestigious campus of Yale University.

While each of these three events can be linked to specific dates, their sociological significance lies in their evolutionary influence as forces following the dates of their initiation. For example, the repeal of Prohibition required a new set of cultural beliefs consistent with the legal availability of alcoholic beverages. The Alcoholics Anonymous modality (which evidently was not named until necessity demanded a title for the "Big Book" in which its principles were codified [Pittman, 1988]) provided what many regarded as a unique and strikingly effective means for transforming the identities and behavior of persons called "alcoholics." The activity at Yale placed the understanding of alcohol and its related phenomena in the scientific laboratory, environs that were markedly different from churches and legislative halls, the primary settings where alcohol had been analyzed, understood, and written about for most of the preceding century.

These three historical events and their sequelae may be viewed in combination as the basis for a "paradigm" central to alcohol-related phenomena that has had substantial influence over the ensuing 50 years. Thomas

Kuhn (1962), a historian whose work enjoys a spectacularly diverse set of citations, put the term *paradigm* into widespread scholarly use as a means of describing the organization of scientific specialties over time. Paradigms constitute ways of viewing phenomena that are encumbered by numerous assumptions that are often framed in unique language and require specific research methodologies.

Kuhn's fundamental argument, which is eminently sociological, is that a dominant paradigm can constitute a complex set of organizational forces that may be at least partially independent of objective reality; indeed, the essence of paradigms is that their construction may preclude from consideration certain concepts or logical research strategies. A given paradigm may be ultimately displaced by a new one, but this typically requires a relatively long-term accumulation of observed "anomalies" that are not explained by the paradigm or that challenge its fundamental validity. For the most part, Kuhn's examples are internal to the "hard" sciences, although social scientists have readily borrowed the paradigm idea in order to describe lesser degrees of paradigm development or varying degrees of "pre-paradigm" status of the social sciences (Lodahl & Gordon, 1972; Ritzer, 1975). Of particular significance are the organizational and intellectual investments that sustain paradigms.

There can be little doubt that events of the 1930s provided a new way of looking at alcohol that at least partially fulfills the criteria of a paradigm, one we can loosely refer to as the medicalized conception of alcohol problems (Roman & Blum, 1991; Schneider, 1978). Avoiding a detailed definition of this conception, which is readily available in myriad sources, the fire and smoke that have come to surround it within the scholarly community might be used as indicators that it is a paradigmatic structure around which anomalies have accumulated (Fingarette, 1988; Peele, 1989). Without doubt, this loosely defined medical model certainly has been and remains an influential if not the dominant model in organizing and directing scientific research, exemplified at the time of this writing by the initial reports of the discovery of the chromosome location of the genetic material inferred to have a causal relationship to chronic alcoholism.

But unlike most of Kuhn's examples, this scientific paradigm has substantial origins and support in the "folk" therapy of Alcoholics Anonymous. Further, its dominance has been deeply embedded in a complex of political and organizational interests that have limited the influence of competing models of understanding (Roizen, 1987; Wiener, 1981). Thus, there are tensions that add to the dynamism of alcohol-related activities as an exciting arena within which to work. These tensions are found among the dominant patterns of scientific research, the implications of this research for practice and policy, and the ever-present "anomalies" that at least partially challenge the dominant paradigm (for example, spontaneous remission of substantial proportions of problem drinkers, evidence of controlled drinking among alcoholics, and apparent influences of alcohol pricing on indices of rates of alcoholism).

This abbreviated description provides a context within which to consider the role and the influence of sociological theory and method on the understanding of alcohol-related phenomena. Within this context, sociological influence and input have been present since the early days of the ascendancy of the medicalized paradigm. This presence, initially in the work and person of Selden Bacon, is a curious phenomenon in itself.

While the Laboratory of Applied Physiology was endeavoring to define issues of alcohol and physiology as legitimate topics for scientific investigation, it is noteworthy that social science became part of this endeavor in the early 1940s, well before the academic community had allowed much credence to the scientific status of sociology. This impact began with the addition of Bacon to the staff of the laboratory. Bacon was a Yale-trained sociologist specializing in crime and criminal justice studies, with an appointment as an instructor in the Department of Sociology.

In a recent interview (*British Journal of Addiction,* 1985), Bacon explains the ascendancy of sociological influence in these early years on the basis of a 1943 state-commissioned study in Connecticut to find why employable people were going to jail and thus being lost to the war effort. Bacon had been doing research on the police, and thus was selected to lead this manpower study.

Although not explicit, it appears that the connection with the laboratory came about through suspicion that jailings were linked to alcohol abuse, which indeed was shown to be the case. This led to two years of support from the state for outpatient alcoholism clinics in Hartford and New Haven; the support was to be administered by the laboratory, with Bacon assuming part of this responsibility. While this explanation of the manpower study accounts for Bacon's initial appointment, it does not explain why a hard science center was willing to engage itself in social science pursuits. To explain the welcoming attitude toward a sociologist, one is tempted to look to E.M. Jellinek, who himself can be labeled as an interdisciplinary scholar of the first order, and who had been appointed to the staff of the laboratory several years earlier.

Thus, one might observe that Bacon quickly proved his worth to the laboratory by conducting a study with social science methods that was quickly transformed into application, which in turn generated public support. That the laboratory, with its stated goal of physiological research, would accept the administration of alcoholism clinics suggests a degree of organizational hunger and receptivity to political savvy. Bacon appeared to have demonstrated such through accepting a role in administering the clinics and through his election, shortly thereafter, as chairman of the Connecticut State Commission on Alcoholism. Through demonstrated substantive interest in alcohol problems, research capabilities, and political skills, it appears that Bacon rather quickly became accepted as a member of the laboratory staff.

This gives pause and raises the question as to whether it was Bacon or sociology that was the primary carrier of this influence into alcohol studies. With due respect, the academic entrepreneurship evident in attracting Bacon to the laboratory cannot be ignored. Further, for those who have known him personally, Bacon brings a style to his scholarly work that can only be described, albeit tritely, as charismatic. One must conclude that both the man and the discipline were of equal importance for this to occur.

Bacon's work and the initial reception to it within the budding specialty of alcohol studies represented a dramatic launching for the influence and involvement of social scientific scholarship. Tracing what has occurred in "alcohol social science" since the 1940s in a truly comprehensive way is a major and important project that remains to be done, although Bacon (1984) has offered a rather pessimistic overview of what has followed upon his initial efforts. For present purposes I skip over most of that history and instead offer a rather subjective view on where alcohol social science is today, then return to what I would regard its halcyon days of the 1960s, and finally look toward what the future might hold.

The Contemporary Scene

For the broad field of alcohol studies in the United States, sociocultural, political, and organizational features of the 1990s can be taken to represent the fulfillment of wishes and dreams that ebbed and flowed from the 1930s through the 1980s. There are many objective indicators that the scholar of today whose work is focused on alcohol issues need not be marginal in the sense of being identified with matters that are subject to intense societal conflict and ambivalence, or which are largely ignored in the allocation of societal resources.

The following is a brief catalog of items that describe the status of alcohol issues in the United States in the 1990s:

- There is doubtless widespread societal recognition of the significance and importance of alcohol-related issues, reflected in both public attitudes and through the pronouncements of leaders in both the public and private sectors.
- Across the public and within leadership circles there is a widespread attitude of cooperative and constructive approaches to deal with the problems that emanate from the widespread use of beverage alcohol.

- There is at least an overtly cooperative attitude between the constituencies engaged in dealing with some of the consequences of alcohol consumption and the producers and distributors of alcoholic beverages.
- While age-limits for legal consumption of alcohol have been raised, federal taxes increased, and health-related warning labels mandated on alcoholic beverage containers, beverage alcohol is likely as available and accessible to consumers now as at any other time in American history.
- To a considerable extent, approaches to alcohol problems are lodged in the relatively high-status social institutions of medicine and public health.
- Support for dealing with alcohol problems has been institutionalized in health insurance coverage for treatment interventions.
- There has been massive growth in the numbers of individuals who work in some aspect of alcohol problem prevention or intervention.
- An organizational patron for alcohol studies is institutionalized within the federal government. Substantial financial support for both biomedical and psychosocial research on alcohol issues is supplied through the National Institute on Alcohol Abuse and Alcoholism (NIAAA). Together with the National Institute on Drug Abuse and the National Institute of Mental Health, the NIAAA participates with the National Institute of Health in a peer review system for grant applications. Within the medical research establishment, it is clear that alcohol studies have genuinely "arrived."

One could take each of the above descriptions of the status of alcohol issues in the 1990s and, through comparisons with the past, demonstrate the changes that have occurred. Yet from the vantage point of social scientific research, there are nagging feelings that something is not quite right. There are, and there must be, distinctive differences from past eras when alcohol issues languished in sociocultural apathy or at the various times when they have been embedded in organizational conflict. There is a sense that a vitality in "alcohol sociology" has been lost, along with the sources of that vitality.

At least two major changes have occurred. First, there has been a separation between the social scientists engaged in research on alcohol-related phenomena and the social movements that surround the "claims-making" activities by which particular strategies to deal with alcohol problems gain ascendancy. Second, there has been a decrease in the apparent solidarity of the sociological scholars engaged in alcohol studies.

At a somewhat higher level of abstraction, these two observations may be tied to two other phenomena that affect both the strength and the solidarity of the sociological study of alcohol problems. The first of these "second-order" phenomena is centered on the continuing lack of agreement on the proper focus for alcohol studies, an issue that is intensely foreshadowed in Bacon's 1943 essay. This lack of agreement is reflected in a lack of research continuity and replication. At the same time, the phenomena associated with drinking appear to be changing with the increased presence and attention to other drugs within American culture.

The second of the second-order phenomena is the declining dominance of the medicalized approach to alcohol issues among social scientists who are engaged in addressing these issues. As is detailed in Chapter 11, the "engine" of scientific inquiry that legitimized much of the post-Prohibition paradigm centered on medicalized conceptions of alcohol problems is no longer the exclusive property of social scientists primarily focused on problems, but is now shared with others who do not concur with many of the assumptions and implications of the medicalized approach.

The Halcyon Years?

Before turning to further analyses of these issues, it may be useful to look at a particular segment of the past. In examining the apparent solidarity of sociologists engaged in alcohol studies, an interesting point of contrast is a volume produced nearly 30 years ago under the auspices of the Society for the Study of Social Problems (SSSP). While not an overt theme of *Society, Culture, and Drinking Patterns* (Pittman & Snyder, 1962), the medicalized concept was evident as a guiding assumption across many of the chapters but was neither stridently supported nor strongly criticized.

The tone of those portions of *Society, Culture, and Drinking Patterns* that bear upon the medicalized concept may be described as facilitative rather than skeptical of the continuing diffusion and expansion of this interpretation of alcohol abuse and alcoholism. It may be inferred from this volume that the medicalized concept at that time was not a target for potshots by sociocultural analysts. It seems instead to have been a subtle yet visible rallying point for cooperation and support between researchers and practitioners. It was likely the basis for solidarity between social scientists and biological scientists working in the field of alcohol studies.

As a cultural artifact descriptive of intellectual orientations at a particular point in time, *Society, Culture, and Drinking Patterns* may be the best social scientific representative of the genre of the "modern" approach to alcohol-related phenomena. Within both practice and research in the alcohol field, the designation of "modern" was not uncommon in the 1950s and 1960s. Pains were still being taken to emphasize simultaneously (1) the enlightenment and progress brought by the "scientific" approach to alcohol issues and (2) the dissociation of this contemporary scholarly work from what were obliquely caricatured as the narrow perspectives and failed strategies of the Temperance and Prohibition era. For better or worse, the central totem in this modern and scientific approach was the medicalized concept of alcohol problems and alcoholism.

A partial exception to the perspective on the medicalized concept presented throughout *Society, Culture, and Drinking Patterns* is the chapter by Seeley (1962). This is a landmark in anticipating the implications of medicalization and the complexities of the multilevel political arena that would come to surround alcohol problems in the years following the book's publication (Cahalan, 1987; Wiener, 1981). Seeley's chapter may be seen as a charter for the extensive literature on sociopolitical aspects of the medicalized concept of alcohol problems that was to emerge over the next quarter-century. Nonetheless, in 1962 Seeley's observations were hardly viewed by the volume's editors as a central or thematic contribution. His chapter was included in a final section of the volume that comprised somewhat of a potpourri of essays only partially directed toward the section theme of "systems of control."

Since 1962 there has been a tremendous growth of societal and social science interest in alcohol problems, probably far exceeding the imaginations of those who contributed to *Society, Culture, and Drinking Patterns.* Alcohol issues have been caught up in a power surge of the societal "attention cycle" to social problems. In the arenas of research, policy formation, and public interest alcohol issues have, from some points of view, achieved the level of attention that has been heartily strived for by the modern movement's leadership for the past half-century.

While there are multiple indices to measure the actual extent to which the medicalized conception of alcohol problems has been socially and culturally accepted in American society, developments through the 1970s and 1980s led to this conception becoming by 1990 the principal "governing image" of alcohol problems in American culture and society (Room, 1978). This cultural support does not stem from the medicalized concept per se, but from the implied means of its discovery and elaboration: To borrow from Freud, the medicalized approach to alcohol problems has traveled on the "royal road" of science to institutional recognition of its value as the primary approach to understanding and dealing with alcohol abuse and alcoholism. Further, the cultural acceptance of the medicalized approach and its "scientific essence" may be seen as the indirect reason for the remarkable increase in the availability of funding for research on alcohol problems over the past 20 years.

Such support has not been limited to biomedical studies, despite the perception by many of a perceived imbalance in that direction. By being more or less aligned with this medical/scientific model of alcohol problems, social scientists successfully entered what were first the tents but soon became the halls of Public Health Service-based funding for alcohol-related research. However, this road has become increasingly well traveled; a broadened application of science as the vehicle for the advocacy of other approaches has contributed to changes in the influence of the medicalized concept of alcohol problems.

Returning to *Society, Culture, and Drinking Patterns,* it may also be argued that the legitimizing of the involvement of social scientists and social scientific analyses has fostered changes in the influence of the medicalized conception of alcohol problems. Beyond critiques of the medicalized conception of alcohol problems, social science-based analyses have at one and the same time affirmed the salience of sociocultural approaches to alcohol problems while contributing to conceptual and operational murkiness and confusion. This scene contrasts with the remarkable simplicity of concepts and strategies which were apparent when the medical-scientific approach began its ascendancy on the shoulders of Alcoholics Anonymous more than 50 years ago.

It is clear that "unity" in the social scientific subcommunity focused on alcohol issues has never been realized except perhaps for very brief periods. Such unity is much less of a fact in 1991 than it was 30 years ago when the chapters for *Society, Culture, and Drinking Patterns* were being drafted. That period was perhaps a watershed time for common or at least complementary concerns among social scientists studying alcohol problems. The vigor of that enterprise has multiple indicators: the scope of the 1962 volume, the number of original chapters that the editors were able to solicit, the fact that the volume was a conscious product of a more or less collective effort fostered by the Society for the Study of Social Problems, and the subsequent momentum that led to the production of a narrower volume under the same auspices (and involving many of the same authors) eight years later (Maddox, 1970).

After 1970 there is scant evidence of the growth or even the maintenance of this collective interest. We have not seen the rise of a specialty association centered on social science and alcohol. There is no alcohol specialty journal within which the paradigms and methods of social science research can be represented without competing with biomedical and evaluative research. Participation by social scientists in the Research Society on Alcoholism (the major scientific organization centered on alcohol studies) is absolutely minimal. The Committee on Drinking Behavior of the SSSP, which was the primary source of much of the 1962 materials, was transformed in the mid-1970s into a section on drinking and drugs. Research presentations centered on alcohol have actually declined within this organization's annual meetings and official publications over the past three decades.

What are the sources of this apparent decline and disintegration? Among the transformations that occurred between the 1960s and the present are the following:

1. The sheer growth of alcohol studies through the relatively massive infusion of funds and resources has affected the extent to which unity and commonality with social science could be reasonably expected to persist. To an extent, the relatively small community of alcohol social scientists in the 1950s and 1960s could draw senses of solidarity from their minority status, the relative absence of resources to support their work, and (according to Robert Straus) even a perceived sense of shared stigma generated through scientific association with a stigmatized set of behaviors. Growth in resources and numbers of scientists has eliminated size and minority status as bases for unity.

2. The rising dominance of psychological perspectives in the spectrum of behavioral science research on alcohol issues has displaced sociological perspectives and even suggested the relegation of social science to areas such as program evaluation rather than basic research. This is clearly linked with the increased support available through the NIAAA. The psychological perspective is certainly not uniform, but draws substantial attention to etiological issues, which have generally been secondary in sociological analyses. Further, many psychologists are strongly sympathetic with the notions of the genetic transmission of alcohol problems and other biomedical perspectives that have characterized the renewal of the biological paradigm within psychiatric research and literature. To an extent, the status of psychological research on alcohol has been enhanced by the association with the broad model of "heritability," with the centrality of notions such as "expectancies" in the prediction of problematic drinking behavior, and with the ever-increasing prominence of psychotherapy in alcoholism treatment (Roman & Blum, 1991).

Thus, while there appears to have been some degree of parity in the representation of social and psychological scientists in alcohol studies in the 1960s, this balance has slipped heavily in favor of psychologists by the 1990s. Further, Bacon's occupancy through the 1960s and until the mid-1970s of the premier position of director of the Center of Alcohol Studies at Yale and Rutgers was a means for maintaining central representation of the sociological perspective.

Finally, recent decades have been marked by what some would regard as decline and fragmentation within sociology (Smelser, 1988). Graduate programs have not been magnets for the best and the brightest. The large cohort of sociologists trained under post-World War II G.I. Bill support have passed into retirement. Thus, the numerous forces that have affected the fate of the parent discipline may be reflected in the fate of alcohol sociology. Despite a recent review that finds that alcohol studies offer incredible opportunities for a vast range of sociological analyses (Bucholz & Robins, 1989), there is little evidence that this magnetism is contributing to significant growth in the numbers of sociologists engaged in some aspect of alcohol studies.

3. Through the 1970s, there has been a massive growth in literature that is both directly and indirectly supportive of the "social constructionist" perspective on alcohol issues, using both historical and contemporary data. In other words, many analyses have shown the political and organizational forces that affected the rise to dominance of the medicalized perspective on alcohol problems. Closely aligned are perspectives from conflict theory that view social constructions of social problems as power struggles wherein one group seeks to dominate another through impugning deviant or stigmatic conditions, and then undertaking the management of these conditions (Conrad & Schneider, 1980). These analyses have been accompanied by a vast array of critiques of the disease concept and the medicalized perspective. This shifts the basis for unity among social scientists adopting such perspectives from sharing their paradigm-based discoveries to sharing their critiques of the objective realities surrounding alcohol issues. One would surmise that survival as critics depends more upon producing new and unique accounts rather than sharing and agreeing upon common critiques. Thus, in terms of sustaining a subspecialty such as alcohol sociology, it may be argued that social constructionism and conflict theory as shared perspectives may be, over the long term, fragmenting rather than unifying forces.

4. These trends may have served to undermine a perspective that was once at least a moderately unifying force for sociologists in alcohol studies. They also have affected the credibility of social scientists' involvements with the practitioner community, to the extent that the community hostilely rejects any challenge or criticism of the disease concepts of alcoholism and alcohol problems (Madsen, 1988). Thus, it may be observed that there is a split between those social scientists who are aligned and work with the practitioner community and those who take a more purely academic perspective on alcohol problems, their etiology, and their management.

Without doubt, across the range of academics working in alcohol studies prior to 1970 one could find distinctive intimacy between researchers and practitioners and sympathy with those affected by alcohol problems. Such closeness was both a precursor and a byproduct of the necessity of gaining entry for data collection in an era when research funding was minimal. The charisma that drew and sustained the commitments of recovering alcoholics spilled over onto many of these academics. To a large extent, this lying down with the natives carried many risks for scientific objectivity. The production of good scholarship under such conditions was a genuine art form that has never been objectively evaluated. Such crossovers between the academy and the "real world" have been largely lost as alcohol studies have grown as an interest area, research funding has become more bountiful, and alcohol interventions have become embedded in entrepreneurial growth and in large-scale bureaucracies.

The Future

What is the future for social science studies of alcohol-related phenomena? Few signs suggest a reorientation toward Bacon's 1943 agenda in which there would be a somewhat balanced interest across the range of drinking behaviors as well as problematic drinking behaviors. If anything, the current phase in the cycle of social problem interest, together with the increasingly accepted dominance of biomedical models of interpretation and intervention, suggests even more pressure away from attention to "normal" and socially constructive drinking behavior.

The future for sociological work is not dismal, however. It may be important to recollect a brief but dramatic nightmare of the early 1980s. Many have already forgotten that members of the emerging Reagan administration attempted to end support of research by social scientists that was provided by agencies under the Public Health Service umbrella. While it may have had some brief impacts, the goal of excluding social science work from these sources of support did not succeed. This nightmare, which conceivably could have ended financial support through the National Institute on Alcohol Abuse and Alcoholism for social science research on alcohol-related phenomenon, did not turn into reality. There is indeed an infrastructure of support based on perceived need for social science research in the loosely defined arena of behavioral problems. It may be observed that the National Institute of Mental Health, whose funding of research centered on urban social problems other than mental illness was the principal stimulus for the attempted "embargo" of funding, has apparently narrowed its scope of social science research support to studies that are focused on some aspect of psychiatric disorder. But social science research in alcohol studies seems rather robustly supported, particularly if one makes comparisons to the pre-Reagan years.

It should be noted that many in the social sciences would disagree with the assertion of a strong level of current support, but such gloomy assessments are rarely well informed and are usually based on a narrow understanding of how the funding mechanisms work. Scholars who do not apply for federal funding to support research do not receive it. Further, a look at the balance between biomedical and psychosocial research that is funded would seem to indicate that social science work receives short shrift. However, such a review must consider the makeup of the pool of applications and the relative sizes of the scientific subcommunities from which applications might be generated. From a range of informal observations and loose counts, I would suggest that there is relatively equitable attention to social science work for which support has been requested, keeping in mind that social science funding applications are more likely to be plagued by methodological ambiguities than those in biomedical research. However, as has been mentioned, there is no evidence of any groundswell in the growth of sociologists' involvement in alcohol studies. Thus, while funds are available for social science research, the relatively small numbers of sociologists with interests in this topic does not augur for significant future growth in this research enterprise.

The future is unclear. From this writer's perspective, the future offers exciting stimulation for social scientists involved in alcohol studies. There appears to be an increasing sensitivity within sociology for adapting to role behaviors required by involvement in research that bears on practice and is policy-related. Perhaps a renewal of the earlier collaborations between sociologists and alcohol problem-directed practitioners is a possibility. At the same time, there is evidence of increasing numbers of social scientists with few if any skills in dealing with data collection in the real world. Stimulating scholarly interest and investments of energy in primary data collections, which are essential for a better understanding of alcohol-related issues, is dampened by the widespread acceptance within sociology of knowledge-building primarily on the basis of secondary analysis of data collected by others for other purposes.

Conclusion

In concluding this introduction, I call upon one who would surely have been the contributor of a chapter to this volume had he not been taken prematurely by death several years ago. Ira Cisin was a significant pioneer in the initiation of nationwide survey data collections on American drinking practices, data bases that continue to be collected and offer potential for many of the analyses suggested in Bacon's 1943 agenda. Cisin was a participant in a 1977 symposium at Rutgers that led to the publication of a volume (Keller, 1979) that has been relatively neglected but is an extremely important source of insights for understanding alcohol studies from a sociology of science perspective. At that symposium, in what would probably be regarded as a keynote or plenary address, Bacon (1979) outlined the advantages of a distinctive academic discipline centered on alcohol studies and tentatively labeled it alcohology. This essay in many ways carries the themes from the 1943 essay and can be fruitfully read in conjunction with the earlier work.

Bacon's typically sweeping comments were followed by several commentators, including Cisin (1979), who was constructively cautious about

Bacon's suggestions. In his observations, based very much on the real world of 1977, Cisin used the theme of "liberation" to describe some severe and practical problems that then stood in the way of the development of an alcohology, much as they continue in 1990 to stand in the way of the development of a sociology of alcohol-related phenomena that can become much more of a magnet for sociological interest than it has been in the past.

The following are several of Cisin's suggestions (with some added paraphrasing) concerning the forces from which alcohol-related scholarship must become liberated:

1. The jargon of "alcoholism," "alcohol abuse," "responsible drinking," "problem drinking," and other such concepts must be effectively and consensually defined and operationalized if research is truly to move forward. Clearly, one cannot utilize data in theory testing or theory construction with concepts that are ambiguously defined, or whose definitions are grounded in political considerations.
2. Alcohol studies must be "liberated from justifying our existence in irrelevant ways" (p. 31) in the political arena. Accepting the primacy of the social-problem significance of a phenomenon directs research primarily toward the political-problem issues rather than toward good quality science. Such emphases distract attention and emphasis from an objective scientific approach before a research question can get "out of the gate." Cisin uses a delightful analog: "Jupiter is a very long distance from our planet; the exact number of miles is hardly important." Such is the contemporary case with defining the relative rank of alcohol usage patterns as causes of death, injury, grief, crime, and other dimensions of social mayhem. Special examples include the efforts to estimate the economic costs of alcohol problems (without adding to the equation their benefits in terms of the work created by the myriad of consequences!) and the more recent (and continuing) shootout over whether alcohol or illegal drug problems are "worse."
3. Alcohol studies must be liberated from dogma, such as alcoholism is really an illness, that abstinence is the only route to recovery, and that all alcohol problems can be classified on a single quantitative dimension of "early, middle and late." Some who are initially attracted to alcohol studies are dismayed by some social scientists' uncritical acceptance of these dogmas, but may be even more bewildered by those researchers who privately reject these dogmas but who embed the dogmatic assumptions in their scientific work.

It is striking that these barriers to the attractiveness of alcohol studies have persisted with little change over the years since they were published. There is no doubt that they were also evident long before Cisin wrote them down. Perhaps this tenacity points more toward the need for acknowledgment and accommodation of these apparent barriers rather than toward hoping that social scientists will somehow "buck up" and join this research specialty despite them.

Acknowledgment

Partial support from Research Grant R01-AA-07218 and Training Grant T32-AA-07473 from the National Institute on Alcohol Abuse and Alcoholism during preparation of this manuscript is gratefully acknowledged.

References

Bacon, S.D. (1979). Alcohol research policy: The need for an independent, phenomenologically oriented field. In M. Keller (Ed.), Research priorities on alcohol. *Journal of Studies on Alcohol,* Supplement 8.

Bacon, S.D. (1984). Alcohol issues and social science. *Journal of Drug Issues, 14,* 7–29.

Beauchamp, D. (1979). *Beyond alcoholism.* Philadelphia: Temple University Press.

British Journal of Addiction. (1985). Conversation with Selden D. Bacon. *British Journal of Addiction, 80,* 115–120.

Bucholz, K., & Robins, L. (1989). Sociological research on alcohol use, problems, and policy. *Annual Review of Sociology, 15,* 163–186.

Cahalan, D. (1987). *Understanding America's drinking problem: How to combat the hazards of alcohol.* San Francisco: Jossey-Bass.

Cisin, I.H. (1979). From morass to discipline in one grand leap. In M. Keller (Ed.), Research priorities on alcohol. *Journal of Studies on Alcohol,* Supplement 8.

Conrad, P., & Schneider, J. (1980) *Deviance and medicalization: From badness to sickness.* St. Louis: Mosby.

Fingarette, H. (1988). *Heavy drinking: The myth of alcoholism as a disease.* Berkeley: University of California Press.

Keller, M. (Ed.). (1979). Research priorities on alcohol. *Journal of Studies on Alcohol,* Supplement 8.

Kuhn, T.S. (1962). *The structure of scientific revolutions.* Chicago: University of Chicago Press.

Lodahl, J.B., & Gordon, G. (1972). The structure of scientific fields and the functioning of university graduate departments. *American Sociological Review, 37,* 57–72.

Maddox, G. (Ed.) (1970). *The domesticated drug: Drinking among collegians.* New Haven, CT: College and University Press.

Madsen, W. (1988). *Defending the disease: From facts to Fingarette.* Akron, OH: Wilson, Brown and Co.

Peele, S. (1989). *The diseasing of America.* Lexington, MA: D.C. Heath.

Pittman, D., & Snyder, C. (Eds.). (1962). *Society, culture, and drinking patterns.* New York: John Wiley and Sons.

Pittman, W. (1988). *AA: The way it began.* Seattle: Glen Abbey Books.

Ritzer, G. (1975). *Sociology: A multiple paradigm science.* Boston: Allyn and Bacon.

Roizen, R. (1987). The great controlled drinking controversy. In M. Galanter (Ed.), *Recent developments in alcoholism,* Vol. 5 (pp. 245–279). New York: Plenum Press.

Roman, P. (1988). The disease concept of alcoholism: Sociocultural and organizational bases of support. *Drugs and Society, 2,* 5–32.

Roman, P.M., & Blum, T.C. (1991). The medicalized conception of alcohol problems: Causes and consequences of murkiness. In D. Pittman & H. White (Eds.), *Society, culture, and drinking patterns reexamined.* New Brunswick, NJ: Rutgers Center of Alcohol Studies.

Room, R. (1978). *The governing images of alcohol problems.* Unpublished doctoral dissertation, University of California at Berkeley.

Schneider, J. (1978). Deviant drinking as disease: Alcoholism as a social accomplishment. *Social Problems, 25,* 361–372.

Seeley, J.R. (1962). Alcoholism is a disease: Implications for social policy. In D. Pittman & C. Snyder (Eds.), *Society, culture and drinking patterns* (pp. 586–593). New York: John Wiley and Sons.

Smelser, N. (1988). Introduction. In N.J. Smelser (Ed.), *Handbook of sociology.* Newbury Park, CA: Sage Publications

Wiener, C. (1981). *The politics of alcoholism.* New Brunswick, NJ: Transaction Books.

CHAPTER 1

Sociology and the Problems of Alcohol: Foundations for a Sociologic Study of Drinking Behavior

SELDEN D. BACON

The Scientific Approach to Social Problems

The scientific approach to a social problem, be it poverty, crime, disease, inebriety, or any other, is one of the last to be utilized by the society. The findings of scientific research are accepted by the public only after long delay, and their application to the problem may be only partial. The scientific investigation will be directed, in all likelihood, at particular, not general, aspects of the problem, and will depend on highly specialized techniques for the solution of highly specialized problems.

The scientific attitude, in contrast to that of an aroused public, is both cautious and humble. It does not angrily look for "who's to blame" and rush to the attack; nor does it quietly hide its head in the sand and, through euphemism and pretense that there is no problem, attempt to avoid the whole situation. Science begins by attempting to define the area in which the problem is located. There follows analysis of the various problems making up the whole, for science is both sufficiently humble and well trained to realize that no long-lasting social dilemma of wide incidence can be solved in one step. Then, after the selection of a particular problem, relevant data are collected, explanatory hypotheses are developed, and the testing of these is initiated. New questions are posed, hypotheses are refined, and more and more testing under a variety of conditions is carried out. With skill, diligence, and luck, the problem may be solved—not the total social problem which stimulated interest at first, but the specific problem which the scientists abstracted.

This process takes men, money, materials, and time, especially the last. The larger problem, meanwhile, is as prevalent and pressing as it ever was. Even when the scientists solve their particular problems, the society

Reprinted with permission from the *Quarterly Journal of Studies on Alcohol, 4*, 402–445, 1943.

may be as disrupted and may be suffering as much as previously from the total problem. Before turning to the subject of alcohol it will not be profitless to consider briefly the approach of science to another serious social problem, that of venereal disease, more particularly the problem of syphilis. Successes and failures of science in this field present some analogies to what has already occurred in the attempt to gain control over problems of alcohol and broadens the scope of imagination in looking forward to future developments.

Science entered the war against syphilis after many other approaches, all popular in struggles to conquer social problems, had been used: running away, isolating or killing persons obviously afflicted, blaming traditional enemies and scapegoats, ordering other people to "do something," hiding the problem, pretending it wasn't so, even trying to laugh it off, and searching and hoping for miracles. All these may be noteworthy and laudable activities in some situations, but for purposes of meeting a real problem, whether venereal disease or unemployment or crime or inebriety, they are obviously inefficient.

When the scientists entered this field, they did not even attempt to eradicate the disease. They were aware that such an achievement was far in the future. They attempted to define the disease, to find the specific entity or entities resulting in the observed physiological symptoms, to discover a technique of identifying its presence, and to develop a means for eliminating the causal entity without otherwise damaging the human body. Their success in this highly ambitious program forms one of the great chapters of science; they achieved every one of their aims. But they did not solve the problem of syphilis.

Years have passed, decades have passed since these particular problems were solved, but syphilis is still a major social problem, although some improvement in the situation is beginning to appear. The scientists who discovered *spirochaeta pallida* and Salvarsan, who have since refined the techniques of identification and therapy, did not approach the psychological, legal, economic, familial, or moral problems involved. Nor did they use techniques or have the training which would have led to solutions in those fields. The public in the United States was attacking all these problems by the subtle method of denying existence of the disease, a method quite incompatible with the approach of the scientists, as indeed are all the popular methods mentioned above.

This recital is not intended to imply that science does not provide efficient techniques for dealing with social problems. The example is presented to show that the scientific approach is selective; in the case of this venereal disease, the scientists employed biological and chemical techniques and they achieved biological and chemical successes. If anything, their achievement is a great spur to scientific work in the psychological and social aspects of the problem. The popular techniques listed never work. The scientific technique has worked where it has been applied. It seems to be indicated that an application of the scientific method in other aspects of the problem would be well worthwhile.

The reasons why this has not been done are manifold. First, all the objections to science in general are appreciably enhanced when the object of study is human motivation or human behavior. Science threatens our nearest and dearest beliefs, attitudes, and pleasant conceits as they relate to our philosophies of family, government, property, and religion. Second, the scientific method originated and has had its greatest development in the pursuit of knowledge about strictly material phenomena; it is not mere chance which brought about the use of chemical and biological techniques in the scientific approach to the problem of syphilis and also to the problem of alcohol. As a result, not only the public but also many scientists are of the opinion that science can be used only in treating material phenomena; science is often considered to be a procedure tied up with machines and test tubes. Third, science is slow, lacks glamour, promises nothing; it cannot compare to a whirlwind campaign by a philanthropist, a district attorney, or an evangelist. Fourth, the function of attending to such matters is assigned, in public opinion, to certain prestigeful groups whom our society has endowed with special competence in these matters (legislators, politicians, ministers, columnists, artists, and a variety of amateur meddlers), and who want no trespassing on their preserves by the scientist, a new type of meddler. It must be admitted, too, that this newcomer can be as dogmatic, arrogant, and tactless as any of the others.

Yet without the scientific approach to social problems, the hope of solution, the hope of avoiding the ills, the pains, the waste, the conflict, and the destruction involved, appears slight indeed. For our final problem is not what the spirochete or alcohol does to the body, nor of what the structure and operation of such physiological and physical entities may consist, but the relation of these entities to individual and group life. That is why syphilis and alcohol are social problems. That is why public effort is expended on these subjects rather than on barnacles, as studied by Darwin, or fruitflies, subject of the most intensive scientific investigation. From the physiologic and chemical and biologic viewpoints these last are just as significant as alcohol or spirochetes, but they are not of great significance socially. If they should be found to cause human beings great pain or if they should become of great social value, then time and money would be poured into an analysis of their structure and processes and relationships. It is the individual and social aspects of phenomena which make them the object of pressing attention. The individual and social aspects, then, must be understood if there is to be control of the total situation.

We may take it for granted that a real problem is involved in the utilization of alcohol as a beverage by human beings. Such a conclusion appears inevitable, whatever criteria may be applied.

Furthermore, it is assumed that a scientific attitude will produce more efficient results than any of the other approaches utilized hitherto. Pleas, cajolery, dictatorship, flight, ignorance, pretense, verbal and physical attacks, wishful thinking, dreams, and abject suffering seem to have had little effect. To some, science may not seem the best or even a hopeful attempt; but those who earnestly desire a solution of the problem will welcome the effort of the scientist.

Thus far the participation of science in the alcohol problem has been limited, for the most part, to the utilization of two or three techniques which apply to narrow segments of the problem. It has been realized that a true conception of a problem of alcohol is far too complex and unwieldy to be attacked by any one research method. It has also been realized that available knowledge, until the last 25 or 30 years, has been not only slight, but obscured by much misinformation and misconception, both reinforced by considerable emotionalism. The latter realization is, in itself, a forward step. It is not an advance which has been fully appreciated by the general public or even by all those who are deeply stimulated by the problem. There are still many who believe that they know everything about this simple problem, which, they are convinced, can be solved if only one or another form of action is taken.

With the understanding that they should approach the problems of alcohol rather than the problem of alcohol, scientific workers have, first of all, investigated the physiological aspects of the problem: what the nature of alcohol is, how it affects bodily processes and structures, how it is affected by the body. In the course of these studies a great deal of fallacious theory was eliminated, a large body of tested data was collected and classified, and an understanding of the physiological properties of alcohol in relation to the human system has been attained which is far, far more applicable than that of earlier times. This development has been necessary for an understanding of the total problem, has been of greatest use in the diagnosis and treatment of diseases related to alcohol, and has, of course, added to the universe of general physiological knowledge and theory. It has not appreciably changed the overall situation in regard to the social problem of alcohol. Nor, until other disciplines advance equally far, would this be expected.

Another scientific avenue of approach, closely related to the physiologic, has been that of pyschology. Here the emphasis has been on the effects of certain quantities of alcohol, taken under certain conditions, on the nervous and, consequently, the physiological potentialities of the animal organism, including the human. Here, again, fallacious assumptions of the past were cleared away, verified data were accumulated and classified, and a body of theory was developed which not only helped to present a clearer picture of the person with alcohol in his system but, also, to refine the knowledge and understanding of psychological processes as a whole. Some of this knowledge has been utilized in deciding legal questions, in identifying excessive drinking, and so forth. It has not, however, explained the motivation for drinking, the learning of drinking habits, the reasons why some people drink more than others, the effect of punishments on drinking, and so on, although some progress has been made in this last field through the study of conditioning techniques.

Allied to the psychologic approach has been that of the psychiatrists and psychoanalysts, although their primary interest has been therapy rather than research. They have studied the development of the psychotic drinker and the neurotic drinker. The psychiatrists have been interested especially in a classification of personality types among drinkers and in the incidence of drinking among certain categories of abnormal personality. The analysts have emphasized the origins of the psychological problems of abnormal drinkers. Both groups have considered data which are treated by students of social phenomena, but, quite understandably, only from the viewpoint of specific individuals. For students of society, the individuals so described represent such a tiny and exotic portion of the whole community that the resulting generalizations have not been sociologically informative, although they have posed questions and suggested hypotheses of great interest.

Professional sociologists have so far done little or no original work in this field. The subject is always mentioned (usually a whole chapter is given to it) in textbooks on social problems, but these consist almost wholly of reviews of the physiological, psychiatric, and psychological findings. Some sociologic generalizations may or may not be derived from these, and may be combined with generalizations rising from studies of other pathologic social groups. As Queen and Gruener (1940) point out several times, there is a wide arena for sociologic study in the phenomenon of drinking behavior, but it is practically untouched at this time.

The Function of Sociology in the Scientific Study of the Problems of Alcohol

What aspects of the total problem of alcohol present situations, dilemmas, and questions which can be properly submitted to the techniques, experience, knowledge, and logic of the sociologic method? Sociology deals primarily with groups of people, their attitudes, their patterns of behavior, and their material equipment. Therefore, in contrast to physiologists who study the nature of alcohol and its effects on the organs of the body, and in contrast to the psychologists who study the effects of the substance on the nervous system and consequent action-potentialities of animal life, the sociologist would consider the drinking behavior. As can readily be seen, this is a totally different world from that observed by the two disciplines mentioned. For them the alcohol can be admitted to the body by a hypodermic needle, in a capsule, or through a rubber tube. Much of their research can be carried out with dogs and rats. Whether a human subject is Jew or Gentile, rich or poor, married or single, a divinity student or an agnostic plumber, makes little or no difference.

The sociologist is interested in the customs of drinking, the relationship between these customs and other customs, the way in which drinking habits are learned, the social controls of this sort of behavior, and those institutions of society through which such control issues. The sociologist wishes to know the social categories in which much or little or no drinking occurs, he seeks correlations of amount and type of drinking with occupational, marital, nationality, religious, and other statuses. More importantly, he poses the broad questions: What are the social rules concerned with drinking? What are the pressures for and against this practice? How does this behavioral pattern jibe with other institutions and folkways? The sociologist is interested in changing patterns of drinking and in their relation to other changes in the society. The sociologist studies the effect of no drinking, some drinking, or excessive drinking on groups, attitudes, and behavior. As drunkenness may result in punitive, preventive, or therapeutic measures, the sociologist observes, classifies, analyzes, and compares these activities.

Rather than questioning the process by which John Doe, an abnormal drinker, came to have his particular personality; rather than studying the effect of alcohol on metabolism or the liver, the sociologist wants to explain such questions as why excessive drinking is so common among Protestants but rare among Jews; what sociologic phenomena are correlated with the low incidence of inebriety among women and the high incidence among men; what is the significance of the abnormal marital status of the common police-court inebriate; what are the social forces in a therapeutic agency such as Alcoholics Anonymous which underlie its considerable degree of success; how do drinking practices and attitudes vary with social classes; what is the effect of varying methods of control (religion, schooling, legal fiat, prison sentences, etc.) on drinking habits of different categories of persons; and other questions of this nature. The answers to these questions must be sought, however, not as ends in themselves but rather for the generalizations which may be derived from them concerning the total problem.

A fair degree of optimism accompanies the proposed research. In the first place, there seems little reason to believe that there is a physiological need for alcohol, or that drinking stems from an inherited craving. Nor does the drinking of alcohol appear as a specific response determined by the physiologic-psychologic nature of man. Certainly no virus carries alcoholic behavior to the human system; inebriety has an onset and development totally different from that of syphilis or yellow fever. Unless convincing evidence to the contrary can be discovered, and there is no sign of any such discovery at present, it may be assumed that drinking[1] behavior is subject to the same mode of analysis as any other form of behavior, whether it be table manners, football, marriage, or earning a living.

A factor which has delayed and discouraged an adequate analysis of drinking behavior has been the failure to recognize the relation of inebriety to all other forms of drinking. Consequently, there has been a failure to orient properly the abnormal phenomena which hitherto have dominated all studies in this field. This exotic fraction of drinking behavior has attracted all the attention, just as the comet or shooting star elicits more comment than do the millions of "ordinary" stars. In the average citizen this imbalance of interest is not blameworthy. It is perhaps true that most sciences had their genesis in the observation and analysis of the immediately painful and the extraordinary, but that stage should by now have been passed. The entire field of social science may be freely criticized on this score; in many instances it may still be found gazing in starry-eyed wonder at the occasional volcanoes, emeralds, and icebergs, to analogize from geology, when it has a gigantic earthcrust as its field of inquiry. The exotic and the pathologic are useful fields of scientific inquiry, but they have their limits. One could study American society by examining the millionaires, or the unemployables, or both, but the result would surely be extraordinary.

This sort of erroneous or, perhaps, naïve approach has appeared in studies of drinking. Nor is this true only of studies of drinking behavior by persons not scientifically trained. It is also clearly evident in scientific studies of drinkers, almost all of which have concerned themselves with wealthy alcoholics, psychotic alcoholics, or alcoholic felons. Inebriates, however, are but a minor percentage of drinkers. Haggard and Jellinek (1942) estimate that out of some 44,000,000 persons in the United States who drink, perhaps 600,000 can be called abnormal drinkers. Of these 600,000 perhaps 10 percent would be psychotic drinkers, certainly not more than 2 percent would be in the class who enter private sanitaria. At best, a highly selected 1 percent of all drinkers have been studied; obviously, this is too biased a sample from which to generalize about drinking behavior. It constitutes a basis of study far more limited than would be that of millionaires and unemployables in a study of United States society.

The study of the exotic or pathologic members of a total group is not to be scorned, nor should the students of such types of persons be dissuaded; but such studies must be oriented within the total field and the generalizations limited to the segment studied. Studies of drunkards arrested by the police, although not so limited as those mentioned above, are still concerned largely with the pathologic portion of all drinkers. For, of the 44,000,000 Americans who use alcoholic beverages, not one-half, not one-fifth, probably not even one-tenth, are ever arrested by the police.

Until the drinking behavior of a representative sample of the drinking population is observed, analyzed, and described, characterizations of the tiny proportion of abnormal drinkers are likely to be as biased and as fallacious as the studies of Lombroso about criminals, studies of the insane during the eighteenth and nineteenth centuries, and comments about the lower-class poor issued by wealthy, upper-class philanthropists of almost any age.

The need for general orientation of problem drinking behavior within the setting of all drinking behavior, including abstinence and antidrinking activity, is one of the major reasons for the types of study here proposed.

As to the general validity of the sociologic approach to a field of this or of similar nature, any argument presented here would be either gratuitous or hopelessly inadequate. To those aware of the nature of society and of social process, structure, and function, nothing profitable could be added in any foreshortened description. To those opposed to or even to those merely unaware of this field of knowledge, the determining influence on human behavior of customs, sanctions, socially interpreted environment, automatic adjustment, and the like would remain an obscure, esoteric discipline. All that can be said is that mystic, whimsical, contradictory, unexplained, and unverifiable explanations of any sort of human behavior are unacceptable to a scientific approach. The unexplained, if so labeled, is an exception to this dictum.

In the field of studies of human behavior there is sometimes a question about the scope of sociology as opposed to the scope of psychology. Actually the relationship of the two fields is more a division of labor than a conflict of interests. There are certain aspects of behavior, however, which belong more obviously in one field than in the other. The forms of behavior, for example, are largely a sociologic matter, while the living through of one of these forms by John Jones, the intensity and sensitivity of his participation, belong rather in the field of psychology. More specifically, John's drinking alcoholic beverages, the amount, type, time, place, and surrounding situation, and the attitude of friends, family and the police are largely determined sociologically; but within this larger determination is John's particular degree and quality of acceptance or rejection of these sociogenetic forms. In any event, John will not drink like a Zulu or an Austrian or a Japanese; in fact, he will not drink like a New Yorker or a Californian or a ditch digger or a Yale man or a Kentucky mountaineer, unless he is or has been in socially significant contact with such a group.

The study to be proposed here will emphasize heavily the sociological aspects of drinking behavior. No study having the purpose, scope, or character of that here proposed is to be found in the literature. Social surveys of drunkenness have been made in many communities, but these may be uniformly characterized as covering the pathologic aspects only, and even then presenting little more than figures on arrests or statements as to the evil of drunkenness, often with suggestions of model reforms. The early study of Charles Booth (1902), covering London at the close of the last century, is sufficiently superior to be considered an exception. Specific studies of the alcohol problem are not common and are usually pointed at particular aspects of abnormal drinking. The reports of the Committee of Fifty, which also appeared at the end of the last century, are probably worthy of special note since they represent a realistic attempt at an objective study, and since ample time, although rather limited funds, was made available. The studies made by groups with predetermined emotional attitudes can be considered as having only occasional value on discrete observations. In almost all these reports attention is given wholly to abnormal drinking; they are studies of the exotic fraction of the whole field of drinking behavior.

The sociologic approach to the problems of alcohol, then, is not in conflict with the research of other students. It has a problem totally different from that of the physiologist or the physiological psychologist. It has objectives more similar to those of the psychiatrist, but its techniques, its scope, and its emphases form a contrast to those of psychiatry; even so, it may present results complementing the findings of psychiatrists. Naturally, without a physical world, without the physiological and psychological entity "man," there would be no groups, no human behavior, no attitudes, no artifacts, no alcoholic beverages, no social problem. All advances in the understanding of the relevant physical components and the physiological and psychological nature of man are of the greatest pertinence and necessity for efficient study of the social factors involved.

It should be added that to the sociologist the center of the problems of alcohol appears more nearly sociological, or sociopsychological, in nature than physiological or biological. And it is believed that, although knowledge of the physiological and biological factors is essential for the attainment of greater understanding and for the increased possibility of control which that understanding will permit, the problems of alcohol are intrinsically social in character and will submit to rational solution only after the social elements, processes, and relationships have been analyzed with all the honesty, humility, and zeal the scientific method allows.

Framework for a Sociologic Study of Drinking Behavior

Before presenting concrete proposals for a sociologic study of drinking behavior, it seems advisable to set down some general statements concerning the observation and classification of behavioral data. Since this particular field has hardly been touched, no detailed system can be offered. Nor would this be desirable in view of the danger of predetermining results. Broad lines of inquiry concerning behavior in general, however, are available, and this experience should be utilized. Furthermore, sociologic and anthropologic studies of total societies present a general orientation of drinking behavior in its relation to all behavior, so that legitimate avenues of exploratory approach, at least, are also at hand.

The human consumption of alcohol is probably not the core or center of any major pattern of activities in the sense that food, shelter, sexual activity, child rearing, or war can be the core of a large and complex totality of organized and interdependent patterns of behavior. This is not to deny that at certain times and places the imbibing of alcohol is the central activity, or that certain economic, recreational, and perhaps other activities are primarily centered around the production, distribution, or consumption of alcoholic beverages. It does imply that these are specialized, particular phenomena, and that, except in rare cases, such as those described by Bunzel (1940), they are not pervasive or dominant activities in the total social scene.

Although drinking behavior cannot qualify as a major human institution (as the family, government, religion, economic organization), it is, like all human behavior, institutionalized. As such it may be expected to show certain components; components characteristic not only of the totality of such behavior but also of each segment of drinking behavior; components that should be noted in the observation and recording of all such data. A listing of these components follows.[2]

Function

The function of any given form of behavior may be divided into two aspects: an individual function and a social function. Expressed very simply, the first of these merely infers that some need is met by activity. A basic or a secondary drive of the individual is satisfied.[3] The behavior will soon disappear if this is not the case (the psychologic correlate is the principle of extinction). The function is very likely to be multiple in nature. The function must be ascertained by a trained observer. The importance of any particular function of a given act may change; for example, the function of a mace may change from that of striking enemies on the head to that of giving prestige to certain assemblies; that of a ship's lantern, from giving light to lending "quaintness" to a "tea shoppe." The social function will be the attainment of a social need; it might be merely the function of lending prestige to an organization the basic function of which may be far different from that of the act concerned. People cannot live except in groups; keeping the group extant may demand activities which are painful to each individual, but necessary to the life of all; for example, traffic rules. Function can be very deceptive, difficult to discover. One of the reasons for this difficulty can be found in the next component to be discussed.

Rationale or Charter

Every behavior pattern is equipped, implicitly or explicitly, with an explanatory, justifying idea. It is possible for this explanation to jibe perfectly with function; possible, but extremely unlikely. The whole field of law is a good example of charter. The introductory statements of purposes of organizations furnish examples. The correct, ideal, respectable explanations of conduct or of organizations come under the same heading. When the ethnologist asks the native why cross-cousins do or do not marry, he will probably get "charter" for his answer. Mythology, old saws, ethical systems, abstracted norms, all help to form charter. There is no need for this charter material to be logical. It is quite possible that to an outside observer it will be both internally inconsistent and at comparable odds with the "real" function of the activity. It is often contradicted, even in the actor's mind, by his own activity, but, and this is a very significant item, it is an important factor in every activity; that is, it influences the carrying out of the activity.

In considering drinking behavior, function and charter should be sharply differentiated. It requires no great social sensitivity to appreciate the lack of a well-integrated philosophy concerning the drinking of alcoholic beverages in our society. People can easily be embarrassed by questioning which touches on their attitudes, the training of their children, and their own behavior on this score. They will give explanations which, to the well-informed, are patently false; moreover, they will back up their justifications with a good deal of fervor. They will say, for instance, that they drink because it is good for their health, because of the delightful taste, because they want to (this is much more realistic as an answer, but hardly leads to greater understanding), or because they must be hospitable. There is also a good deal of combative reasoning centering around the argument that whether there is a reason for drinking or not, no one can order the speaker not to drink. Emphasizing further the insecurity of any society-wide belief and policy, is the occasionally furtive character of drinking. "Sneaking a quick one" is a phrase characteristic of this sort of behavior.

The presence of anxiety and uncertainty in justification of drinking behavior makes it difficult to discover the real functions of drinking by direct questioning. The opponents of such behavior sometimes deny any real function; the drinkers tend to be aggressive, insincere, or irrelevant. To conclude, however, that there is no real function, socially or individually, would be to fly in the face of history and reason. A practice which has flourished in almost all societies in almost all times cannot be characterized as of no consequence. That some drinking is not socially functional is quite clear, e.g., solitary, morning drinking. That some drinking is not individually functional is equally obvious. By observing the drinking behavior, its concomitants and later consequences, a keener understanding can be obtained, resulting in a more discriminating and more widely oriented perception of why people behave in this fashion. This, in turn, can lead to the more efficient control of drinking which is held to be dysfunctional. Blind attacks on all drinking seem to most persons to be inefficient. No control at all is equally inefficient. Discrimination and selectivity, both for understanding and for control, are imperative. A study of function and of charter will lead in this direction.

The mythology, folklore, and superstitions about the powers and functions of alcohol form part of the charter of drinking behavior. That much of this material can be attacked or ridiculed on scientific grounds is perhaps true; it is nonetheless relevant and significant that such beliefs are powerful reinforcements to behavior, whether that of the prohibitionist or the excessive drinker. More than that, they are an integral part of the tradition handed on to rising generations, and thus operate as instigators of behavior. The experience of a "good" education is no particular bar to the belief in fallacious concepts; highly trained lawyers and doctors may believe in the stimulating, body warming, and medicinal qualities of alcohol. Belief that drinking is a sign of vigorous manhood may be a spur to action outweighing the deterrent effects of statistics about accidents, poverty, or disease which may be invoked against the habit.

There is a wealth of charter material concerning the drinking of alcoholic beverages. It can be found in maxims, in poems, in plays, in advertising, in songs, in books on etiquette, and innumerable other sources. It is a constant source of behavior, of rationalization, and of indoctrination. The part it plays in the total pattern of drinking behavior is of great significance. The theoretical background and announced concepts of organized religion, of law, of social work agencies, of recreation groups, of organizations of producers and distributors of alcoholic beverages, and of temperance and teetotal groups, as well as the attitudes of groups only secondarily concerned, play an important role which must be studied if the motivations behind both drinking and abstinence, and if the scene in which drinking occurs, are to be competently portrayed.

The Way of Activity

This refers to the technique of achieving the function, e.g., the way one lifts one's hat, the way one courts a fiancée, makes a shoe, ties a rope, or addresses a letter. Carefully collected and classified information concerning the ways in which people drink is a sine qua non for the study of drinking behavior. Drinking behavior is not limited to the lordly sipping of liqueurs by the haut monde, rowdy singing and stein waving by college boys, and rolling in the gutter on the part of the "depressed third of a nation." There are all sorts and kinds of drinking, and there are all sorts of behavior consequent to such modes. Both the type of drinking and consequent activities are related to various social strata—occupation, age, sex, nationality, and other categories. To consider all drinking as the same kind of behavior is an affront to intelligence as well as to the senses. But what are useful classifications of this sort of activity? One can go from the extreme of considering each individual act as a unique experience, to the opposite of characterizing all such acts as sins. Neither is a useful concept for scientific purposes. Collection of a large mass of data on drinking usages will allow classification for various relevant purposes. Until such a collection is achieved, however, verification of hypotheses and insurance of adequate scope in the positing of such hypotheses are not possible.

Material Apparatus

In drinking behavior, this would refer to bottles, casks, tables, rails, spigots, glasses, aprons, checks, etc. Material apparatus is an essential part of behavior. It can set limits to action; it can, through association, stimulate action. Material apparatus in the drinking situation can be most influential. The old Chicago saloon, for example, was furnished with chairs, newspapers, card tables, free lunch facilities, and, in a few instances, even with pianos, gymnasia, and handball courts (Koren, 1899). A bar without chairs, without space, without free lunch or other attractions may well speed drinking, increase the amount consumed ("buy drinks or get out"), and thus assist in completely changing the practices and atmosphere of the older saloon. Material apparatus is also an important factor, in our society, in that it can be so expensive and so durable that it is maintained long after the original need has passed, with the result that old, possibly otherwise maladjusted, practices survive with it.

Material apparatus is also significant in drinking behavior in that certain articles come to have a value of their own. Cocktail shakers, old kegs, particular drinking glasses, and other items become significant symbols for particular categories of persons in the society. They may enhance the values of drinking. They may serve as symbols of evil to antidrinking groups. In our society the attempt to better one's social position by material purchase and by conspicuous consumption has been a practice of long standing. Material apparatus relevant to drinking behavior, such as chromium soda containers, home bars, antique decanters, and brandy inhalers, are examples of drinking articles which may have a secondary, perhaps even a primary, importance because of their social class prestige or "snobbery value." Although not as important as function or charter in the field of drinking phenomena, material apparatus is a significant characteristic deserving full attention.

Place and Time

Many behavioral patterns are most commonly, sometimes exclusively, acted out at particular places; they often represent the optimum behavior in some places, are merely acceptable in others, and may be utterly forbidden in still others. The same statements may be made regarding time. It is worth noting that what may be considered abnormal, even criminal or insane behavior in one temporal or spatial setting may be socially acceptable in another. When the behavior considered is of a settled, socially integrated character, no great problems appear because of these distinctions; but when, as in the case of drinking in our society, the behavior is not stable and well-oriented to the rest of culture, problems may be generated. For example, drinking activities generally tolerated on "moral holidays," such as Mardi Gras and New Year's Eve, may be exercised on more and more occasions; behavior once limited to the bar or saloon may be extended to the country club, the professional meeting, and the political convention.

So important is the place and time discrimination in drinking behavior that one type of deviation from the cultural norm is almost sufficient in itself to mark a person as an alcohol addict, namely, the practice of morning drinking. Solitary drinking, also, is sufficiently "out of place" to raise the suspicion of neurotic elements in the drinker's personality makeup. It is to be noted, however, that morning drinking under socially specified conditions, such as a wedding breakfast or a hunt breakfast, may carry no taint of addiction whatsoever.

Functionaries

These are the persons who have specialized functions to perform in carrying out the particular pattern of behavior. In religion, for example, the priests or ministers, choir boys, vestrymen, and musicians are examples of functionaries. Such persons have an interest in the behavioral pattern apart from and greater than that of the ordinary participant. They serve as foci of maintaining interest, sustaining old ways, stabilizing personal relationships within the institution, and carrying out changes in technique or policy. The saloonkeeper, barmaid, liquor salesman, and bottle manufacturer are all functionaries to a greater or lesser degree in the drinking complex. Officers and writers and speakers of temperance and prohibition groups are functionaries of antidrinking or controlled-drinking groups.

The social class, regional, religious, age, sex, educational, nationality background, and other characteristics of the functionaries may be of great importance in the determination of behavior, especially in the matter of emphasis or de-emphasis of certain attitudes and activities. The entrance of a quasicriminal group into the business of distributing liquor, or the assumption of control by a particular political or religious group within a temperance or prohibition movement, form dramatic examples of the influence of functionaries in the whole field of behavior related to the drinking of alcoholic beverages.

The Personnel or Participants

To repeat the example from religion given above, the church members would come under this heading. There can be many classes of participants, based on all sorts of criteria (wealth, education, experience, sex, age, kinship); the different classes may have different ways, privileges, and so forth. The distinction between functionary and personnel may become clouded; the host at home drinking parties, for instance, may be a borderline case.

Membership in a group may be formal, accompanied by ritual and contract, or it may be informal, as in a friendship group. Not to be overlooked are categories of persons barred from membership in one association or another. Participation in associations is an important subject for the student of drinking not only because of the influence brought to bear on drinking behavior by the fact of membership in particular organizations (Poteat, 1943), but also because of the relationship between abnormal drinking and general nonparticipation in groups (Bacon & Roth, 1943).

The Rules

These involve implicit or explicit norms concerning the behavior, covering all the phases mentioned thus far. There are right and wrong places to drink, times to drink, people to drink with, types and amounts of alcoholic beverages to imbibe, subjects to talk about and activities to pursue while drinking, ways of mixing drinks, methods of imbibing (fast, slowly, with straws, inhaling of bouquet, holding glass, mug, or bottle, etc.), clothing to wear at drinking parties, ritual phrases and body movements, and so forth. These rules vary with different groups. In any given group they may vary in time, in response to both individual desires and social changes in the total society.

These norms are not irrelevant minutiae in the field of drinking behavior, subjects of potential interest only to the collector of exotic folklore. For the sociologist interested in the problems of alcohol, these rules carry highly significant implications for determining the incidence and defining the limits of socially accepted drinking behavior. Furthermore, rules may act of themselves as motivations toward behavior. Knowledge of the properties and protocol of drinking is undoubtedly regarded in many strata of our society as a graceful social asset. To know that one should savor, not gulp, or to be able to detect the subtle difference between a Mouquard 1927 and a Montvoisier 1932 is considered by many the mark of a cultured gentleman. The fact that the Joneses were unaware that the wine should not be served ice cold with the soup course is a satisfying experience of no mean quality to the Browns, who have less money, perhaps, but whose ancestors had achieved greater prominence.

The Sanctions

These include the rewards and punishments connected with the activity under consideration. The tacit sanctions are more difficult to detect, and are probably quantitatively and qualitatively more important, than the fairly obvious sanctions of a formal nature. The negative sanctions of overt character have been the subject of a great deal of attention in relation to the drinking habits of specific individuals as well as of large groups. Pleading and scolding, fining and imprisonment, discharge from occupation and ostracism by neighborhoods and cliques have all been tried repeatedly. In some cases they may have had the desired effects. In the great majority of recorded instances they have failed. It is an important question why these techniques, successful in many other situations, have shown the poorest results for the control of drinking, particularly excessive drinking.

The social rewards for drinking behavior require to be studied not only with wide scope and keen discrimination, but also with all the objectivity that the scientist can summon to his aid. To approach this subject with a predetermined scorn or animus, an approach not unknown in this field, could lead only to meager results and to an underestimation of the forces which are at work. That there are great rewards for drinking in our society can hardly be denied; being a genial and lavish host, being a connoisseur of wines or whiskies, being able "to hold your liquor," are obviously rewarding states of affairs. Drinking can lead to the easing of possibly tense relationships between casual acquaintances as well as between parties to business or professional agreements. Drinking is closely associated with many pleasant occasions and situations and, in addition to any direct reward it may hold in itself, is reinforced as a source of pleasure because of this association. As a depressant, alcohol may be individually rewarding to the harassed or anxious person. Drinking is a component of many rituals which may be deeply satisfying to the individuals participating in them. It is not impossible that the individual in this last instance may consciously dislike the taste and afteraffects of the alcohol but still derive great pleasure from the ritual as a whole and be strongly opposed to any changes in the forms and procedures involved.

Equally important in the study of drinking behavior are the rewarding and punishing sanctions for abstinence.

The Sanctioning Agents

These are the roles or social positions through which punishment or reward issues. Occasionally there may be roles specifically concerned with the control of drinking behavior; the tithingman of Boston in the eighteenth century might be listed as an example. It has already been pointed out, however, that drinking is rarely the core of any major pattern of activities, and we may therefore expect to find the chief sanctions issuing from roles which are primarily established in other institutions—the father, the minister or priest, the teacher, the employer, and so forth. This raises a question significant to those desirous of controlling drinking habits: What sanctioning agents will be most efficient in activating the proposed norms? That egregious errors have been committed on this score is common knowledge. Whether or not the observer wishes the norms activated is, of course, a different matter.

The 10 component parts of behavior patterns here described could undoubtedly be elaborated into further parts. On the other hand, even the present listing would command some repetition; for example, rules and charter may overlap considerably, and the same is unquestionably true of sanction and function. Since the present purpose is an application of sociology to a particular problem, rather than an elaboration or modification of sociologic theory, further discussion would appear to be out of place.

The subjects described herein may be regarded as component parts of all social behavior. These parts become activated in different constellations and to varying degrees through the operation of certain sociologically significant aspects of behavior. These latter aspects, too, should be observed and recorded with the collection of all data. They are described under the seven headings that follow.

Stratification. The particular activity which is discussed may be called a folkway, a term developed by Sumner. Both Sumner and Malinowski, whose analyses of culture have been freely utilized in this report, were primarily concerned with preliterate groups. In dealing with large, industrialized, urbanized societies, particularly those which attained their size through accretion from diverse societies, the task of describing and classifying behavior becomes more complex than it was in dealing with relatively small, homogeneous tribes. One of the primary differences between the preliterate and the large, urban society is the highly important phenomenon of subsocieties. The people of New York City and the people of the Idaho region do not form distinct societies; they share many aspects of a common culture, they occasionally intermarry, they are economically interdependent, politically joined, linguistically identical at least for the main language, and have, in many ways, a common tradition. On the other hand, to consider them all as of one society in the way that the Polar Eskimo or the Nama Hottentots or the Samoyed compose discrete societies, would obviously be shortsighted. There are marked differences even in basic values, in daily routine, in occupational characteristics, and so forth. The process of developing subsocieties may be labeled stratification. The criteria for such stratification may be found, chiefly, in divergencies in folkways. Studies of folkways, therefore, whether of drinking or of marriage, must be oriented within particular strata of the society.

Stratification is present in all societies; it was observed by Sumner, by Malinowski, and by other students of preliterate societies. With the enlargement of human groupings from hundreds and thousands to hundreds of thousands and millions, and with the many changes that have occurred in what is generally referred to as the Industrial Revolution, this stratification has expanded and intensified enormously. It has resulted in an increasing unreality of statements of abstracted norms of behavior when applied to whole societies, although it must be admitted that the employment of loose averages in describing social phenomena seems as popular as ever. Any descriptions of the drinking habits of "Americans," for instance, are likely to be not only meaningless but also misleading. This applies, also, to the great majority of behavioral patterns. Statistics on the sale of alcoholic beverages in the United States are an example of sociologically meaningless or misleading descriptions of behavior. This does not mean that such statistics may not be useful for other purposes. For instance, the overall figures for the United States show that in 1940 the per capita consumption of alcohol was distributed among the various beverages as follows: beer, 0.54; wine, 0.12; distilled spirits, 0.49 gallons. But the corresponding figures for the District of Columbia are: beer, 0.76; wine, 0.14; distilled spirits, 1.92. Although these are all higher than the national average, one figure is only 16 percent higher, another is 40 percent higher, and the third is almost 400 percent higher. The corresponding figures for Iowa are: beer, 0.47; wine, 0.01; distilled spirits, 0.35. These are all lower than the national average, but again in entirely different proportions; the beer drinking is quite similar to the national average, and the drinking of distilled spirits is about 70 percent of the national average, but wine drinking is only 8 percent of the national average. There is, moreover, no reason to believe that such figures refer to the whole population of either Iowa or the District of Columbia, and no reason to assume that the same categories of population are concerned in the two regions. Patterns of behavior in the United States show decided geographic variation. That this is as true of drinking, as it is of murder, the use of beauty shops, the playing of contract bridge, religious affiliation, or types of jobs, is apparent from the analyses of Jellinek (1942) and Jolliffe and Jellinek (1942).

Stratification has been described thus far as a regional matter only, not because this is the most important distinction—it definitely is not—but because it is an obvious one. Stratification in behavior is almost always obvious between the sexes and between different age groups. Differences in occupation, financial status, race, nationality, religion, social class, education, and rural or urban residence may also be significant for the particular folkway under consideration. Significance can only be determined by observation; if the two or more groups do the act and think about the act in the same way, then, for that act, the otherwise significant stratification is not significant. Some minor folkways may be closely similar throughout many strata of the population, such as, perhaps, the use of bed pillows, cigarette smoking, or telling time.

Social stratification, then, is a phenomenon that must be taken into account in any study of human behavior. Generalizations about psychotic drinkers, wealthy drinkers, Irish drinkers, or student drinkers cannot be considered applicable to all drinkers. Nor can the drinking behavior of any one group or category be held as a norm for our complex, highly diversified society.

The distinction between folkways and mores. It has been pointed out previously that drinking behavior does not form the core of any basic human activity. It is therefore not expected to discover anything that could be regarded specifically as drinking mores in the correct usage of the term. An exception to this might well appear in a mos against drinking in certain subsocieties of the country. Folkways may be defined as the usual, expected, and accepted ways of behavior in any society. They are handed on from one generation to the next, are considered wise, "natural," lucky, human, God-given, and so forth. They generally developed over a long period of time through a process of trial and error and unconscious adaptation. Through contact and conflict with each other, they tend to become consistent, mutually compatible. They may be described in terms of the component parts previously listed.

The mores are a particular type of folkway. They are emotionally highly significant. Societal welfare is believed to be dependent upon their existence. Punishment for infraction is likely to be prompt and severe. Mores are difficult to define with any great exactitude; they are perhaps most aptly exemplified by certain powerful "don'ts" in any society. They may be said to be the accumulated learning of past generations, derived from the most repetitious and seriously dangerous variations in human behavior and from the most successful adjustments to these dangerous situations. Murder, unrestricted sexual activity, economic parasitism, failure to protect the younger generation, unrestrained aggression in the in-group—these are examples of areas in behavior which had to be firmly controlled if a society was to exist. Positive activities and negative controls in these areas form the body of the mores. Mores are so strongly indoctrinated in the young, penalties for violation are so great, and failure to adhere to the mores is so realistically painful, that any change in the mores requires much time. Mores develop automatically, just as folkways do. Like folkways, they are societal, not universal phenomena. Although all societies had to adjust to the problems mentioned above, they did not all reach the same adjustments.

Two further characteristics of the mores are of especial significance for the study of drinking behavior: (1) mores are institutionally pervasive; and (2) all members of the society act as sanctioning agents of the mores. To illustrate the first of these, folkways may often be applicable only in specific situations; there are certain ways (also time and place, sanction, material apparatus, etc.) common to the factory, to the home, or to the dance hall, common to eating, card playing, or worshipping, which are limited to those situations. Mores, however, are applicable in all situations. When particular folkways come into conflict with mores, the folkways will change. Drinking folkways as such do not run counter to the mores. But behavior consequent to drinking, especially heavy drinking, is very likely to run directly counter to the mores. The proponents of prohibition have often attempted to persuade the public that the drinking and the consequent behavior are the same. The public, or large portions of it, has refused to believe this, has refused to react to drinking on a mos level, and continues to regard it as a folkway.

The fact that mores are sanctioned by all members of the society whereas folkways are sanctioned by those holding particular roles may be illustrated by a few homely examples. There are good ways and bad ways of driving a car, referring to smooth changing of gears, efficient use of the "choke," steady pace, and so on. Young boys or servants or lower-class persons given a ride are not expected to criticize the driving of the ordinary adult. If they do attempt to punish the "poor" driver, they are very likely to be punished themselves. The behavior is in the realm of folkways. The driver's social equal may criticize, but he is likely to refrain. The driver's social superior may criticize freely; the boy, driving his father in the latter's car, had best mind his "p's and q's." If the ranking of the two persons involved is uncertain or competitive, as in the case, let us say, of a man and his mother-in-law, there may be a refusal to accept the criticism and even aggressive counterattack. When mores are involved, however, the situation is different. Although problems arise immediately when a person of lower status punishes or attempts to punish his superior for violation of the mores, nevertheless this action will generally be supported by others. A son can attack his father for abusing his mother or his brothers and sisters. A servant can punish his employer, directly or indirectly, for criminal or perverse activity. Even the stranger can bring pressure to bear on the immoral member of the community.

In bringing punitive or restrictive sanctions to bear on drinking behavior, this discrimination has often been overlooked. Lower-class groups (so considered by the objects of their activities) have attempted to punish and restrain upper-class groups. Persons of low status have been made functionaries to enforce new restrictions on groups of high-status persons. Much of this activity has been undertaken because the proponents considered all drinking phenomena on the same level as perversion or anarchy. This attitude has frequently brought about results directly contrary to their purposes.

Although fine discrimination between folkways and mores is eventually a subjective matter, the differentiation for practical purposes is not difficult. Nor is the distinction of fads from folkways too difficult. Just as mores are very powerful folkways, so fads are very weak folkways. Fads do not center around basic social activities, are not heavily sanctioned by the major agencies, and show great variation. Types of alcoholic drinks can rise and fall much as do styles in women's hats. Certain forms of ritual drinking, such as toasts and bachelor dinners; some recreational drinking, such as Saturday-night parties and festival occasions; and perhaps some theoretically medicinal drinking, are more on the order of folkways, subject, of course, to the stratification limitations previously discussed. The only emotionally powerful, all-pervasive, heavily sanctioned, and conceptually socially beneficent custom concerned with drinking in America would be the taboo on any and all drinking, this being limited to certain subgroups of our population, and the more general restriction on drinking for those in certain age groups.

In the study of drinking behavior, as in the study of any behavior, the distinction between mos and folkway must be kept in mind. The specificity and variability of drinking behavior per se must be sharply distinguished from the effects of drinking on other behavioral patterns which may be diffuse and permanent in nature. Too emotional feelings about drinking behavior should not be expected to be on the same level as feelings about immoral or antimoral behavior which might occur after heavy drinking. The sanctions brought to bear with success upon actions and actors violating mores cannot be expected to succeed or even to be efficiently activated in the attempt to control drinking behavior.

Discussion of this point emphasizes the need for the study of drinking as a part of human behavior, in contrast to the study of the behavior of inebriates alone. These discriminations seem less significant in the consideration of alcohol addicts. To understand the problems of alcohol, however, they are essential.

Flexibility and rigidity of folkways. In establishing the presence of particular folkways, there is often a laudable intention on the part of the observer to state his observations with a high degree of exactitude. There is, perhaps, a certain amount of emulation or competition between the social scientist and the physiologic scientist, especially in the mind of the former, which may lead to this oversimplification and possible overexactitude of description of folkways. The nature of the phenomenon does not lend itself to sharp distinctions and black-and-white definitions. Folkways may be imperative, highly desirable, desirable, permitted, permitted on occasion, or, if not in one of these categories, they should be described on some such sliding scale. This characteristic should be fully considered in the discussion of rules, charter, sanctions, and sanctioning agents. (Also, the relative importance of the particular act should be noted; it may occur within five seconds and not be repeated for a week; it may be merely an adjunct to much more important activities, in the mind of either the actor or the observer, or both.) This subject is especially emphasized because observers of culture have too often reported that some specific form of behavior was characteristic of a particular group, without considering the emphasis laid upon the way (behavior) in action, explanation, offense, etc. There is a vast difference as to quantitative incidence in contemporary American society between driving vehicles on the right side of the road and getting one's shoes shined. In most situations there is probably no great significance to either habit. Taking one drink in a lifetime, however, may have tremendous significance for the specified person.

Relations with other institutions. As a pattern of behavior the drinking of alcoholic beverages would be a very complex subject in this aspect since it is an accompaniment to so many groups, activities, and situations which, in themselves, have nothing to do with drinking in a direct sense; for example, marriages, funerals, christenings (all the rites du passage), recreational gatherings of almost all sorts, and rituals and assemblies connected with a multitude of diverse interests.

A special subject under this heading, although complex enough to demand separate treatment, will be merely mentioned; that is, the incompatibility of the behavior under consideration with other institutionalized modes of behavior.

Social change. Folkways are dissimilar to physical traits in that they are subject to rapid change (even mores, or such basic structures as the family, may change in less than a century). Students of culture who have spent most of their time with the reports of ethnologists may unconsciously consider customs as rather dry, static phenomena; one reads that the Cheyenne had such-and-such a myth, finds references to that fact in other dissertations, and so forth. It must not be forgotten that, actually, a Mr. Blank reported that a native told him of such a myth on a certain date. Ten years earlier, or later, the myth might have been used in different contexts, emphasizing different characters, values, problems, etc.; it might be considered only a quaint rarity 40 years later, or it might have become a dominant means of passing on an important message to youth. Students of culture who have spent most of their time considering the modern scene may go to the opposite extreme and think that everything is in change all the time. Both views are unbalanced. Change occurs at different tempos, under different conditions, and with varying results, depending on many different variables. It is of great importance for the student of behavior to relate the folkway in which he is interested to its own history and to historic phenomena in general.

Social change is a difficult matter to describe, since the change may occur in some of the aspects of a folkway, but not in others. Thus, the function may change, but the rationale may continue; the ways may change, but the rules may not; and so on. Further, the given folkway may not change, but the related institutions may change. One is more likely to observe rate of change than absolute change or no change at all. The fact of stratification must also be considered here; the change may take place among men, but not among women; among younger persons, but not among older; among upper social class persons, but not in lower classes; in New England, but not in Texas; among professionals and business executives, but not among small tradespeople. Furthermore, several changes may be taking place at once, and some of these may be incompatible. Taking our lead from Mumford (1938) we may say that the following aspects of change in a folkway may be noted: the archaic, the recessive, the dominant, the insurgent, and the experimental. In any pattern of related behavior traits, we are likely to find all these aspects if our base of observation is sufficiently wide. Any particular habit may very possibly show a history in which it starts with the last-named and is, today, in the archaic. The observer of drinking habits must be very careful to see whether the particular aspect he is describing is in one of these categories; this may well depend on the particular actor or explainer of acts who is the source of information.

The teaching and learning of the folkway. That attitudes, roles, behavioral patterns, and, in fact, all social characteristics are acquired by the individual through social means, not by physical inheritance, is a principle accepted today by practically every student of the social sciences. This in no way denies or minimizes the fact that a physiologic and psychologic being is both the target and the instrument of such learning.

The form and function of the characteristics learned, the scene in which learning takes place, the sequence of experiences, the statuses of those teaching the ways and of those learning, the relationship of the various ways to each other and of the various teaching agents, the rewards and punishments employed, the relation of early learning to more mature experience, all of these aspects of the teaching and learning of the folkways have formed an area of special interest to the sociologist. The general label of "socialization process" has been used to describe this field.

Much of the teaching and learning concerned with ways of drinking, and the charter which accompanies such ways, proceeds as normally and regularly as does that concerned with other ways of life. This is more readily perceived in nations other than America, perhaps, but even in some of our more recently arrived groups of foreign origin it is obvious that drinking customs and beliefs are handed on to the rising generation without perturbation, without anxious and uncertain application of sanctions, without threat and denunciation by agents of an outraged God. To the recently arrived visitor from southern Europe, for example, the emotional tension and organizational activities surrounding drinking practices are somewhat bewildering.

Among some American groups there would appear to be fairly stable social adjustments concerning the teaching and learning of habits concerned with drinking alcoholic beverages. It is equally true, however, that among many groups no such stability exists. The charter material is often conflicting and disorganized; the learning often starts only in late adolescence, a period in which the usually powerful social roles of father, mother, teacher, minister, and adult neighbor are at low ebb; the learning may often occur in places and at times not controlled by the ordinary agents of sanction; introduction to such habits may be attained under the guidance of persons of lower social status than the neophyte; early participation and activity in drinking behavior may well be colored by an aggressive, even antisocial, atmosphere; various sanctioning agents may be in conflict as to the correct norms involved; a wide choice of activity and interpretation may be left to the young individual concerned.

Here, then, is an obvious area of social maladjustment. It is to be carefully noted, however, that there is nothing inherently maladjusted, that this is no universal, predestined and unavoidable evil. A fairly uniform, stable, non-anxiety-provoking adjustment can be made and can be handed on to the rising generation without the shock, the aggression, the undisciplined participation, and the individual disorientation which sometimes, perhaps often, occurs in our society. To the sociologist this disorganized scene presents a challenge. Carefully orienting the pathologic within the larger sphere, he analyzes, compares, synthesizes, and sets up postulates to explain the behavior involved: what are the significant factors connected with this malfunctioning in the socialization process; with what conditions are they correlated; to what effects do they lead; what may be learned about socialization in general from the particular case in hand.

Whether normal or abnormal, organized or disorganized, the teaching and learning of drinking behavior is an essential field of research for the student of the problems of alcohol. If our understanding of the sociologic "whys" of drinking is to be enhanced, the process by which individuals in the various strata of our society take on drinking customs must be thoroughly portrayed.

Individual participation and interpretation. The fact that a given individual may adhere to the folkways but may do so with great effort, with loathing, with private interpretations, or with other modifying attitudes, has been discussed at some length by several social psychologists. It is probable that the apparently diffuse state of charter, sanctions, and ways characteristic of drinking behavior in our society allows more variety of private interpretation than would otherwise occur. Bunzel (1940) has noted that even in a society where there is practically unanimous approval of continued and excessive drinking, there is secret objection and only simulated adherence to the custom on the part of some of the young.

Orientation for the Study of Drinking Behavior

General Remarks

Granted the validity of a behavioral approach to the problems of alcohol, and granted an acceptable discipline for the observation and analysis of behavior, there still remain the questions, what behavior should be studied, where, in how great detail, in what manner, and so forth. For it is impossible to study all drinking activity by all methods of study.

One could chose a specific component of behavior—for example, charter; and a specific group or classification of persons—let us say, Negroes, or women; and a particular source of data, as life histories, or novels, or court records. One could study any of these, and it would probably be worthwhile on a descriptive level; there is very little information on the subject as such. One could study drinking societies or antidrinking societies. One could study drinking customs connected with marriage, conventions, college boys, death, or prostitution. Another approach would be to study drinking and its relation to various psychological tensions. A whole dictionary of drinking terms, drinking slang, or even of the usual linguistic errors made by intoxicated persons is not impossible.

The criticism of such approaches at the present time, as previously indicated, is that they are too specialized. At the very outset it must be realized that no general picture of drinking habits is available. Until such a description is available, study of one classification of drinking habits is not advisable. Intense study of specific types would better follow or accompany a general survey, not precede it.

Such approaches are also too segmented from the sociologic viewpoint. By this it is inferred that any such study would involve the lifting of particular behavioral patterns out of the cultural and social context. Present always is the question, "How far can one go in leaving out related phenomena in order to highlight the particular object being described?" The keyword in this question is "related." Relationships apart from drinking habits can be covered fairly well by reference to standard studies (of family life to job, of religion to education, of recreation to standard of living). Unfortunately, relationship of one aspect of drinking behavior to another aspect of drinking behavior, or of drinking behavior to other, nondrinking patterns of behavior, cannot be covered by reference; they have not yet been studied. It is not impossible, then, that a particular act might be considered of great importance in Negro drinking, or in the control of women's drinking, when broader study would show that the behavior involved is common to the whole field. Perhaps the converse is an even more probable mistake: some behavior considered as ordinary drinking might actually be limited to the particular aspect or group under study.

Another weakness of such specialized fields or types of study is that, in all likelihood, they would be difficult or impossible to standardize. Unless specific studies can be correlated with or complementary to other knowledge, their utility is limited to stimulation of interest or to description of the matter in hand. This weakness is perhaps illustrated by the current state of studies of city life. There are many different approaches; some are undoubtedly excellent researches, but they lack a common denominator. The historical approach of Mumford, the artistic approach of Farrell, the financial-political-administration approach of surveys by taxpayers' associations, crime surveys, studies of minority groups, official directories, health studies, novels about social problems—all these, and others as well, have made contributions; but they are almost impossible to combine, they cannot be used to evaluate a sociologic hypothesis. As a result, verification of theories must depend on one study or type of study, usually the study from which the hypothesis evolved, or there will be no verification at all. The only other possibility is a duplication of the field work of another reporter, couched in terms relevant to the hypothesis. Since community surveys are expensive in men, time, and money, this can rarely be done.

Standardization is not to be confused with censorship and direction, forcing all studies into a predetermined form and automatically eliminating some forms of research. It infers no such negative action. It implies that, whenever research in the field is to take place, other workers are interested in certain types of problems, and it would be helpful to them if certain information, basic to them and probably to many other workers, were collected. Standardization may call for extra work; it does not deny any work.

Nor would such requests be made only of those interested in the field of drinking behavior. The sort of information required is relevant to any general study of behavior. That it has been omitted, or only mentioned casually, or observed merely through impression, may be due in part to fear of treading on the toes of informants, but it is partly due to oversight. Some of our famous community studies imply that the drinking of alcoholic beverages is unknown in the American town. A recent study of the recreational folkways of students at a large American college, for example, did not mention at all the drinking of alcoholic beverages. Whether this be ethically determined "social science" or carelessness, is beside the point. Standardization will enable verification of results, will make specific studies far more significant, will open up new channels of investigation.

In order to attain some of the values made possible by standardization, it is held that a general orientation study is preferable to a series of researches, special in method and special in type of data observed. This in no way forestalls or denies the value of specialized studies. It is hoped that they will increase in number and in value because of the wider, more general approach.

Study of Drinking Behavior in the Community

The considerations mentioned above point to a general study as more advisable at the outset. This does not mean a study of the drinking habits of all the people in the world or of all the people in America. It means that a fair sample of all the relevant facts about drinking behavior in a relatively discrete social group should be covered. Such a group would be one which is geographically unique, which is identified as a distinct social entity by both members and nonmembers, and which presents in microcosm most of the major aspects of the larger society. This would mean that the group would include families bringing up their children, workers in a variety of industrial, commercial, and professional occupations, and organized religious activities with representation by more than one sect; it would include governmental activities, and would be characterized by some of the usual stratification (economic, social, nationality background, etc.) to be found in the whole society. In other words, it would be a community. A purely industrial, purely residential or purely recreational community would mean a specialized study undertaken for specified theoretic purposes. The "typical community" is, of course, a figment of the imagination; this does not deny, however, that there are communities which are more or less atypical. It is quite possible that one of the more typical communities would have practically no Negroes, or would have an age distribution that was askew in some category. This distinction should be described, and its possible effects on drinking behavior or problems must be considered; but its existence would not negate the choice of a community for this sort of study. Recognition of the danger of overgeneralization is, in itself, a potent safeguard.

Aside from avoiding the atypical, certain characteristics should be avoided or recommended in selecting a community for the survey of drinking behavior:

1. It should be a community characteristic of a sociologically determined region.
2. It should fall into one of the categories urban, rurban, or rural.
3. It should be a community subjected to an acceptable sociologic survey at a relatively recent date.
4. It should be a community which has not been subjected to any violent changes in the immediate past.
5. It should be a community which has been liberally documented (newspapers, historical societies, governmental records, surveys, etc.).
6. It should be a community of sufficient age to have attained an independent tradition.
7. It should be a community large enough to illustrate fully various types of the activity to be studied; e.g., it should have adequate representation of such obvious focal points of drinking activity as hotels, night clubs, taverns, private clubs, and typical events such as wakes, weddings, conventions.[4]
8. It should be a community large enough to digest such a survey without concurrent imbalance of the activity surveyed.

9. It should be a community convenient for those entrusted with the study.

Communities meeting these requirements are in no way difficult to find, although the recommendation that there should have been an acceptable social survey of recent date might appear a stumbling block. However, this is not the case. In the Middle Atlantic and Southern New England regions, for example, studies have been made recently of New York, Newburyport, the several New England communities covered by C.C. Zimmermann, the Connecticut towns described by N.L. Whetten and his collaborators, and also New Haven, Atlantic City, and Lowell, to mention a few that come readily to mind. Reference would uncover many more community studies in the same area. Moreover, special studies, such as those on income distribution and consumer habits, have been made for many cities. This situation can be duplicated in many other regions.

The Historical Data

Following the selection of a particular community, a general statement as to its present makeup and its historical development is essential. Without the perspective and orientation to be derived from such an introduction, unfortunate blunders of emphasis, paradox, and trend can occur.

Historical sources containing material peculiarly relevant to drinking behavior should be sought industriously and with keenly critical appreciation. Not only is such material scattered and rare, but that which can be discovered is likely to be of a highly contentious or perhaps unconsciously biased nature. The legal sources will be more abundant than others, but their imbalance on the side of the abnormal, the tabooed, and the criminal, as well as the rationalizing nature and political fluctuation of law in general, render them subject to marked limitations. Reports of temperance, religious, and other societies with decided views on the ethics of drinking are a useful source, but require always to be labeled as such. Surveys of vice, crime, drinking, poverty, and the like are not uncommon in any sizable community. Special studies of immigrant groups represented in the community should be carefully examined for relevant descriptions and analyses. Industrial and government reports on production and distribution can furnish suggestive information. In addition to legislative and judicial reports, taxing, licensing, and inspecting agencies often have maintained records over long periods of years. Reports of administrative officers such as the mayor, comptroller, etc., are usually on file. Political campaign speeches are often available. Sermons and tracts of the past are often well preserved, sometimes even better than official records. Travelers' comments are often surprisingly detailed and relevant. Novels written by members of the community, collections of letters, wills, and similar data may be maintained by local libraries and historical societies. Books privately printed by local antiquarians may be difficult to discover, but can repay the search. Diaries, parish records, institutional records, doctors' records, etc., may also cast light on the drinking behavior of earlier days. Newspapers are of especial value because of their continuity; the advertisements, descriptions of social events, cooking hints, and general reporting can all be of great relevance. Novels of the time, whether or not concerned with the community, can be of real value in the presentation of insights and in unconscious reflection of the values, groups, and material culture of the time; that such works generally are concerned with a particular social class is no bar to their validity; that their comment on particular cultural phenomena is accidental or casual is, on the whole, an asset rather than a liability. Many other sources of an historical nature would undoubtedly be discovered: materia medica, pictures, warehouse receipts, recipe books, manuals on manners, social histories, material on drinking from fairy tales and anecdotes of heroes, reports of military men, ships' logs, all come to mind.

The field to cover is vast. Yet this is not the major purpose or field of the proposed study. The historical study, therefore, despite its necessity, despite its undoubted attractions of antiquarianism, esoterica, humor, charm, and incidental drama must be oriented within its limited place, must be so disciplined as to allow rapid, concentrated reference on specific relevancies. This means that a predetermined classification system is required. Furthermore, contrary perhaps to ideal patterns of inductive inquiry, a certain rigidity of classification will be necesary. There is not the time to mull over the mass of historical sources from which data on drinking behavior may be obtained. Nor would such data alone be adequate for a complete picture of the drinking folkways of any particular period.

No specific classification can be presented here. From the preceding description of the component parts of behavior and of the aspects from which they must be viewed, a general understanding of the main headings and of the chief needs for cross-reference can be achieved. A decimal filing system such as that of the Index to Cultural Materials, or a punch card system with bibliographical emphasis, are both efficient and tested techniques. The development of the system most useful for the present purpose will depend on further study and on experience with the material to be classified.

The general purposes, however, can be outlined. A body of data is necessary. Every time that particular information is desired it should not be necessary to cover the entire literature to obtain the relevant facts. Not only does this take an inordinate amount of time in itself but it demands continual reinvestigation of the same sources. Adequate classification of a large body of data allows comparison and verification which would otherwise be almost impossible. Adequate classification means flexible classification, allowing abstraction of data of diverse nature, and subsequent merging of two or more types without destruction of the system. The system should also be flexible in that it should allow for future elaboration or addition of new major headings. Uniformity in system, accessibility to any qualified person, and easy manipulation of filing forms are all requisite.

Such an historical-cultural collection would be made for the selected community and, later on, for all other communities chosen for study, whether as representative of different regions or of different types of communities. Such a collection would not only be unique in its contribution to the greater understanding of the problem of alcohol, it would also form a major contribution to the development of historical and social analysis.

Data collected in contemporary surveys would also be incorporated in the file. This does not require a restricted type of investigation. As long as the descriptive classification is broad and flexible, the research need not be confined. It is quite possible that certain types of study would not fit, as types, into such a scheme at all—the life history, for example. This in no way precludes use of the life-history technique.

Such historical studies do not have to be made all at one moment. They need never be considered as finished; evidence concerning drinking in historic Boston or New Orleans or Miami can be added to the files at any time, or a new community can be described. It may be that a regional file or even a national file would be useful. Naturally, there is no intention of avoiding non-American communities.

The Techniques of Observation of Behavior

Introductory

One technique of study has already been indicated, that of documentary study. It has also been pointed out that available data are casually scattered in diverse sources, are often artistic (not bound by the discipline of scientific observation), are heavily weighted by material on the abnormal, are all too likely to be contentious and heavily (often admittedly) biased, or, in all too many of the categories suggested by the analysis of behavior, are almost nonexistent. This means that the major work will at first consist of the collection of data. The scope of such a collection has been indicated in the two previous sections on analysis of behavior and the historical background.

The most common techniques include, in addition to historical documentation, studies of current documentation, field studies, participant observation, organizational surveys, informant interviews, case studies, life histories, and questionnaires. The use of most of these will be advised.

Current Documentary Sources

One of the criteria for the selection of a community for the purpose of surveying drinking behavior was that it should have been recently subjected to a satisfactory sociologic survey. The inclusion of socially relevant data and the orientation of drinking behavior within larger social and behavioral patterns are essential. There is also documentary evidence concerning the current state of affairs in regard to alcoholic beverages in governmental, journalistic, commercial, religious, and other sources.

The Field Study

This refers to the direct observation and reporting of the behavior involved. Only certain aspects of drinking activity are open to this approach. As a technique it has been used for the saloon fairly often (e.g., Committee of Fifty, Springfield Survey). The observer may report on the location, architecture and furnishings, functionaries, personnel, amount and types of liquor consumed, prices, general character of behavior, etc. Personal contact with those observed is not essential. A schedule of the data to be observed is imperative. The same technique can be used for some types of ritual drinking, for conventions of temperance societies, for festival and orgiastic drinking, for drinking in commercial establishments.

This technique cannot apply to either family drinking or illicit drinking. This technique cannot directly study the function of drinking in the case of individuals. It is limited to only the most obvious aspects of sanctions. The same is true of the relationship of drinking behavior to other institutions and habits.

Field surveying is valuable as a first step in the observation of behavior. For one thing, it is possible to make mistakes of omission or commission without permanently damaging the source of data. A good informant cannot be pestered by "retakes" due to the investigator's errors; the same is true of case studies or organizational studies. A saloon or a railway dining car or a convention, however, is open to repeated studies with a minimal chance of offending the group surveyed. For a second reason, information, problems, aspects, hypotheses, and standardization can evolve out of the field survey which may determine or influence the course of further studies. It must be remembered that the study of drinking behavior is practically an innovation; all the information desired cannot be detailed or even outlined before observation occurs. A third reason for commencing a study of drinking behavior with the field technique is that it has been used, although rather superficially, in the past; there is thus some guide to action. The saloon, or the tavern or bar, would appear the most valuable object of such a survey, since it is less specialized, less spasmodic, and more widely distributed than any of the other phenomena open to this form of study.

Participant Observation

Both the field survey and the participant observation technique may be labeled as "uncontrolled observation." A certain amount of impressionism, unconscious bias and emphasis, and even unconscious selectivity of data to be observed are almost sure to accompany these methods. In view of the present developmental status of social science, this is hardly surprising. The checking and rechecking of observations, the formulating of more and more precise schedules, and the definition of relevant units of behavior will all help to overcome these liabilities.

If these faults are likely to appear in the field survey technique, they are almost certain to be present in participant observation. Nevertheless, the assets of this means of observation far outweigh the liabilities. The following remarks on the subject are quoted from Young (1939):

> The participant observer, generally using non-controlled observation, is a scientific student who lives in and shares in the life of the group which he is studying. It will be recalled, for example, that Charles Booth had used non-participant observation in his extensive studies of the life and labor of the people of London. Nels Anderson, author of the famous study of *The Hobo,* . . . was an intimate participant observer of the life of the hobo on the road, in the "jungle," in lodging houses, at Hobohemia, at work, at Hobo College in Chicago, and under other circumstances. He closely identified himself with the life of the hobo for an extended period of time and gained a great deal of insight into the inner life and processes of hobo life which would have been almost impossible had he not been able to eliminate social and mental distance through intimate participation.
>
> Robert S. and Helen M. Lynd (*Middletown,* p. 506) write that members of their staff—during the study of Middletown—lived in apartments or rooms in private households:
>
> "In every possible way they shared the life of the city, making friends and assuming local ties and obligations as would any other residents of Middletown. In this way a large measure of spontaneity was obtained and the 'bug on a pin' aspect reduced. Staff members might dine one night with the head of a large manufacturing plant and on the next with a labor leader or a day laborer. Week in and week out they attended churches, school assemblies and classes, court sessions, political rallies, labor meetings, civic club luncheons, missionary meetings, lectures, annual dinners, card parties, and so on. At the end of the

> period they had access to a kind of information which would have been entirely inaccessible at the outset." . . .
>
> Participant observation enables one to penetrate behind the thinking, feeling, and acting of the group. It facilitates the "sensing" and prepares the learning of the social atmosphere, the total social setting, the inter-relations between the single members and the whole group. Moreover, participant observation has its psychological values: it tends to accustom the group to the observer until it accepts him wholeheartedly and incorporates him more or less as a member. He thus gains the rapproachement which is almost indispensable for more intimate case studies and interviews later on.

The three illustrations presented by Young are all pertinent to the proposed study. Charles Booth (1902) actually considered some aspects of the drinking behavior of Londoners, but largely in an impersonal way; the reader is given a picture of what went on in a pub, but there is not much understanding of the *how, why, who,* and *what of it.* There is no warmth of realistic appreciation of living people. Quite unconsciously the sympathetic reader tends to cover the bare skeleton of description presented by Booth with blood, skin, and personality, all derived from his own imagination.

Anderson (1923) showed that participant observation on the level of a disorganized group was feasible and useful. The Lynds (1929/1937) did the same with a total community. In a study of drinking behavior, both types are relevant; disorganized drinkers and "normal" drinkers should both be studied by this technique. Participating with the disorganized drinker should be considerably easier than with the hobo or almost any other pathologic group; at least, this would be true at times and in places of drinking. In the opening study of the drinking place this technique would be immediately utilized, at the very least for experimental purposes.

To what extent such an approach would be efficient, practical, or more efficient than other means of studying drinking habits in situations other than the tavern or bar, is uncertain. It might be found that use of informants is more practicable. The Lynds were studying total behavior, which made such a technique advisable; since drinking behavior is subsidiary in the lives of most people, it might hardly repay such extensive study.

Survey of Organizations

This refers to the study of certain organizations which are closely related to the phenomena of drinking behavior. There are, for example, the temperance groups. There are therapeutic agencies, such as Alcoholics

Anonymous, clinics, and mental hospitals. There are the flophouses, jails, and police departments. There are even scientific organizations concerned primarily with alcohol. There are also groups which have a subsidiary interest in drinking and the consequent problems, particularly courts, churches, and members of the medical and allied professions.

There can be studies of the institutions which produce and distribute alcoholic beverages. Such studies would differ from the previous two in that formal organizations would be the object of observation; their participation, or at least their willing coöperation, is required. Furthermore, study would be directed not at drinking behavior itself, but at the organization and behavior of those groups as they are related to the drinking habits of others. Some of these organizations may have been studied previously, so that no new surveying would be required. Orientation of such studies into the world of drinking behavior would probably be necessary.

The Informant

Probably as important as any other technique in gaining information about so-called "normal" drinking habits will be that of selecting and questioning informants. As pointed out above, the spasmodic character of "normal" drinking may render the participant-observation technique impractical. One might have to fulfill the role for two years in a single family to observe perhaps two dozen incidents of drinking. In order to acquire a picture of the drinking habits of middle-class, second-generation Italians, for example, one could obtain far more material in far less time from various informants drawn from that category. It is not to be denied that as the evidence comes from a second-hand rather than a first-hand observation, it decreases in value. This can be partially counteracted, however, by use of more than one informant, by careful selection of informants, by careful preparation for the interview.

Informants need not be limited to the subject of drinking habits of people of their own group. There are many persons in the community who may well be mines of information on drinking by others. Social workers, policemen, priests, precinct politicians, firemen, old residents, school teachers, and a host of individuals of all sorts may be sources of valuable data. This will be true for the bar and tavern as well as for the family and private club.

Careful selection of those to be interviewed has been proposed. It should be noted that the sociologist has an advantage over the ethnologist or the opinion-poller in that his subjects are not practically forced on him by linguistic, financial, or other extraneous factors as may be the case with the former, and that they are not picked "at random" as with the latter. Careful preparation of the interview means the full comprehension of the questions to which answers are desired, and sympathetic appreciation of the person interviewed. It has been pointed out already that interviewing should not be undertaken until schedules of problems and desired data are in a fairly advanced stage; interviewees should not be wasted or bothered by inadequate or inept questioning.

Social Case Work Studies

This approach is represented by those brief, often stylized, reports which describe the problem aspects of a person in a social problem category; such analyses are made to assist in the melioration of the particular condition, be it delinquency, unemployment, nonsupport, or any other. Due to their administrative and meliorative purpose, reports of this type are not envisaged as a very helpful technique in achieving the broader, more widely oriented purposes of the sort of study here proposed. Although the study of excessive drinkers is relevant, more intensive and more extensive analysis will be required than is usually presented in the social case work study.

The Life History

The life history is probably the scientific counterpart of the social case record. It has been used generally in describing abnormal persons or persons representative of disorganized groups. As far as the drinking behavior of a "normal" person is concerned, a life history would seem a long, expensive, highly elaborate road around Robin Hood's barn. Drinking behavior takes up so minor a part of most people's lives that such broad and such intense study of the total life pattern and total life-pattern development is disproportionate to the goals proposed. This does not imply that the life history of the "normal" drinker is irrelevant. On the contrary, it is essential to the understanding of drinking behavior. This does not mean, however, that the proposed survey should utilize the technique. Chemistry might be an undiscovered field, essential to the understanding of alcohol problems, but it would not follow that its study should be developed by the present undertaking. The study of life histories of "normal" people living in "normal" groups can be enthusiastically approved by students of the sociologic and psychologic aspects of the alcohol problem, but it is hardly their responsibility or privilege to undertake this task. They can, however, comb the existing literature for any insights or data that are available, can request students interested in the subject to document those aspects related to drinking, and can develop postulates which such histories might verify or disprove.

The life history of the inebriate is a different matter. Here the phenomenon of drinking has become almost a dominant factor in the person's life whether measured by time, money, associations, or effect on nondrinking aspects of his existence. True, this is the study of the abnormal which, without orientation, can be both mistaken and misleading. This danger, however, does not bar such studies; it does call for great care in generalization and for strong statements concerning the limitations of interpretation. Fortunately, the misuse of such material has been so flagrant that the spur of harsh criticism is never too remote.

Scientific approaches to the life history have been made through the individual's own story, through interviewing the subject and concurrent interview of many persons and groups with whom the subject has had significant contact, and through psychoanalysis. All these techniques have obvious deficiencies as well as advantages. All demand considerable expenditure of time, effort, and money. However, the possibility of attaining two or three such reports on persons who have long resided in the community selected for the study should be earnestly considered. Perhaps all three methods could be used.

The model of all such approaches was presented by Thomas and Znaniecki (1927). In that full-length study of the Polish peasant in Europe and America, following a description of the social history of the Poles from 1850 to 1910, a description of the situation of Poles in American cities after the migration, and a sociologic analysis of the major factors involved, there appears a full life history of one Pole who lived through the whole process. The addition of this last volume gives a realistic insight into the more formal and abstract treatment of the sociologic-historical analysis; it brings the various parts of the analysis into an understandable whole, serving both as a realistic aid to perspective and as a synthesis of the total operation. Very much the same function is expected of life history studies of inebriates, although they will function thus for the arena of abnormal drinking, not for all drinking. Naturally, as is potentially the function of all studies of the pathologic, new questions, new orientation, additional insights, specialized techniques, and, always desirable, new funds of relevant knowledge, can be acquired for the greater understanding of the whole field in question.

The Questionnaire

The questionnaire has become exceedingly popular in recent years. It appears to be a rapid and scientific method of acquiring up-to-date knowledge on almost any social matter. Appearances are often deceiving.

The questionnaire method is probably best adapted for acquiring data which are to be used in the quantitative analysis of concretely defined phenomena. It may also be used as a means of acquiring simple information on a large scale. In any event, it is preferably a technique of the later or last stages of an investigation.

Quantitative analysis may or may not be desirable. In establishing the existence of certain processes, or the reality of certain groups, quantitative analysis may be at best a subordinate, elaborative technique; the questions "how often" or "how many" or "accompanied with what" may not be particularly relevant.

Unless concretely definable matters are concerned, the value of questionnaires will have to remain unknown. Age, marital status, number of children, place of birth, and similar data are fairly easily defined. Attitudes about drinking, about laws, about interpersonal relationships, and the like, are not definable for questionnaire purposes. One can obtain "overall impressions" from the answers, but to subject them to quantitative analysis involves assumptions that render the results of questionable reliability.

The use of questionnaires to obtain simple information is perhaps desirable as one among several techniques by which such data may be acquired. Interview, observation, and indirect sources should surely be used with the questionnaire method. Questions concerning material objects, however, are probably answered with an establishable degree of honesty and accuracy. For instance, answers to the questions "do you drink Crown Cola, Coca Cola, Choc Cola or __________ (fill in name) Cola; one, two, three, _____ times a day; _____ days a week; at home, at drug stores, at club" are probably of measurable reliability. However, observation, interview, and reports of drug stores, wholesalers, government inspection agencies, and the like might well correct the questionnaire results. Unfortunately, it is not too likely that the present survey would want simple information of this sort, unless it referred to alcoholic beverages. Information referring to personal use of alcoholic beverages, however, often carries too much affect to be allowed within the category of simplicity.

When a great deal of information has been acquired by other means, it is possible (1) that the establishment of relationships of different patterns of behavior, or of particular behavioral patterns to particular groups, would be satisfied only by quantitative analysis, and (2) that the amount of data acquired by interview or observation would be insufficient for this purpose. The questionnaire method could then be pressed into service. It would appear advisable to use this technique for particular rather than for general purposes, as a subsidiary rather than as a basic method, and toward the close rather than at the beginning of the collection of data.

Problems, Analysis, Exposition

The discussion thus far has dealt with (1) the scientific approach to social problems; (2) the sociological approach to problems of alcohol; (3) the framework of a sociologic approach to drinking behavior; (4) the orientation in space and time of such studies; and (5) the techniques of collecting data.

It might appear that a more specific definition of problems should have been included, perhaps should have preceded much of this discussion. Such a procedure has been avoided deliberately. Some general questions have been suggested and a large area for study has been indicated. In addition, the comparison of the sociologic to the physiological and psychiatric methods, as well as the statements concerning the elements and aspects of behavior, were based on theoretic assumptions and on findings in related fields of inquiry. These form scientific problems by themselves. The framing of specific questions, the exact formulation of hypotheses, the construction of methods for testing such hypotheses, and the form of exposition are all better left, it is felt, to natural development out of and in conjunction with the observation and recording of materials.

There is no scarcity of theories about alcohol and its relation to human behavior and society. There is a real scarcity of sociologically relevant information. With a sufficient body of attested data which are amenable to comparison and synthesis, theoretic formulations can be evolved which will be of far greater utility than those arising from the speculative postulates of the armchair philosopher. Both practically and scientifically, the use of alcohol and the consequent behavior, individual and social, present problems which demand investigation. The first step in the sociologic approach to those problems is the careful observation and collection of relevant data. The present report has outlined the field to be investigated and the techniques that may be employed.

Notes

1. Unless otherwise specified, "drinking" always will refer to the drinking of alcoholic beverages.
2. The broad and general statement which follows cannot be attributed to any one person, certainly not to the present author, although he has presented it in his own language. The purpose of the statement is to introduce certain sociologic techniques and concepts to those students of the problems of alcohol who are without training or experience in the field of sociology, and who are interested in the contributions it may make. The influence of W.G. Sumner and A.G. Keller probably predominates. The writings of Malinowski, and more particularly the seminars he conducted in the years immediately preceding his death, are drawn upon heavily; his article (1941/1942) in the *Sigma Xi Quarterly* is particularly relevant. The "Outline of Cultural Materials," a hand-

book and an index for the collection and codification of cultural data by Murdock and his associates (1938) of the Institute of Human Relations at Yale University, has been used freely. More important than any publication has been the writer's association with these authors for a period of several years and also with many others, including, particularly, Professors M.R. Davie, J.W.M. Whiting, E.M. Jellinek, J. Dollard, and C.S. Ford.

3. One of the few culturally relevant treatments of the use of alcohol is centered around the problem of its societal and individual functions; this is the excellent cross-cultural comparison and analysis by Horton (1943).
4. For the rural American community this poses a special problem.

References

Anderson, N. (1923). *The hobo.* Chicago: University of Chicago Press.

Bacon, S.D., & Roth, F. (1943). *Drunkenness in wartime Connecticut.* Hartford: Connecticut War Council.

Booth, C. (1902). *Life and labor of the people in London* (Vol. 8) London: Macmillan & Co.

Bunzel, R. (1940). *Psychiatry, 3,* 361.

Haggard, H.W., & Jellinek, E.M. (1942). *Alcohol explored.* New York: Doubleday, Doran & Co.

Horton, D. (1943). *Quarterly Journal of Studies on Alcohol, 4,* 199.

Jellinek, E.M. (1942). *Quarterly Journal of Studies on Alcohol, 3,* 267.

Jolliffe, N., & Jellinek, E.M. (1942). Effects of alcohol on the individual. In E.M. Jellinek (Ed.), *Alcohol addiction and chronic alcoholism* (Vol. I, chap. 6).

Koren, J. (1899). *Economic aspects of the liquor problem.* Boston: Houghton, Mifflin & Co.

Lynd, R.S., & Lynd, H.M. (1929/1937). *Middletown.* New York: Harcourt, Brace; *Middletown in Transition.* New York: Harcourt, Brace.

Malinowski, B. (1941/1942). *Sigma Xi Quarterly, 29,* 182; *30,* 66.

Mumford, L. (1938). *The culture of cities.* New York: Harcourt, Brace.

Murdock, G.P., and others. (1938). *Outline of cultural materials.* New Haven: Institute of Human Relations.

Poteat, W.H. (1943). *Quarterly Journal of Studies on Alcohol, 4,* 195.

Queen, S.A., & Gruener, J.R. (1940). *Social pathology.* New York: T.Y. Crowell Co.

Thomas, W.I., & Znaniecki, F. (1927). *The Polish peasant in Europe and America.* New York: Knopf.

Young, P.V. (1939). *Scientific social surveys and research.* New York: Prentice-Hall.

CHAPTER 2

Competing Paradigms in Biomedical and Social Science Alcohol Research: The 1940s Through the 1980s

Kaye Middleton Fillmore

There is marked conflict in the contemporary scientific community with respect to explanations of alcohol problems. On the one hand are etiological explanations of a biological nature, specifically those hypothesizing a biological link for alcohol problems between parents and children. On the other hand are explanations that propose multiple processes of an environmental nature, hypothesized to contribute to the emergence, persistence, and decline of multiple alcohol problems on both the individual and aggregate levels. These differences in explanation bear the stamp of distinct disciplinary approaches to the study of drinking practices and problems, biomedical versus social science.

It has been stated elsewhere that the contemporary polarization in these explanatory motifs has the effect of producing such fierce competition between the two broad disciplinary approaches that it has precluded development and testing of competing interdisciplinary hypotheses (Fillmore & Sigvardsson, 1988). Furthermore, this polarization has minimized consideration of either the actual or potential ethical and social consequences of single bodies of scientific explanation that may be adopted as social policy.

Are the tensions reflected in the contemporary polarization of these explanatory motifs new, or did they have their origin in the early years of the modern alcoholism movement? In this chapter, the emergence and decline of explanatory motifs are described for the 40 years since the beginnings of the modern alcoholism movement with the hope of identifying the origins of the current polarization.

Six documents are used as basic source materials to describe the explanatory foci in alcohol research from 1942 to 1987. These documents represent the formal statements of various "gray eminences" in the field that have emerged time and again as offerings for the course of future research. They are analyzed to discern similarities and differences in research direction over time. The first two consist of the research priorities for the field outlined by E.M. Jellinek in 1942 (Jellinek, 1942), and a letter, published the following year, from Bacon to Jellinek describing his view of the place of social science in the field (Bacon, 1943). These two documents provide a picture of the differences in the two major disciplinary perspectives, biomedical science and social science, during the early years of the modern alcoholism movement. The third is a review of the field by scholars at the Center of Alcohol Studies published in the *Quarterly Journal of Studies on Alcohol* in 1952 (Bacon, 1952) appraising the alcohol field at its "10th anniversary." Reports of research directions and accomplishments from physiology, social science, and treatment are included. For the late 1970s and early 1980s, two reports are used—one on research priorities from a group of scholars assembled at the Center of Alcohol Studies (Keller, 1979) and the other an Institute of Medicine report on research recommendations for the alcohol field (Institute of Medicine of the National Academy of Sciences [IOM/NAS], 1980). For the late 1980s, another Institute of Medicine report on research recommendations (IOM/NAS, 1987) is used.

The principal recommendations in these documents are categorized with respect to research concerning (a) the etiology of alcoholism; (b) the study of a wide range of drinking practices, alcohol problems included; (c) the influence of alcohol/alcoholism on society; (d) the influence of alcohol on the body; and (e) the effects of treatment. For each topic area, emerging and declining interests in the biomedical and social science fields are traced with respect to their potential for strain and tension.

The Early 1940s: The New Scientific Approach to the Study of Alcohol Problems

Descriptions of the early alcoholism movement are abundant, advancing the notion that a joint effort by lay people and scientists was undertaken to promote public acceptance of both the scientific approach to the problems of alcohol and the humanitarian treatment of alcoholics (Conrad & Schneider, 1980; Johnson, 1973; Room, 1978). Undoubtedly, a number of the scientists, then members of the faculty of Yale's Laboratory of Applied Physiology (later to be called the Center of Alcohol Studies for some of the laboratory's activities; it is presently located at Rutgers University), were important players in this fledgling movement.

There was a critical dilemma in simultaneously advancing a social movement and scientific work on the alcohol topic. Lay persons working on behalf of the movement advanced scientific claims that were sometimes at odds with state-of-the-art scientific knowledge. For example,

Marty Mann, a lay spokesperson for the movement through her leadership of what was to become the National Council on Alcoholism, claimed a physiological basis for alcoholism, maintaining that this knowledge had been available for more than a century (Johnson, 1973). On the other hand, Yale scientists, such as Jellinek and Haggard, were clearly aware that a physiological basis for alcoholism (or any other basis, for that matter) had not yet been discovered. Regardless of their objective appraisal of the state-of-the-art knowledge, modern historians of the alcohol movement have advanced the notion that the scientific inquiry of the day reinforced the ideology of the social movement (Herd, 1984). As Bacon, a member of the Yale research team, retrospectively put it in 1979, the disease concept of alcoholism was, for scientists of that time, to be "the wheelhorse of change" but not the change itself:

> Alcoholism has been the great cult excitement of the last 34 years. I was one of the people who helped build up the cult, in the early days before I got thrown out when I said, "Well, that wasn't quite what I meant." Alcoholism was to be the wheelhorse of change; it was not to be the change. It was taken over.... Recently the cult has become so powerful, 1960–1972, that it has taken over traffic problems and youth problems, and they are all called alcoholism, which of course is a lot of nonsense if the word is to refer to a disease-like entity of some sort or a life disorder of some sort. Alcohol problems are much, much bigger than that. (Bacon, 1979)

The "great cult excitement," to use Bacon's phrase, dominated the research priorities expressed by Jellinek in 1942 for the emerging field. The substance of the research, Jellinek said, should be "primarily relevant to the central problem of alcohol, which is the problem of the origins and development of addiction and other forms of abnormal drinking." Thus, Jellinek's priorities for research in the field focused on physiological experiments, the role of personality, and the influence of social factors, all directly related to the etiology of alcoholism (see Table 1). These directions were in marked contrast to the temperance paradigm of alcohol as a risk factor leading to health and social consequences.

However, the research in the field had not yet pointed to any single etiological basis for alcoholism, whether it be physiological, psychological, or social. Bowman and Jellinek's (1942) review of the literature regarding etiological explanations of alcoholism concluded that while the "terrain" of alcohol problems had been well explored, the etiology had not. Existing physiological theories were not, according to them, true etiological theories at all but rather *descriptions* of dependence or addiction. Allergic theories, then popular with lay promoters of the medical model, were dismissed as theories of etiology. Further, theories of person-

TABLE 1
Etiology of Alcoholism

1942	1952	1980	1987
The influence of social factors on inebriety (includes cross-cultural comparisons of societies with varying degrees of inebriety, incidence of inebriety in differing historical periods and among specific groups plus correlation between inebriety and trade policy, taxation, etc.) (J)	Physiological "defects" (includes allergy, which has no support; adrenal hypofunction, which has no support; and nutritional deficiency, yet unknown as etiological) (L&G)	Biochemical and genetic issues (includes studies of genetic variation in metabolism and development of animal models)	Biochemistry of ethanol metabolism (ultimate goals = heritable aspects of behavior + individual susceptibility to medical consequences)
Role of personality in abnormal drinking (J)	Use of the social problems approach "to underscore the inadequacies of earlier definitions and impressions of the nature of 'the alcoholic' " (B)	Clinical and epidemiologic (including genetic) issues (includes natural history of alcohol-related diseases, epidemiological and longitudinal studies of risk factors for multiple dependent variables including alcohol abuse and other disorders)	Heritable determinants of risk (includes interdisciplinary longitudinal studies delineating clinical subtypes of alcoholics + studies of genetic markers)
Physiological experiments that touch on tolerance and habituation as central to understanding origins of inebriety (J)		Research on prevention (includes studies with objective of early identification of those at risk: longitudinal, genetic, biochemical, and psychological screening plus environmental elements: family structure, media, work which deter/promote alcohol abuse)	Animal models (includes studies with genetic implications)
			Neurosciences (includes studies on cellular level of tolerance and dependence plus genetically defined response to alcohol)
			Psychopathology related to alcohol abuse (includes biologic and social factors associated with those psychopathologies that have been associated with alcohol abuse)

J = Jellinek, 1942; B = Bacon, 1952; L&G = Lester and Greenberg, 1952.

ality, while "much more satisfactory and this in spite of their incompleteness, indefiniteness and other failings," had been "somewhat overstated." Psychiatric theories (for instance, those concerned with homosexuality or suicide) "contribute to the knowledge of the terrain on which alcohol addiction grows but not to the knowledge of the decisive factors." Social theories (for example, poverty, the accessibility and inexpensiveness of alcohol, social ambivalence and cultural theories) had not received sufficient scrutiny in the scientific literature "although they are prominently discussed in propagandistically oriented writings."

Theories of constitution and heredity, which had enjoyed prominence in the early part of the twentieth century, were not currently popular: "at least in the scientific literature, there are nowadays no representatives of the view that alcohol addiction itself can be directly inherited" (Bowman & Jellinek, 1942). While the reviewers recognized the higher incidence of alcoholism among those with alcoholic parents, they maintained that:

> All one can say is that persons with such hereditary liability or with such dispositions have a greater probability of succumbing to the risks of addiction. A full realization of this is important from the standpoint of therapy and mental hygiene. As Norman Kerr said, "The idea of hereditary disease is gratifying to the patient but makes fatalism which stultifies all effort to overcome it." (Bowman & Jellinek, 1942)

This important conclusion was made even clearer in the *Quarterly Journal*'s fifth lay supplement titled "Alcohol, Heredity and Germ Damage":

> Is the craving for alcohol inherited? . . . our answer would correctly be "no, it is not inherited." We could make this answer because abnormal drinking and the craving for alcohol are acquired traits and acquired traits are not inherited. If, however, we phrased our question another way it would perhaps express more nearly what the reader has in mind on the subject and the answer would be different. To the question, are the children of alcoholics more apt to become alcoholics themselves than are the children of temperate parents, the answer is definitely "yes." (*Quarterly Journal of Studies on Alcohol,* 1941)

The authors of this document, presumably the scholars at the Yale center at that time, also enumerated the reasons for this relationship. Included were poor home environment, child neglect, setting an example of excessive drinking, and, interestingly, a higher incidence of excessive drinking in families where mental disorders and abnormal personalities *are* inherited traits.

Jellinek posed no priority research questions for the study of a wide range of drinking problems and practices (see Table 2). Rather, the ques-

TABLE 2
Study of a Wide Range of Drinking, Alcohol Problems Included

1942	1952	1980	1987
Study of the problems of alcohol embedded in the larger study of the customs and social controls of drinking. Study not confined to the "exotic" (e.g., alcohol problems). Function, rational, activity material apparatus, place and time, functionaries, rules, social change, etc., are the foci of the behavioral approach (B)	Etiology of a social-psychological nature that involves description of the phenomenon of drinking per se and translating the description of both the phenomenon and the problem manifestations into generalized statements. Emphasis on behavioral studies and drinking among specific demographic/age groups (B)	Psychosocial issues (includes effects of price, changing social values, policies on consumption and problems, plus identity of situational factors that affect at-risk drinking)	Social determinants of risk (includes longitudinal studies of general population, studies of beliefs, norms, natural experiments, policy intervention)

B = Bacon.

tions recommended came from Bacon, a young sociologist on the Yale staff, in a paper titled "Sociology and the Problems of Alcohol" (Bacon, 1943), still a classic in the field more than 40 years later. Bacon maintained that to confine the study of alcohol to the "exotic"—that is, the tiny population of alcoholics—would not yield adequate social policy recommendations from the scientific establishment. Rather, the inquiry should be of *all drinking,* with studies ranging from the individual to the social level. Further, society's reactions to drinking were to be included in this range of inquiries (see Table 3).

> The sociologist is interested in the customs of drinking, the relationship between these customs and other customs, the way in which drinking habits are learned, the social controls of this sort of behavior, and those institutions of society through which such control issues. The sociologist wishes to know the social categories in which much or little or no drinking occurs, he seeks correlations of amount and type of drinking with occupational, marital, nationality, religious, and other statuses. More importantly, he poses the broad questions: What are the societal functions served by the drinking of alcoholic beverages? What are the social rules concerned with drinking? What are the pressures for

TABLE 3
Influence of Alcohol/Alcoholism of Society

1942	1952	1980	1987
Societal reaction as a part of studies on all drinking (B)	Societal reaction as a part of studies on all drinking (B)	Research on economic costs on society from alcoholism and alcohol abuse plus studies of fiscal relationship between alcohol beverage industry and governmental entities	Quantifying the impact (surveys of drinking behavior, prevalence estimates on alcohol's role in accidents, morbidity, mortality, etc.) Social consequences of alcohol abuse (includes studies of impact of drinking on children, family, violence, drug abuse)

B = Bacon, 1943, 1952.

> and against this practice? How does this behavioral pattern jibe with other institutions and folkways? The sociologist is interested in changing patterns of drinking and in their relation to other changes in the society. The sociologist studies the effect of no drinking, some drinking, or excessive drinking on groups, attitudes, and behavior. As drunkenness may result in punitive, preventive, or therapeutic measures, the sociologist observes, classifies, analyzes, and compares these activities. (Bacon, 1943)

While Bacon's thesis was not peculiar to *social* science and while he was also reacting to the differences between *applied* science versus *academic* science, the paper also signified the divergence in orientation of biomedical and social science. This divergence has been interpreted by those interested in this history as reflecting a disparity in emphasis and conceptualization of the dependent variable under scrutiny. Early on in the movement, the biomedical approach viewed the dependent variable as "alcoholism," a progressive disease entity, whereas the social science approach viewed the dependent variable as *all* behavior related to alcohol use, including multiple problems with multiple origins and the effect of alcohol on the society per se.

It is thought that the temporal closeness of the nascent alcoholism movement to the repeal of Prohibition served to direct the interest of alcohol scholars in general and biomedical scholars in particular to the study of the etiology of the disease alone rather than the effect of alcohol on society (see, for example, Room 1978). This has been hypothesized as a reaction to the Prohibition movement's emphasis on the role of alcohol consumption as a risk factor leading to social and health consequences. In a personal communication to the present author, Bacon has recently restated the position of those working on behalf of the alcoholism movement in the 1940s:

> The major theme of the time was clearly of both negative and also a positive nature. The negative was a denial of the Temperance-Prohibitionist-Dry position that the only approach was a primitive religious and law enforcement orientation and that *no other position was possible....* The positive approach had one general and one very specific message. The general position was, to the effect, that a scientifically oriented mode of study was available and useful (and very, very different from the religious-political approach of the past). The specific approach was to the effect that (a) what was rather vaguely (at first) called alcoholism was a major societal problem in the alcohol area and (b) it could be and should be treated as a *public health problem.*

Jellinek's early outline of the field emphasized that studies of etiology took precedence over studies concerned with alcohol's effect on the body. Although he did call for research to investigate the direct chronic effects of alcohol on the nervous system (see Table 4), there was no men-

TABLE 4
Influences of Alcohol on the Body

1942	1952	1980	1987
Investigation of the possible direct chronic effects of alcohol on the nervous system (J)	Absorption, distribution, excretion, and metabolism of alcohol (L&G) Effects of alcohol on the body (e.g., on nervous system, emotions, cholesterol content) (L&G)	Neuropharmacological issues (includes studies of effects of alcohol on cell membranes and studies neurotransmitters)	Medical consequences of alcohol abuse (includes genetic and prospective studies of diseases attributable to alcohol abuse) Consequences of drinking on performance (includes studies on dose and motor action)

J = Jellinek, 1942; L&G = Lester and Greenberg, 1952.

TABLE 5
Effects of Treatment

1942	1952	1980	1987
Establishment of clinics staffed by at least one psychiatrist to assign treatment on the basis of origin and form of habits plus follow-up studies re: different treatments. Plus differentiation between alcoholics and schizophrenics (J)	Accurate diagnosis (medical, psychiatric, social), evaluation of acute alcohol intoxication, medication, condition-reflex therapy, disulfiram, etc. (L)	Research on treatment (includes efficacy studies re: interaction of patient characteristics and treatment settings, approaches and goals plus assessment of factors influencing access to treatment)	Not yet published

L = Jellinek, 1942; L = Lolli, 1952.

tion of studies on alcohol's effect on the liver. Herd (1984) has suggested that although there was actually an epidemic of cirrhosis occurring in some portions of the U.S. population between 1941 and 1967, the relationship between cirrhosis and alcohol consumption was all but ignored in epidemiological investigations until the 1970s. In fact, this relationship was neglected until Lieber's (1975) work was performed.[1]

The emphases on the effects of treatment for alcoholism are displayed in Table 5. Jellinek recommended creating clinics in which a member of the medical community would assign treatment "on the basis of origin and form of habits." In addition, an emphasis on differentiating between alcoholics and schizophrenics probably antedated later interest in distinguishing between "primary" and "secondary" alcoholics.

Although the seeds were sown in the 1940s for the polarities in research orientation in the alcohol field, as documented by the recommendations of Jellinek versus those of Bacon, there is little evidence in the literature at that time to suggest that differences between the approaches matched the extent of the breach that exists today. In fact, evidence suggests that, in many respects, these two "camps" represented a united front in their efforts to advance research in the field. The more "unified" approach by those trained in the broadly diverse disciplines of the biomedical and social sciences can perhaps be best understood by their relative isolation from the mainstream of the medical and academic establishments such that those scientists interested in the problems of alcohol were in the business of "creating" their own field of study. In fact, the

Center of Alcohol Studies (the organization growing out of the Yale Laboratory of Applied Physiology) was conceived as an interdisciplinary research unit from its inception. And, although Jellinek confined his research priorities to the testing of etiological hypotheses, he also held great hopes for social factors, among others, to explain the development of alcoholism.

The 10th Anniversary of the Modern Alcoholism Movement

In 1952, members of the scholarly community at the Center of Alcohol Studies published a "10th anniversary" review of the field. Noteworthy is that this review was titled "Alcoholism, 1941–1951: A Survey of Activities in Research, Education, and Therapy," although Bacon's introduction to the survey clearly embedded the notion of alcoholism in the broader study of alcohol problems. Bacon could report the success that the alcoholism movement had enjoyed over the past decade:

> The problem of alcoholism has been the one receiving the lion's share of attention by research workers, by action programs, and by scientific and popular writers. Since 1940 many research projects and many publications have emerged, many hospitals and clinics have been created, new legislation—federal, state, and local—has been passed, voluntary agencies have grown up the country over, and education within and without the school system has undergone a profound change. (Bacon, 1952)

But even with this abundance of new research projects, in addition to the political victories of the still young alcoholism movement, the reports of the reviewers on the biomedical work of the decade did not yet point to breakthroughs in understanding the etiology of alcoholism. Lester and Greenberg (1952) reviewed three major etiological theories that were thought to discriminate between individuals having "inherent physiological 'defects' ": allergic hypersensitivity, adrenal hypofunction, and relative nutritional deficiency (see Table 1). The first two, they maintained, had no support; the third, although doubtful, was still under investigation. They concluded: "At present, certainly, we are far from an answer in physiological terms for the genesis of addiction to alcohol."

The major contributions in the biological area during the decade of the 1940s primarily related to descriptions of alcohol in absorption, distribution, excretion, and metabolism (see Table 4). Lester and Greenberg recognized the lack of research on "the physiological economy of the body" such that they could report that "little is known of either the beneficent or deleterious effects of alcohol on the body." They called for further work in exploration of the hangover effect, the mechanisms entailed in the production of fatty liver and cirrhosis, and alcohol tolerance.

A report from Bacon (1952) on the contributions of social science research during the decade illustrated his concern with the continued concentration on the "exotic" form of drinker, the alcoholic, among the social science/alcohol studies community. However, he pointed out that the contributions of this orientation (the social problems approach that focused almost entirely on alcoholics and alcoholism) had served to highlight the fact that there was not one problem classification, but many (see Table 1).

> The major value of the social-problem approach thus far has been to underscore the inadequacies of earlier definitions and impressions of the nature of "the alcoholic".... Earlier writers would present theories explaining the causes of alcoholism and indicating appropriate therapies or techniques of prevention which were based almost entirely upon observation of special types of alcoholics, far from representative of the total alcoholic population.... The increasing amount of data made available by the disciplined students of the social-problem approach has made it clear that many of these conflicting viewpoints can be resolved on the grounds that the authorities were discussing different types of problems rather than achieving different answers for the same problem. (Bacon, 1952)

This statement represents a long line of social research to follow that would painstakingly, if somewhat naively, test the assumptions of the disease concept of alcoholism. It would conclude that Jellinek's hypotheses regarding a progressive, unaltered entity were difficult to document in both alcoholic and general populations (Jackson, 1957, 1958; Room, 1966; Seeley, 1959).

Bacon also highlighted the fact that there was no consensus on the definition of alcoholism, the exotic dependent variable of inquiry, which a number of studies were attempting to predict. His recognition of this basic scientific problem in the field in the 1950s has not been resolved to date. Further, this lack of consensus has probably contributed to the breach between the two broad disciplinary perspectives.

> ... it can be reported that an effective and generally acceptable definition of alcoholism is still being sought by those in psychiatry, public health, law, insurance, the Veterans Administration, police, education, personnel work, and many others. In the absence of satisfactory definitions of problem drinker, pathological drinker or excessive drinking, it is not surprising that a variety of definitions of the problem are competing, that there is controversy both over the nature of the problem and also over the proper area for therapeutic and preventive research or action, and that there is flat disagreement concerning the magnitude of the problem. (Bacon, 1952)

By pointing out the inability of the alcohol field to arrive at a consensus on the definition of their object of study and by pointing out that the inquiries of social scientists had found that a unitary entity called "alcoholism" was, in fact, a multitude of problems, Bacon's assessment of the contributions of the social problems approach to the etiology of alcoholism was prophetic and helped to undermine the notion.

According to Bacon, research he had called for in 1943, that of basic social science or behavioral research on *all* drinking, had received relatively little attention over the past decade. He noted as exceptions the studies of Jewish drinking (Bales, 1946; Glad, 1947; Snyder, 1958), of Italian Americans, a study of college students then in the field (Straus & Bacon, 1953), a tavern study, and an anthropology of drinking (Horton, 1943).

Bacon attributed the lack of research in the behavioral arena to the relative youth of the field in general as well as to its essential threat to the beverage industry and to temperance organizations. But he projected that the behavioral study of alcohol problems had the potential to develop hypotheses and test them, a potential that would far surpass research performed in the laboratory for the simple reason that the very essence of the dependent variables was behavioral.[2]

Lolli (1952) reviewed the progress of the 1940s in the treatment of alcohol addiction (see Table 5). In contrast to the present day, a time in which the therapeutic modalities of Alcoholics Anonymous have escalated to dominate the treatment terrain, Lolli, like Jellinek before him, envisioned a future in which there would be a diversity of treatments for the alcoholic condition from which to choose.

> There are obviously many different and equally successful ways to treat addiction to alcohol. There probably is only one way to understand it and this way should also supply us with insight into the mechanisms by which different approaches can be equally successful. (Lolli, 1952)

The hope for the discovery of the etiological factor in the disease thus was still very much in the picture of those in the biomedical "camp." In this sense, the importance of diagnosis, on the medical, psychiatric, and social levels, was emphasized with the objective of diverting individuals to the appropriate treatment regime.

The 10th anniversary of the modern alcoholism movement was, it seems, celebrated with a sense of bleakness in both the biomedical and the social science camps. Jellinek's research priorities for locating etiological explanations of alcoholism, whether biological, social, or psychological, had not been fulfilled. Indeed, the inclusion of social science research in the form of a "social problems approach," which was to "serve" the disease concept, seemed instead to be undermining its foundations. Last,

Bacon's 1943 statement of research priorities on the *totality* of drinking behaviors from a broader social science perspective was suffering from neglect, allegedly due to potential threats from political entities.

Research Priorities in the Early 1980s

Between the 1950s and early 1980s, a great deal of change occurred in science in general and in alcohol-related science in particular. The creation of the National Institute on Alcohol Abuse and Alcoholism in 1970 solidified funding to alcohol researchers, young professionals were attracted to the field (Wiener, 1981), and alcoholism treatment expanded. In general, science became "big science" with "big money" (primarily from federal sources) and with advanced technology to perform complex projects (for instance, computers became available and accessible to most researchers). This was, in contrast to the earlier years of the alcoholism movement when funding was severely limited, when the number of scholars in the field was relatively low and when researchers were confined to the computer card counter-sorter (at best) to analyze their data.

In 1977, a group of prestigious scholars in the alcohol field met under the aegis of the Rutgers Center of Alcohol Studies to discuss research priorities (Keller, 1979). Papers and responses on the topics of alcohol studies as a discipline, psychology, biology, and social science were presented. Although this was a conference focusing on "research priorities," no concise listing of the priorities emerged, but the discussion around the various research propositions gives insight into the issues at hand.

In 1980, the Institute of Medicine of the National Academy of Sciences also issued a report titled *Alcoholism and Related Problems: Opportunities for Research* (IOM/NAS, 1980), which explicitly laid out research priorities in biochemical and genetic research, neuropharmacological research, clinical and epidemiologic (including genetic) research, psychosocial issues, prevention issues, treatment issues, and the cost of alcoholism and alcohol abuse. In the report, three of the seven topic areas dealt with issues on the etiology of alcoholism and the priorities within each of these three topic areas (biochemical and genetic, clinical and epidemiologic, and prevention) included some mention of genetic factors (see Table 1).

The critical priority for the biochemical and genetic area related to the study of metabolic pathways of alcohol, which, in turn, raised questions about the genetic factors in these pathways. The report on clinical and epidemiological research emphasized the need for more definitional uniformity in epidemiological research (conforming to clinical criteria) and outlined a number of areas (including genetic) that "can inform and improve clinical practice" (IOM/NAS, 1980). The first priority listed in the prevention section included early identification of target populations at risk, utilizing a number of independent variables, among them genetic factors.

The overwhelming interest in genetic explanations of alcoholism was a far cry from research priorities of the 1940s and 1950s when such explanations had been all but dismissed. Of even more interest, the reliance on these explanatory variables at the Rutgers meeting, only several years before the IOM/NAS report was published, was both cautious and minimal.

At the Rutgers meeting in 1977, Kissin presented the biological perspective. With reference to Goodwin's adoptee study (Goodwin & Guze, 1974) as well as the work of Cruz-Coke (1964) on color blindness and other genetic factors, he said that "these studies suggest hereditary correlates of alcoholism more than they do hereditary etiological factors as is the case in the Goodwin studies" (p. 149). In the discussion, Yedy Israel, also a member of the biomedical community, presented his cautious interpretation of these results:

> We already know that the children of alcoholics have four times the frequency of alcoholism than does the rest of the population. Nevertheless, I ask myself, but what would we gain by knowing this? What are we doing with the sons and daughters of alcoholics? Do we have any socially accepted intervention? I don't know, and therefore I ask myself, is this a priority? I think it would definitely be a priority the moment I would know what to do with it. (Keller, 1979)

Edwards, summing up the meeting, offered the following statement regarding genetics as an etiological factor:

> I am interested in Don Goodwin's work on genetics but he doesn't have to account for all the variance. I don't think we should roll in the aisles with rejoicing because we have discovered a small part of the variance and can call it genetic. We should be concerned about larger parts of the variance which manifestly are not genetic. We should ask those questions not as the mandarin, but how does one have any chance of interdicting the nongenetic transmission of this sort of self-hurting behavior? What are the processes which mediate this: processes of modeling, processes of attitude formation, processes of identity. (Keller, 1979)

What occurred in these short years to transform the etiological explanations from the biomedical community? One of the critical factors was the 1977 publication by Li (1977) on the enzymology of alcohol metabolism among humans. Li found that the isozyme pattern of alcohol dehydrogenase differed in sex and genetic characteristics. In other areas (clinical/epidemiologic and prevention), the work of Goodwin on the hypothesized genetic link between biological fathers of adoptive sons seemed to be gaining acceptance. In the prevention chapter of the IOM/NAS document, this new momentum was defined as coming from several disciplines:

> These studies [twin studies and adoptee studies] are not viewed as conclusive because the sample sizes are small and because there have not been many opportunities to repeat the studies with different population samples. However, in conjunction with clear-cut animal data, it is sufficiently persuasive that there has been a stimulus to search for a specific gene or combination of genes that will place an individual at high risk for alcoholism. (IOM/NAS, 1980)

This collection of studies signaled, for the first time in 40 years of research, a promising etiological hypothesis from the biomedical area. This "new" etiological emphasis is, however, reminiscent of an earlier period in scientific history (at the turn of the 20th century) when these theories dominated etiological explanations (Bynum, 1984).

With regard to social science approaches, there was an appeal to move beyond the descriptive studies that had dominated the field. Nonetheless, the tone of the 1980 IOM/NAS document reflected the recommendations forwarded by Bacon almost 40 years earlier with emphases on multiple designs (process-oriented studies, longitudinal studies, naturalistic studies and "natural experiments"). Studies of factors behind changes in per capita consumption were highlighted. Studies of populations "at risk" were closest to "etiological" questions under consideration with respect to situational or environmental factors that supported or restrained at-risk drinking.

Examination of the chapter on psychosocial approaches suggests that there was no unifying theoretical emphasis in the social science area. Rather, there was a large body of work that was loosely knit together in the description of human behavior from divergent methodologies and viewpoints, some from the individual level, some from the societal level. That there was not a central theme in the psychosocial chapter may reflect the fact that social scientists (with the exception of what Bacon called social problems scientists) did not see this as their charge.

Reflecting the continuing divergent biomedical and social sciences approaches to the dependent variables under study, Room, in the 1979 conference, rather defensively stated the position of social science in the alcohol field:

> . . . perceptions by others of the nature and use of social science research tend to stop at the images of competent technician or advance scout or management consultant. Social scientists, on the other hand, tend to define the nature and worth of their work in quite other terms, in terms of the theoretical contributions it makes or reconceptualizations it suggests. (Keller, 1979)

While some social scientists were continuing to fulfill the role that Bacon had noted in 1952, namely "underscoring the inadequacies of earlier definitions and impression of the nature of 'the alcoholic,'" it was becoming increasingly clear that the theoretical contributions of social scientists studying the *spectrum* of drinking behavior were going relatively unnoticed by the remainder of the field.

A prolonged unresolved problem in the field was evident in the Rutgers and IOM/NAS reports. There was continued disunity across disciplines with regard to definitions of alcohol problems. In this case, recommendations were made to bring epidemiological research closer to that of the clinical picture.

A major change in the social science research priorities occurred between the 1940s and early 1980s. Jellinek (1942) had suggested that the research models to take precedence were those in which alcoholism was the dependent variable and that studies using alcoholism as an independent variable were valuable only for publicity reasons. By contrast, the 1980 IOM/NAS report devoted an entire chapter to the economic costs of alcoholism and alcohol abuse for the society as a whole. The chapter outlined and recommended studies concerned with the relationships between the alcohol beverage industry and government, which had the pragmatic social policy potential of raising alcohol taxes to prevent alcohol-related problems.

Prevention measures on a number of levels, including the societal, were addressed in another chapter. Studies of social policy that would control per capital consumption were recommended among others. The "new" emphasis, drawing from the work of Canadian researchers (de Lint & Schmidt, 1968) on the distribution of alcohol consumption in the population (politically reflective of a much earlier era in American alcohol-related politics which maximized *all* the problems of alcohol with regard to their toll on society), has led some to call those who perform these studies and advocate these measures as "neo-temperance."

The IOM/NAS recommendations for research priorities in the treatment area echoed in part Lolli's earlier review in the 1950s. Lolli had emphasized the possibility of a panorama of alcoholism treatment options. But while the IOM/NAS reviewers addressed the issue of matching patients to different treatment modalities, only two major alternatives seemed available, those that emphasized outcomes of nonabstinence or "controlled drinking" versus those that emphasized outcomes of abstinence. Otherwise, the considerations of treatment variability were concerned with structural characteristics such as length of care, inpatient versus outpatient care, or the qualifications of the therapists. Obviously, Lolli's expectations of a grand panorama of treatment alternatives had not been realized. However, during the early 1980s there was a rise in conceptual "sub-divisions" of alcoholics (for example, primary versus secondary alcoholics or court-referred versus self-referred alcoholics) which allegedly required "treatment matching" to maximize successful outcome (whether defined as abstinence or "contolled drinking").

In sum, the differences between the biomedical and social science approaches were widening by 1980. The differences were reflected in divergent dependent variables, alcoholism in the biomedical camp, and, increasingly, per capita consumption in the social science camp. Further, specification of the explanations and "control" of alcohol problems were taking on strongly divergent postures.

On the one hand, the new findings regarding the heritability of alcoholism would later imply means of control through warnings to the children of alcoholics, although biomedical scientists in 1977 had doubted there would be *any kind* of socially acceptable intervention. Importantly, biomedical scientists (clearly by 1980) were beginning to believe that they had the answer to Jellinek's etiological puzzle.

On the other hand, the hypothesized relationships between the per capita consumption and alcohol problems were beginning to imply a means of control of *all* problems of alcohol by reducing the amount of or accessibility to alcohol in the environment. Some social scientists, while not directly addressing Jellinek's etiological puzzle, were suggesting that if the entire population was controlled with regard to the availability or accessiblity of alcohol, the prevalence of chronic addictive drinkers would also be reduced.

Contemporary Research Priorities

In 1987, the Institute of Medicine of the National Academy of Sciences issued its second alcohol-related report titled *Causes and Consequences of Alcohol Problems: An Agenda for Research* (IOM/NAS, 1987). (Treatment and prevention issues were part of a report published after this manuscript was completed.) Ten topic areas were covered. Half of these areas included various aspects of the genetic hypothesis: biochemistry, heritable determinants of risk, animal models, neurosciences, and psychopathology. Cross-cutting the ten topic areas, the committee identified six common themes in the recommendations that were considered "of crucial importance to continued progress in learning about drinking and the problems related to drinking" (IOM/NAS, 1987). Among them were (1) "the need for consistent and precise terminology," which as been an ongoing and unresolved problem of definition in the field from the 1940s to the present; (2) "the power of animal models and genetic approaches," which were only emerging as promising areas for study in the late 1970s but by 1987 were dominating the field; and (3) "the need to integrate biologic and behavioral approaches," allegedly a new emphasis in the field.

The committee also bridged the gap between the 1980 report and the current one by pointing out four important interim research progress examples: (1) membrane research that was allowing "more targeted study of membrane structure and function in relation to basic biologic mechanisms of intoxication, tolerance, and dependence" (IOM/NAS, 1987); (2) advances in describing heterogeneity among alcoholics, particularly with regard to subtypes of genetic predisposition; (3) insights derived from studying the context of drinking and the social expectations of drinking, in particular cognitive and psychomotor functions influenced by expectations of effect yielding hypotheses conducive to biological studies and biobehavioral studies; and (4) research related to the temporal sequencing of alcoholism with other psychopathologies.

The authors stated that the report was guided by a "biobehavioral perspective." By this it was meant research "that will permit study of the interactions of biologic and behavorial variables that contribute to the causes and consequences of alcohol problems" (IOM/NAS, 1987). In this sense, the *spectrum* of drinking behaviors was given secondary importance to the integration of social and biological explanations for behaviors called problems, and the order of research priorities, stated 47 years earlier by Jellinek, was restated and revitalized in this report.

The biobehavioral approach, it was maintained, dealt with such questions as parceling out the variance of genetic factors in the total constellation of alcohol problems; in determining the compatibility or incompatibility between biologically determined factors in variable social environments and among given subgroups; in explaining "social dependence" on alcohol versus biological dependence; and in advancing understanding of the behavioral and biological factors in withdrawal symptoms. In short, it seems that this new perspective focused more on the development of *interactive* interdisciplinary hypotheses. This difference is critical. Even assuming the identical dependent variable of interest (a position not identified with the social science objectives as outlined by Bacon), the biobehavioral approach would *not* propose studies that would pit the assumptions of the social scientist against those of the biologist but rather would propose studies that would determine how much of the variance the biological variables explained, how much the social science variables explained, and an interaction value. Furthermore, it is critical to point out that the model driving the biobehavioral approach was not that of the social sciences (where the spectrum of drinking could conceivably be the dependent *or* the independent variable) but was the traditional biomedical model.

With regard to the heritable risk hypotheses, the authors of the report stated that the synergy of research advances "in defining determinants of alcohol dependence, in statistical methodology, in the recognition of heterogeneity, and in the identification of physiologic markers with familial association" (IOM/NAS, 1987) laid the groundwork for future research in this area. Between 1980, when Goodwin's work was cited in the first IOM/NAS report, and 1987, a Swedish adoption study (Bohman, Sigvardsson, & Cloninger, 1981) generally replicated Goodwin's findings. While some of the results from this important study were published before 1980 when the first IOM/NAS report was issued (Bohman, 1978), the impact of the study on American scientists was not felt until it was "imported" to this country by Cloninger, who collaborated with Bohman and Sigvardsson when he was a visiting professor at Umea University (Cloninger, Bohman, & Sigvardsson, 1981). In addition, the work of Cadoret and colleagues on a smaller adoptee sample was also published in 1978 (Cadoret & Gath, 1978), lending additional credibility to the genetic hypothesis.

The empirical work of Cloninger, Bohman, and Sigvardsson has probably contributed considerably to the emphasis on the "biobehavioral" approach, particularly with regard to the advances noted above in the hypothesized heterogeneity among alcoholics. Their work on genetic predisposition has pointed to two forms of alcoholism, one associated with genetic factors, the other associated with environmental factors. Although these authors and others who espouse the genetic hypothesis favor a biobehavioral approach, the work to date places greater emphasis on the "bio" than on the "behavioral."[3] In fact, it is performed primarily by those in the biomedical camp and may be regarded as analogous to a social scientist attempting to perform biomedical research.

While social science approaches were still very much apparent in the IOM/NAS report (see Tables 3 and 4), the pendulum had quite definitely swung toward studying the etiologies of alcoholism, particularly geared toward the use of genetic hypotheses. Three chapters dealt with the social sciences, but only two of the ten chapters could, without qualification, be identified with Bacon's earlier call for a study of *all* behaviors associated with drinking: one was concerned with quantifying the impact of alcohol on society, the other with the social consequences of alcohol abuse.

The chapter dealing with quantifying the impact of alcohol on society, including alcohol's role in accidents, its interaction with the use of other drugs, its contribution to various causes of death, and its role in the creation of "victims," such as children and spouses (see Table 3), relied on epidemiological or survey research as basic methodological strategies but also recommended further research to promote the understanding of "the biological and social mechanisms leading to differential susceptibility among individuals and subgroups for alcohol-related problems" (IOM/NAS, 1987).

The second chapter was related to the social consequences of alcohol abuse (including studies of alcohol's relation to children, family violence, crime, homelessness, and drug abuse). While more closely aligned to Bacon's call for a study of the spectrum of drinking, it nonetheless evaded it by dealing primarily with "adverse social consequences."

The chapter coming closest to Bacon's call was titled "Social Determinants of Risk," where recommended analyses were based on study, not only of individual variation in drinking behavior but on societal and historical variation as well (see Table 2). Emphasis was placed on the social and historical correlates of drinking in which entire populations shifted and on drinking beliefs and drinking settings. But this chapter paled in comparison to the bulk of the report, which was devoted to the biological or biobehavioral views. Interestingly, the chapter was the last of 12 chapters, perhaps reflecting the composition of the committee (2 of 10 were trained in the social sciences).

The areas of study concerned with the influence of alcohol on the body included medical consequences of abuse (again with discussion of genetic components) and the consequences of drinking on performance. The latter was an area characterized as one providing an interface between the biological and behavioral sciences in which expectations regarding drinking and biological mechanisms, particularly from the electrophysiologic area, interact.

The 1987 report reflects the status of today's alcohol research world—one dominated by biomedical research-based etiological hypothesis that is aimed at the "exotic," namely alcoholism. The efforts to create an integration of the "bio" and the "behavioral" have thus far, in this author's opinion, reflected more of the former and less of the latter and, furthermore, have not emphasized the role of culture and the spectrum of both independent and dependent variables that Bacon originally recommended for the behavioral approach.

Commentary

The Political and Social Influences

No era of scientific inquiry is without its political influences, influences potentially directing the nature of the research questions. At the time of this writing, the politics in 1942 surrounding alcohol-related science have been commented on, which, according to other investigators, influenced the research questions of that era. It is more difficult to comment on the political influences of today, primarily because we are living these hours. But how will those scientists in 2034 see us? Will we be seen as a body of scientists responding to and reinforcing an increasingy conservative conceptualization of human life, an era of biological romanticism?

How will the social forces around us be interpreted? We see today, for instance, a biological determinism on the rise in a social and political world where a wide variety of alcohol problems are increasingly being "treated" under the guise of "alcoholism" and under coerced conditions (conditions sometimes characterized as punishment rather than treatment), where per capita consumption is decreasing, and where a large and relatively understudied social movement, the Adult Children of Alcoholics (perhaps to have the ideological and social impact of that of Alcoholics Anonymous in an earlier era), blames *all* intra- and interpersonal human problems of its members on the alcohol use of parents.

Will the continued focus on disease-like entities be deemed appropriate by our scientific grandchildren under these conditions? Will the almost exclusive focus on etiology be considered a valuable contribution to science? Will our scholarly grandchildren muse, as we do about earlier historical eras, that the scientific community from the mid- to the turn of the 21st century experienced an implosion, rather than an explosion, of hypotheses concerning alcohol-related problems, configured somehow by the social and political events of the time? Will the fact that science for a brief period toyed with the problems of alcohol be regarded as an amusing historical anecdote?

Disciplinary Ascendancy

The ascendancy of one discipline in the alcohol field has been traced in this chapter. The explanation offered by this discipline essentially lies in a predisposition to a condition called alcoholism; scientific efforts to map out this etiology in the near future will explore the environmental and social conditions under which the predisposition is supported or restrained. In the enthusiasm of locating this predisposition, little attention has been paid to the continuing lack of consensus on just precisely what the definition of the condition is. This problem is perhaps most evident by observation of the fact that the alcohol field is continually holding conferences to redefine the concept of alcoholism.

Let us imagine the scientific future, perhaps 10 years hence, of the impact of disciplinary ascendancy in the explanation of a social problem. Let us imagine that a biological marker, perhaps an element that can be easily identified in the blood at birth, is located that, contingent on certain environmental conditions, will predispose some people to alcoholism. Let us now restate some of the research questions we have covered since the 1940s in those terms:

- What are the societal functions served by the drinking of alcoholic beverages for those predisposed versus those not predisposed?
- What treatment regimes will maximize outcome for those alcoholics with and without predisposition?
- How can we develop prevention strategies for the predisposed versus the nonpredisposed?
- What are the economic costs on society from those predisposed to alcoholism?
- What is the temporal sequencing of predisposed alcoholism with other psychopathologies (for example, criminality, antisocial behavior)?
- What are the societal consequences for societies or historical periods having a higher or lower prevalence of the genetically predisposed?

Our imaginary forecasting of scholarly questions when directed by the ascendancy of one disciplinary approach may serve to color and transform research questions from *all* other disciplines (at least research questions for which funding support will be granted) such that the unique paradigms from individual disciplinary approaches will be subordinated to the reigning discipline. This implication is touched on below.

The Objectives of Study and the Notion of a Coordinated Scientific Effort

It has been noted that although there were breaches between the disciplinary efforts in the 1940s, particularly with regard to the dependent variable under study, there was an underlying assumption that if the disciplines worked togther in some joint interdisciplinary effort, the answers to alcohol problems in general would be located. It has also been noted that the most recent recommendations in the field are concerned with an intense interdisciplinary effort to understand the interactions between biological and environment factors that contribute to the incidence of alcoholism. This continuity in the field, although perhaps for quite different purposes and under quite different political conditions, should give us pause.

Frankl maintains that to understand the whole of man the disciplines must integrate their knowledge:

> The challenge is how to attain, how to maintain, and how to restore a unified concept of man in the face of the scattered data, facts and findings supplied by a compartmentalized science of man.... What is dangerous is the attempt of a man who is an expert, say, in the field of biology, to understand and explain

> human beings exclusively in terms of biology. The same is true of psychology and sociology as well. At the moment at which totality is claimed, biology becomes biologism, psychology becomes psychologism, and sociology becomes sociologism. In other words, at that moment science is turned into ideology. (Frankl, 1969, as quoted in Glaser, 1974)

It is suggested here that this recommendation requires more consideration from a number of different perspectives. It requires us to step back and look at the objectives and meanings of science.

On the one hand, there is great benefit in persons from diverse fields in the scientific arena joining together in pursuit of developing an understanding of a number of topic areas, including the study of human beings. After all, social scientists would hardly deny that human behavior is influenced by biological factors any more than biologists would deny that it is influenced by social or environmental factors. Interdisciplinary approaches, of course, have the potential of contributing to the ongoing description of human behavior. These efforts are maximized when the *definition* of the object of inquiry is agreed upon by those representing the diverse disciplines. Furthermore, there is much to be learned from an interdisciplinary perspective that has practical applicability, particularly with regard to the quality of human life, but it is important to point out that the goal of practical applicability is *not* the objective of science. Nonetheless, many people, scientists included, regard the essence of science to be a concerted interdisciplinary effort that will eventually describe the whole of a subject, including "the whole of man," and that description will bear one integrating theoretical stamp.

On the other hand, there are scientific perspectives that diverge from the above. Since *all* is not known about *any* object of study, certainly the study of human beings, it is incumbent on scientists to bring to the inquiry their own unique disciplinary perspectives. This may mean that differing perspectives will produce *competing, if not incompatible, theories* with regard to the object of inquiry, including human behavior. In this respect, the object of inquiry may be conceptualized in different ways and, indeed, the questions raised in reconceptualization from one disciplinary perspective may be provocative, radical, and contrary to other disciplinary perspectives. In this view of science, it is incumbent on scientists from different theoretical perspectives to propose hypotheses that have *no hope* of being integrated into a central theoretical function that can explain the "whole" of any object of scientific inquiry, including the "whole of man."

There are, of course, pitfalls and dangers in both perspectives of science, particularly with regard to the social and ethical consequences of each of these broad outlooks. Because our universe of study concerns a social problem, there is danger in singular explanations of human behavior when placed in the hands of political bodies. Researchers are obliged, at the very least, to acknowledge competing hypotheses such that science will not serve political regimes, regimes that may advocate loss of human rights, if not human life (Fillmore & Sigvardsson, 1988). It might be mentioned as well that it is not unknown for scientists to either seduce or mislead politicians to be advocates of their own single disciplinary perspectives.

There are also pitfalls and dangers in a coordinated scientific effort that seeks one universal answer to one puzzle. Our universe of study does include, but is not confined to, social problems and this, it is believed, is our dilemma as scientists in the alcohol field. The mandate of the National Institute on Alcohol Abuse and Alcoholism and other like bodies throughout the world is not directed toward the objectives of science but toward seeking cures and developing prevention strategies for alcohol problems. The mandate of science includes the derivation of hypotheses that may be *totally at odds* with both the nature of the officially stated puzzle and the coordinated effort to understand that puzzle. If the community of scientists loses sight of this end, it becomes a slave to social policy.

In 1943, Selden Bacon appealed to this new field, in analogy, not to confine its inquiry to the "exotic"—to the emerald, the volcano, or the iceberg—but to the whole earthcrust and to bring to that study a broad range of questions and discipinary approaches. Although Bacon stated the appeal from the mandate of social science, it is restated here as an appeal for the mandate of science as a whole.

Acknowledgments

This work was supported by a National Institute of Alcohol Abuse and Alcoholism Research Scientist Development Award (1KO1 AA 00073-04) and is written in honor of my mentor and friend, Selden D. Bacon. It is hoped that Bacon's influence in my scholarly life has balanced my appreciation for research that studies both the "exotic" and "nonexotic." Should bias be present, it results from Bacon's recommendation that to understand the "exotic" one must understand the "nonexotic": the field of social science "may still be found gazing in starry-eyed wonder at the occasional volcanoes, emeralds, and icebergs, to analogize from geology, when it has a gigantic earthcrust as its field of inquiry" (Bacon, 1943).

Notes

1. This important work influenced the emergence of a contemporary social science environmental explanation for the prevalence of alcohol problems (including cirrhosis) across social groups in the form of the Ledermann (1956)

hypothesis that postulates a relationship between per capita consumption and the prevalence of various alcohol problems.

2. By 1958, however, Trice and Pittman (1958) outlined a rather large collection of social science work that, although widely scattered, spoke to the substantial creation of a social science of alcohol.
3. An article by Reich, Cloninger, Van Eerdewegh, Rice, and Mullaney (1988) attempts to bring together these two fields by documenting and integrating temporal trends in alcoholism and genetic predisposition to them, but the research falls well below the mark with regard to its social science charge for two reasons. First, evaluation of temporal trends of alcoholism in the general population uses noncompatible operational definitions across time, resulting in the conclusion that there has been a major increase in the prevalence of alcoholism from 1968 to 1983, a finding not generally supported by major social science epidemiological investigations. Second, the lack of attention to research that specifies the ages at which spontaneous remission is most likely to occur (as derived from social science investigations) influences the conclusions such that what the authors perceive to be a cohort effect may actually be an age effect. This article laid the groundwork for a conclusion in the 1987 IOM/NAS report that "environmental influence alone cannot be used as the basis for distinguishing between familial and non-familial alcoholism" and justified further research in this area.

References

Bacon, S.D. (1943). Sociology and the problems of alcohol: Foundations for a sociologic study of drinking behavior. *Quarterly Journal of Studies on Alcohol, 4,* 402–445.

Bacon, S.D. (1952). Alcoholism, 1941–1951: A survey of activities in research, education and therapy. I. Introduction. *Quarterly Journal of Studies on Alcohol, 13,* 421–424.

Bacon, S.D. (1952). Alcoholism, 1941–1951: A survey of activities in research, education and therapy. IV. Social science research. *Quarterly Journal of Studies on Alcohol, 13,* 453–460.

Bacon, S.D. (1979). Discussion: Social science. In M. Keller (Ed.), Research priorities on alcohol: Proceedings of a symposium. *Journal of Studies on Alcohol,* Supplement 8, 289–321.

Bales, R.F. (1946). Cultural differences in rates of alcoholism. *Quarterly Journal of Studies on Alcohol, 7,* 480–499.

Bohman, M. (1978). Some genetic aspects of alcoholism and criminality. *Archives of General Psychiatry, 35,* 269–2176.

Bohman, M., Sigvardsson, S., & Cloninger, C.R. (1981). Maternal inheritance of alcohol abuse: Cross-fostering analysis of adopted women. *Archives of General Psychiatry, 38,* 965–969.

Bowman, K.M., & Jellinek, E.M. (1942). Alcohol addiction and its treatment. In E.M. Jellinek (Ed.), *Alcohol addiction and chronic alcoholism.* New Haven: Yale University Press.

Bynum, W.F. (1984). Alcoholism and degeneration in 19th century European medicine and psychiatry. *British Journal of Addiction, 79,* 59–70.

Cadoret, R.J., & Gath, A. (1978). Inheritance of alcoholism in adoptees. *British Journal of Psychiatry, 132,* 252–258.

Cloninger, C.R., Bohman, M., & Sigvardsson, S. (1981). Inheritance of alcohol abuse. *Archives of General Psychiatry, 38,* 861–868.

Conrad, P., & Schneider, J.W. (1980). *Deviance and medicalization: From badness to sickness.* St. Louis: C.V. Mosby.

Cruz-Coke, R. (1964). Colour-blindness and cirrhosis of the liver. *Lancet, 2,* 1064–1065.

de Lint, J., & Schmidt, W. (1968, September). *The distribution of alcohol consumption.* Paper presented at the International Congress on Alcohol and Alcoholism, Washington, DC.

de Lint, J., & Schmidt, W. (1968). The distribution of alcohol consumption in Ontario. *Quarterly Journal of Studies on Alcohol, 29,* 968–973.

Fillmore, K.M., & Sigvardsson, S. (1988). "A meeting of the minds"—A challenge to biomedical and psychosocial scientists on the ethical implications and social consequences of scientific findings in the alcohol field. *British Journal of Addiction, 83,* 609–611.

Frankl, V.E. (1969). *The will to meaning: Foundations and applications of logotherapy.* New York: World Publishing.

Glad, D.D. (1947). Attitudes and experiences of American-Jewish and American-Irish male youth related to differences in adult rates of inebriety. *Quarterly Journal of Studies on Alcohol, 8,* 406–472.

Glaser, F.B. (1974). Medical ethnocentrism and the treatment of addiction. *International Journal of Offender Therapy and Comparative Criminology, 18,* 13–27.

Goodwin, D., & Guze, S.B. (1974). Heredity and alcoholism. In B. Kissin & H. Begleiter (Eds.), *The biology of alcoholism,* Vol. 3: *Clinical Pathology* (pp. 37–52). New York: Plenum.

Herd, D.A. (1984). *Ideology, history and changing models of liver cirrhosis epidemiology.* Paper presented at the Alcohol Epidemiology Section Meetings of the International Council on Alcohol and Addictions, Edinburgh, Scotland.

Horton, D. (1943). The functions of alcohol in primitive societies: A cross-cultural study. *Quarterly Journal of Studies on Alcohol, 4,* 199–320.

Institute of Medicine of the National Academy of Sciences. (1980). *Alcoholism and related problems: Opportunities for research.* Washington, DC: National Academy Press.

Institute of Medicine of the National Academy of Sciences. (1987). *Causes and consequences of alcohol problems: An agenda for research.* Washington, DC: National Academy Press.

Jackson, J. (1957). The definition and measurement of alcoholism: H-technique scales of preoccupation with alcohol and psychological involvement. *Quarterly Journal of Studies on Alcohol, 18,* 240–262.

Jackson, J. (1958). Types of drinking patterns of male alcoholics. *Quarterly Journal of Studies on Alcohol, 19,* 269–302.

Jellinek, E.M. (1942). An outline of basic policies for a research program on problems of alcohol. *Quarterly Journal of Studies on Alcohol, 3,* 103–124.

Johnson, B.H. (1973). *The alcoholism movement in America: A study in cultural innovation.* Unpublished doctoral dissertation, University of Illinois, Urbana-Champaign.

Keller, M. (Ed.). (1979). Research priorities on alcohol: Proceedings of a symposium. *Journal of Studies on Alcohol,* Supplement 8.

Ledermann, S. (1956). *Alcool-Alcoolisme-Alcoolisation: Donnees Scientifiques de Caractere Physiologique, Economique et Social.* Paris: Presses Universitaires de France.

Lester, D., & Greenberg, L.A. (1952). Alcoholism, 1941–1951: A survey of activities in research, education and therapy. III. The status of physiological knowledge. *Quarterly Journal of Studies on Alcohol, 13,* 444–452.

Li, T.K. (1977). Enzymology of human alcohol metabolism. In A. Meister (Ed.), *Advances in enzymology,* Vol. 45 (pp. 427–484). New York: John Wiley and Sons.

Lieber, C.S. (1975). Alcohol and the liver: Transition from metabolic adaptation to tissue injury and cirrhosis. In J.M. Khanna, Y. Israel, & H. Kalant (Eds.), *Alcoholic Liver Pathology* (pp. 1–18). Toronto: Addiction Research Foundation.

Lolli, G. (1952). Alcoholism, 1941–1951: A survey of activities in research, education, and therapy. V. The treatment of alcohol addiction. *Quarterly Journal of Studies on Alcohol, 13,* 461–471.

Quarterly Journal of Studies on Alcohol. (1941). Alcohol, heredity and germ damage, Lay Supplement 5. New Haven, CT: Laboratory of Applied Physiology, Yale University.

Reich, T., Cloninger, R., Van Eerdewegh, P., Rice, J.P., & Mullaney, J. (1988). Secular trends in familial transmission of alcoholism. *Alcoholism: Clinical and experimental Research, 12,* 458–464.

Room, R. (1966). *Notes on "Identifying problem drinkers in a household health survey" by Harold Mulford and Ronald Wilson.* Unpublished manuscript, Social Research Group, Berkeley, California.

Room, R. (1978). *Governing images of alcohol and drug problems: The structure, sources and sequelae of conceptualizations of intractable problems.* Unpublished doctoral dissertation, University of California, Berkeley.

Seeley, J. (1959). The W.H.O. definition of alcoholism. *Quarterly Journal of Studies on Alcohol, 20,* 352–356.

Snyder, C.R. (1958). *Alcohol and the Jews.* New Haven: Yale Center of Alcohol Studies.

Straus, R., & Bacon, S.D. (1953). *Drinking in college.* New Haven: Yale University Press.

Trice, H., & Pittman, D. (1958). Social organization and alcoholism: A review of significant research since 1940. *Social Problems, 5,* 294–307.

Trice, H.M., & Wahl, J.R. (1958). A rank order analysis of the symptoms of alcoholism. *Quarterly Journal of Studies on Alcohol, 19,* 636–648.

Wiener, C. (1981). *The Politics of alcoholism: Building an arena around a social problem.* New Brunswick, NJ: Transaction Books.

CHAPTER 3

The Promise and Problems of Alcohol Sociology

HARRY G. LEVINE

I entered the alcohol field in 1974 as an impoverished graduate student in sociology at the University of California at Berkeley. Knowing nothing about alcohol research, but needing income, I accepted what I thought would be a very short stint as a National Institute of Alcohol Abuse and Alcoholism (NIAAA) predoctoral dissertation fellow. The fellowship was offered by the Social Research Group (SRG) that was affiliated with the School of Public Health at U. C. Berkeley.

Since the early 1960s the SRG had done research on a range of alcohol- and drug-related topics, especially national, state, and regional surveys of drinking practices and problems. Over the years, SRG staff members Genevieve Knupfer, Don Cahalan, Robin Room, Walt Clark, and Ron Roizen identified many of the distinct problems and behaviors resulting from alcohol use. The SRG researchers sought to disaggregate the syndrome of alcoholism and look instead at the different ways that drinking interacts and conflicts with the drinker's life. This focus to their work gave rise to their critiques of the disease concept of alcoholism, to their development of the notion of drinking problems and alcohol problems, and to their collection and analysis of a large and growing body of empirical data on U.S. drinking practices and problems.

All but one of the senior staff had trained as sociologists, and all brought to their work a real interest in the life that surrounded and interacted with drinking. As Room and Roizen would put it, the SRG shop perspective used alcohol as a window or lens for viewing broader questions of society and culture.

In 1977, the SRG was selected as one of the first five federal alcohol research centers and the only one devoted to social science research. By that time I was already on the staff. The new federal grant had opened even further an ongoing debate in the office as to the kinds of research that we as sociologists could and even ought to carry out. As part of that open-ended conversation I wrote a paper pointing out some of the highlights of Bacon's original call for a sociological study of drinking and encouraged us to take up the mission. Like Bacon, I was a relative newcomer to alcohol studies, and, like Bacon, I meant my piece to be provocative. Though I understood some of the enormous constraints of funding, I hoped to stimulate thought about new kinds of research projects.

There were several levels of irony and internal references in what I did.[1] The SRG staff were always reading and talking about the work of Jellinek, Keller, Bacon, and other original members of the Yale and Rutgers centers. SRG staffers poured over each issue of the *Journal of Studies on Alcohol* and they understood themselves as the young turks challenging the elders in the field. In turning to Bacon and quoting him chapter and verse I was engaged in a kind of scriptural argument familiar to my colleagues, and I was using the appropriate holy text. I was also making not very veiled comparisons between our situation at the SRG and the one the Yale center faced in the mid-1940s: Were we going to take Jellinek's path and focus only on the problems of alcohol use, or would we follow the oh so boldly put recommendations of the only sociologist in the original Yale group? Part I of the chapter below is essentially the paper I wrote at that time.

Unlike Bacon, I did not stay around my alcohol center. Shortly after I completed my dissertation on the temperance movement I moved across the country to join the faculty of Queens College of the City University of New York.

In 1980, Robin Room wanted to publish my paper in the *Drinking and Drug Practices Surveyor,* an informal journal and newsletter aimed primarily at alcohol and drug researchers. I used the opportunity to write a section called "second thoughts" in order to broaden the scope of the paper to include anthropological and historical work and to make explicit some things that I had taken for granted. Part II of this chapter is most of that second section.

Shortly after that, Ronald Reagan was elected president. Under the auspices of budget director David Stockman, the Reagan administration proposed a series of budget-cutting measures, including major cuts in social science research. It was in this reactionary and threatening political climate that the Social Research Group changed its name to the Alcohol Research Group. Just prior to publication in 1981 I tacked on an afterword to make some political points addressing the conservative atmosphere of the early Reagan years and to pull back from what now clearly seemed to me to be the overly optimistic and gung-ho tone of the original piece. Part III is that afterword.

Another nine years has passed and I remain more than ever a student of the alcohol field. For me now the interesting question Bacon's paper raises is one I passed over originally because I did not really have an answer: Why wasn't Bacon's agenda followed more, and why has it seemed so hard to put it into effect?

I now approach that question through the lens of the remarkable developments of the last decade. This has included: the rapid growth in public concern with alcohol problems, the amazing spread of the "12th step" movements inspired by Alcoholics Anonymous (AA), the appearance and rapid growth of the movement centered around Adult Children of Alcoholics, the creation of Mothers Against Drunk Drivers and a vigorous antidrunk-driving movement, and the increase with which individuals and groups use the terminology of addiction and compulsive behavior to describe all sorts of activities and problems.

Some observers have rightly interpreted recent developments as the rise of neotemperance sentiment and movements in America. I term this the growing strength of *temperance culture* in America. The irony of Bacon's own work is, I think, that as a crusader and supporter of the alcoholism movement he participated in the creation of sentiment hostile to his own sociological project. In part IV I discuss this point.

I. Selden Bacon's "Foundations for a Sociologic Study of Drinking Behavior" Revisited

In 1943, Selden Bacon, then a young assistant professor of sociology at Yale University, prepared a lengthy memo to E.M. Jellinek, Howard Haggard, and others who had gathered at Yale to study alcohol problems. At the time Bacon was rather periperhal to the scientists and researchers at the alcohol center. Bacon's memo was addressed to the opportunities created by the development of an interdisciplinary alcohol research center at Yale—what the people gathered around Jellinek and Haggard understood as the attempt to work out new post-Prohibition and posttemperance ways of understanding alcohol problems and the place of alcohol in society. They called this perspective the new scientific approach.

Bacon also wrote his memo in response to a paper of Jellinek's, published in the recently created *Quarterly Journal of Studies on Alcohol* (of which Jellinek was the associate and managing editor) and called "Outline of Basic Policies for a Research Program on Problems of Alcohol." Jellinek's purpose in his paper was to "designate those aspects of the problems of alcohol most in need of research." For him, "the central problem of alcohol" was the problem of the origins and development of addiction and of other forms of abnormal drinking (p. 105). Jellinek divided projects into various headings; along with psychiatry, experimental psychology, and research of alcoholic diseases, he included sociology. Indeed, sociology was the first area to be discussed. Under the heading of "A" or high-priority projects, Jellinek listed five specific studies focused on social sources of inebriety or alcohol addiction. Although Jellinek's proposed projects involved conducting a variety of studies on the place of alcohol in society, the only question he allowed as truly legitimate concerned the social factors contributing to alcoholism.

Selden Bacon's long memo was a reply to Jellinek's paper and a critique of it; Bacon offered a developed alternative vision of what sociological research on alcohol could be. Jellinek was relatively unfamiliar with sociology and its orientation. Bacon was a sociologist and he boldly drew upon a broad range of sociological and anthropological approaches and fashioned a lengthy shopping list of questions and issues for alcohol research. Jellinek, it seems, recognized the significance of Bacon's critique for he published the memo virtually as written in the *Quarterly Journal* in September 1943. He also wrote a special introduction to the piece, and subsequently created a monograph series called *Memoirs of the School of Alcohol Studies,* of which Bacon's paper was the first in the series.

Bacon's paper is still a remarkable document. Its questions remain absolutely contemporary and it has not been surpassed as a statement of a distinctly sociological perspective on alcohol. Although some of its phrases and references sound "old fashioned," much of the content of the paper, the questions it raises, and the perspective it offers strike the contemporary reader as new.

In essence, Bacon criticizes previous scientific approaches to studying alcohol for being concerned primarily or exclusively with questions of pathology. Other approaches, including Jellinek's, study alcohol only or principally insofar as it is a problem. Although that might be appropriate for the psychiatrist and physiologist, for the sociologist to study drinking and drunkenness as only a social problem and pathology is to miss the major phenomena—to miss the point of it all. Alcohol and drinking, Bacon said, should be studied as part of normal social life: all the behaviors, ideas, customs, mythologies, patterns, processes, traditions, and institutions surrounding drinking are open for inquiry and examination. Such things should be studied the way the sociologist studies religion or the family or politics—not exclusively for the pathologies and deviances, but in their wholeness, as significant parts of social life.

Part of the reason Bacon's argument sounds so new is because there has not been a body of work developed that self-consciously utilized the perspective he outlines. The "sociologic" approach has not been the one the alcohol field has taken; it has not even been an important minority voice. In the intervening years there have been a number of major studies that fulfill or speak to specific agendas raised by Bacon. But despite the breadth and depth of some of these studies, including a focus on normal drinking and on everyday life, researchers doing this work have by and large presented and justified themselves as problem- and pathology-oriented. Thus, while Jellinek acknowledged the legitimacy of Bacon's position, the pathology-oriented approach Jellinek outlined has remained the dominant one, even the only one, in the field.

A full discussion of why the alcohol field has been dominated by a problem-orientation is, of course, important but beyond the scope of this paper. Bacon discusses the issue at considerable length in a piece in 1976 and makes another call in 1979. Room (1979) also examines some of the issues involved in setting the early alcohol research agendas, while Johnson (1973) discusses the history of the Yale center and the alcoholism movement. Perhaps one important reason why Bacon's particular version of a sociological program was not adopted by more people was because Bacon never systematically developed it in his own work or with students; he himself became diverted from the project. After assuming directorship of the Yale and then Rutgers centers, Bacon was involved in the alcoholism movement as an organizer and as a promoter of the National Council on Alcoholism and of state alcoholism agencies and treatment facilities. When asked in a personal interview what happened in the more than 30 years between his first major call for a sociological approach to studying alcohol and his next one in 1976, Bacon thought for a moment and then suggested that he had gotten "sidetracked."

Whatever the historical explanations may be for the failure of sociologists and other social scientists to develop a nonpathological approach to studying alcohol, it should be possible to begin it now. Indeed, much of a distinct sociological approach already exists half-stated in various pieces of research and analysis and as part of the background assumptions in many writings. But that "sociological" perspective needs to be articulated, developed, elaborated, discussed, examined, and strengthened. Selden Bacon began the task in 1943, providing us with both a tradition and an example. Bacon's paper, "Sociology and the Problems of Alcohol: Foundations for a Sociologic Study of Drinking Behavior," is an excellent place to begin again.

Bacon starts his paper with a few preliminary remarks about science and outlines some of the particular interests and orientations of the sociologist in "the scientific study of the problems of alcohol." He says:

> The sociologist is interested in the customs of drinking, the relationship between these customs and other customs, the way in which drinking habits are learned, the social controls of this sort of behavior, and those institutions of society through which such control issues. The sociologist wishes to know the social categories in which much or little or no drinking occurs, he seeks correlations of amount and type of drinking with occupational, marital, nationality, religious, and other statuses. More importantly, he poses the broad questions:

> What are the societal functions served by the drinking of alcoholic beverages? What are the social rules concerned with drinking? What are the pressures for and against this practice? How does this behavior pattern jibe with other institutions and folkways? (p. 407)

Bacon marks out a broad terrain concerning the relationship of drinking to virtually all other major social activities.

The chief polemical point of Bacon's piece is probably an attack on inquiry and research centered exclusively or primarily on the pathological. Although this critique has rarely been formally developed, its spirit has animated most analyses of general population surveys—it has been the point of such studies. Indeed, survey research is probably the one area in the alcohol field where Bacon's critique has been most appreciated and absorbed; for example, elements of the following echo in the writings of Knupfer, Room, Cahalan, Clark, and Roizen.

> A factor which has delayed and discouraged an adequate analysis of drinking behavior has been the failure to recognize the relation of inebriety to all other forms of drinking. Consequently, there has been a failure to orient properly the abnormal phenomena that hitherto have dominated all studies in this field. This exotic fraction of drinking behavior has attracted all the attention.... It is perhaps true that most sciences had their genesis in the observation and analysis of the immediately painful and the extraordinary, but that state should by now have been passed. The entire field of social science may be freely criticized on this score; in many instances it may still be found gazing in starry-eyed wonder at the occasional volcanoes, emeralds, and icebergs, to analogize from geology, when it has a gigantic earthcrust as its field of inquiry. The exotic and the pathologic are useful fields of scientific inquiry, but they have their limits....
>
> The study of the exotic or pathologic members of a total group is not to be scorned, nor should the students of such types of persons be dissuaded; but such studies must be oriented within the total field and the generalizations limited to the segment studied. Studies of drunkards arrested by the police, although not so limited... are still concerned largely with the pathologic portion of all drinkers. (pp. 408–409)

The vast majority of drinkers, Bacon points out, are not pathological and have never been arrested.

Like all those who eventually go on to do general population studies, Bacon offers a powerful and persuasive rationale for studying the nonpathological, the general population, and the ordinary phenomena of drinking and getting drunk.

> Until the drinking behavior of a representative sample of the drinking population is observed, analyzed, and described, characterizations of the tiny propor-

tion of abnormal drinkers are likely to be as biased and as fallacious as the studies of Lambroso about criminals, studies of the insane during the eighteenth and nineteenth centuries, and comments about the lower-class poor issued by wealthy, upper-class philanthropists of almost any age.

The need for general orientation of problem drinking behavior within the setting of all drinking behavior, including abstinence and antidrinking activity, is one of the major reasons for the type of study here proposed. (p. 410, emphasis added)

After making his critique, Bacon returns to the theme of defining the sociological approach. Bacon titles one major section of his paper, "Framework for a Sociologic Study of Drinking Behavior." In this section the sociological concepts and terms Bacon employs are most outmoded and unfruitful, perhaps because Bacon drew heavily on the approach of William G. Sumner, who for many years had dominated Yale sociology. This conceptual framework, I think, distracted Bacon from the alcohol-related phenomena more than it helped him focus on them. In short, Bacon would have done better, more enduring sociology if he had been less concerned with explicating and classifying things in terms of sociological notions current at the time in his circles. In many ways, the aims of Bacon's proposals are more relevant today than the specific ways he divided up issues and problems. Therefore, I want to simply draw out some of the points and suggestions his paper makes.

Bacon's focus is primarily on drinking construed in as wide a fashion as possible and he offers a number of tasks and guidelines with regard to studying it. He suggests closely observing the varieties of drinking-associated, drinking-related, and drinking-accompanied behaviors. These can be seen as research topics, as ways of asking questions about drinking:

Carefully collected and classified information concerning the ways in which people drink is a *sine qua non* for the study of drinking behavior.... To consider all drinking as the same kind of behavior is an affront to intelligence as well as to the senses. (p. 416)

For example, Bacon urges studying what he calls "the rules" covering drinking:

There are right and wrong places to drink, times to drink, people to drink with, types and amounts of alcoholic beverages to imbibe, subjects to talk about and activities to pursue while drinking, ways of mixing drinks, methods of imbibing (fast, slowly, with straws, inhaling of bouquet, holding glass, mug, or bottle, etc.), clothing to wear at drinking parties, ritual phrases and body movements, and so forth. These rules vary with different groups. (pp. 418–419.)

Another issue we might include in a normative study of drinking Bacon identifies as the social factors that discourage and encourage drinking. In particular, he suggests that the ritualistic aspects of drinking be studied in a "wide scope":

> That there are great rewards for drinking in our society can hardly be denied; being a genial and lavish host, being a connoisseur of wines or whiskies, being able to "hold your liquor," are obviously rewarding states of affairs.... Drinking is closely associated with many pleasant occasions and situations and, in addition to any direct reward it may hold in itself, is reinforced as a source of pleasure because of this association.... Drinking is a component of many rituals that may be deeply satisfying to the individuals participating in them. It is not impossible that the individual in this last instance may consciously dislike the taste and aftereffects of the alcohol but still derive great pleasure from the ritual as a whole and be strongly opposed to any changes in the forms and procedures involved. (p. 420)

In studying or researching these different parts of drinking and social life Bacon points out the lack of homogeneity in the society as a whole and the necessity of paying attention to subcultural differences.

One of the most interesting of Bacon's suggested areas for research and study, still relatively unexplored, he calls the "rationale or charter" of drinking. He has in mind here much of what might be subsumed under studying drinking norms, but what he is describing and talking about would better be called *culture* in the fullest or anthropological sense of the term. Indeed, Bacon refers to anthropology when defining *charter:*

> Mythology, old saws, ethical systems, abstracted norms, all help to form charter. There is no need for this charter material to be logical. It is quite possible that to an outside observer it will be both internally inconsistent and at complete odds with the "real" function.... (p. 414)

Bacon is looking here for the background assumptions, the folklore or traditions and ideas regarding drinking and its effects. Again, the topic is nowadays best captured by the concept of *culture:*

> The mythology, folklore, and superstitions about the powers and functions of alcohol form the charter of drinking behavior.... [I]t is nonetheless relevant and significant that such beliefs are powerful reinforcements to behavior, whether that of the prohibitionist or the excessive drinker. More than that, they are an integral part of the tradition handed on to rising generations, and thus operate as instigators of behavior. The experience of a "good" education is no particular bar to the belief in fallacious concepts; highly trained lawyers and

doctors may believe in the stimulating, body warming, and medicinal qualities of alcohol. Belief that drinking is a sign of vigorous manhood may be a spur to action outweighing the deterrent effects of statistics about accidents, poverty, or disease that may be invoked against the habit. (p. 415)

Bacon continues with a list of fascinating sources of data for these "charter" or cultural elements:

There is a wealth of charter material concerning the drinking of alcoholic beverages. It can be found in maxims, in poems, in plays, in advertising, in songs, in books on etiquette, and innumerable other sources. It is a constant source of behavior, of rationalization, and of indoctrination. The part it plays in the total pattern of drinking behavior is of great significance. The theoretical background and announced concepts of organized religion, of law, of social work agencies, of recreation groups, of organizations of producers and distributors of alcoholic beverages, and of temperance and teetotal groups, as well as the attitudes of groups only secondarily concerned, play an important role that must be studied if the motivations behind both drinking and abstinence, and if the scene in which drinking occurs, are to be competently portrayed. (pp. 415–416)

This analysis of the elements of culture and everyday life, of the influences of various aspects of society on drinking and attitudes, and the use of all kinds of documents to research these questions, is probably the one area that has received the least attention and development in the decades since Bacon first proposed it. It remains an important and fruitful area for inquiry. Bacon, himself, seemed especially sensitive to these sorts of cultural themes and questions. He discusses briefly at one point the importance of what he called "material apparatus" and offers some examples of their relevance for analysis:

In drinking behavior, this would refer to bottles, casks, tables, rails, spigots, glasses, aprons, checks, etc. Material apparatus is an essential part of behavior. It can set limits to action; it can, through association, stimulate action. . . .

Material apparatus is also significant in drinking behavior in that certain articles come to have a value of their own. Cocktail shakers, old kegs, particular drinking glasses, and other items become significant symbols for particular categories of persons in the society. They may enhance the values of drinking. They may serve as symbols of evil to antidrinking groups. In our society the attempt to better one's social position by material purchase and by conspicuous consumption has been a practice of long standing. Material apparatus relevant to drinking behavior, such as chromium soda containers, home bars, antique decanters, and brandy inhalers, are examples of drinking articles that may have a secondary, perhaps even a primary, importance because of their social class prestige or "snobbery value." (pp. 416–417)

This attention to the class- or status-signifying aspects of drink is one of the least explored and probably most important aspects of drinking culture. It is heavily exploited by the alcoholic beverage industries in their advertisements, and it has been part of the defensive response of the beverage industries since at least the mid-19th century: the alcohol industry has long attempted to present itself as a provider of "classiness."

The final section of Bacon's paper contains a proposal for a community study. In the course of outlining the study Bacon reviews again, from a slightly different angle, the sorts of data available and useful to those studying the place of alcohol and alcohol problems in society. Under a heading of "historical data" he compiles an amazing list of data sources that would be appropriate for any sort of sociological study:

> Historical sources containing material peculiarly relevant to drinking behavior should be sought industriously and with keenly critical appreciation.... The legal sources will be more abundant than others, but their imbalance on the side of the abnormal, the tabooed, and the criminal, as well as the rationalizing nature and political fluctuation of law in general, render them subject to marked limitations. Reports of temperance, religious, and other societies with decided views on the ethics of drinking are a useful source, but require always to be labeled as such. Surveys of vice, crime, drinking, poverty, and the like are not uncommon in any sizable community. Special studies of immigrant groups represented in the community should be carefully examined for relevant descriptions and analyses. Industrial and government reports on production and distribution can furnish suggestive information. In addition to legislative and judicial reports, taxing, licensing, and inspecting agencies often have maintained records over long periods of years. Reports of administrative officers such as the mayor, controller, etc., are usually on file. Political campaign speeches are often available. Sermons and tracts of the past are often well preserved, sometimes even better than official records. Travelers' comments are often surprisingly detailed and relevant. Novels written by members of the community, collections of letters, wills, and similar data may be maintained by local libraries and historical societies. Books privately printed by local antiquarians may be difficult to discover, but can repay the search. Diaries, parish records, institutional records, doctors' records, etc., may also cast light on the drinking behavior of earlier days. Newspapers are of especial value because of their continuity; the advertisements, descriptions of social events, cooking hints, and general reporting can all be of great relevance. Novels of the time, whether or not concerned with the community, can be of real value in the presentation of insights and in unconscious reflection of the values, groups, and material culture of the time; that such works generally are concerned with a particular social class is no bar to their validity; that their comment on particular cultural phenomena is accidental or casual is, on the whole, an asset rather than a liability. Many other sources of an historical nature would undoubtedly be discovered: materia medica, pictures, warehouse receipts, recipe

> books, manuals on manners, social histories, material on drinking from fairy tales and anecdotes of heroes, reports of military men, ships' logs, all come to mind. (pp. 434–435)

As Bacon notes, "The field to cover is vast."

The most glaring omission from Bacon's suggestions for study is the institutional context of drinking, especially the two most important actors in the alcohol field: the government or state, including all agencies and bureaus at all levels; and the alcoholic beverage industry, including producers, distributors, lobbies, and advertising and public relations efforts. The government and the industry have been the two most significant institutions shaping drinking and drunkenness: the way drinking enters in our lives; the way it looks; how, when, and where drinking goes on; the kinds of things people drink; the way they drink; and so on. Insofar as the place of drinking in society has been organized, managed, administered, controlled, shaped, and influenced, it has primarily been done by the government and by the industry. In most cases the reasons or the rationale for actions have not been determined by considerations of health, but by economic and political considerations: what is necessary for social order, what is necessary for fiscal considerations, and what is necessary for the industry. Although Bacon mentions institutions such as temperance and public health groups, he stresses insufficiently their importance in shaping drinking patterns.

If a developed list of questions regarding the state or government, the alcohol beverage industry, and other organizations and movements (especially nowadays Alcoholics Anonymous and the alcoholism movement) are added to Bacon's social and cultural questions, we have, I think, a dazzling array of material within the sociological domain.

II. Anthropology, History, and Alcohol Studies: Second Thoughts in 1980

The previous section, first written several years ago, was prepared as a discussion piece for the Social Research Group. Since then I have realized that I had a couple of expectations or hopes for alcohol sociology that were not stated in the paper.

One obstacle to reviving the minimal "sociologic" approach Bacon called for 35 years ago may be the field of sociology itself. Sociology may not be the social scientific or humanistic discipline capable of taking on these questions. It is possible, indeed it seems likely, that anthropologists are at this time the ones most equipped or prepared to engage in such studies. At the minimum, they may be the ones to lead the social sciences and the alcohol field in this program.

One individual who has been speaking in the tradition and for the general aims of the program that Bacon proposed is the anthropologist Dwight Heath. In a number of papers over the last 10 years, Heath has repeatedly urged an approach to the study of alcohol in society that casts as wide a net as possible, considers all sorts of questions, and is not obsessed with pathology.

In 1975 Heath published a very long "Critical Review of Ethnographic Studies of Alcohol Use" in the annual volume put out by the Addiction Research Foundation of Toronto. The paper is a thorough, extremely thoughtful discussion of the vast anthropological literature, its assumptions and orientations. Heath is clearly trying to address the alcohol field; he points out that despite all their studies "anthropologists have generally remained on the fringes" of the alcohol field. Echoing Bacon's complaint to Jellinek, Heath says that is at least partly because in alcohol studies "there is often an obvious—even if not explicit—preoccupation with alcoholism (in a clearly problem-oriented or ameliorative sense) which takes precedence over attempting to understand alcohol in a normal, customary, and value-free sense." Heath suggests that anthropologists have not shared this preoccupation because the focus on problems is at odds with the realities of the societies that anthropologists study:

> *One of the most striking contrasts between the roles and meanings of alcohol in Western and non-Western cultures is the immense emphasis on "social problems" in the former, and a relative absence of such concerns in the latter.* This does not necessarily mean that there is a full consensus about drinking and behavioral consequences in non-Western societies, nor does it mean that drunkenness does not sometimes result in social psychological, economic, or other problems for individuals or groups in such societies. *The evidence does, however, very strongly suggest—and in many instances, explicitly affirm—that alcoholism, addiction, and other pathological manifestations are extremely rare in cross-cultural perspective.* (1975, p. 34, emphasis added)

The gist of Heath's argument is that because of the nature of drinking in non-Western societies, anthropologists have a strong disciplinary investment in studying drinking as part of everyday life. For anthropologists to follow the mainstream of alcohol studies, and focus primarily on the causes and consequences of alcoholism and alcohol problems, would in most cases narrow their field of study down to almost nothing.

Anthropology is concerned with the differences between the West and other cultures, and anthropologists often point out how the modern world is unusual and even unique among cultures. This difference certainly holds true regarding the place of alcohol in society; in fact, drinking appears as almost an ideal case. Heath raises the issue several times in his 1975 article, as if attempting to wake up the alcohol field to the larger world.

> *Alcoholism—even in the general sense of problems associated with drinking—is rare in the vast majority of the societies of the world. One might even go so far as to note that it is almost unknown outside of the mainstream of Western culture, although it is becoming a widespread concomitant of acculturation which often accompanies the impact of modern industrial society.* Another striking feature of the ethnographic literature on alcohol studies, in marked contrast with a major portion of the writings on alcohol in Western society, is the relative lack of concern with social problems, education, prevention, therapy, or even alcoholics and alcoholism as meaningful categories. As already discussed, this is in large part a reflection of the fact that, in cross-cultural perspective, "problem drinking" is very rare, and alcoholism seems to be virtually absent even in many societies where drunkenness is frequent, highly esteemed, and actively sought. (p. 76, emphasis added)

Other anthropologists have addressed these points as well. In his introduction to a fine volume on drinking in a variety of cultures, Mac Marshall (1979) agrees that "the field of anthropology has never held center stage in the inter-disciplinary scientific explorations of alcohol use and abuse that have been conducted in the Western world for a century or more." Marshall also observes that because of the intellectual thrust of their discipline, anthropologists have not been primarily concerned with alcoholism and alcohol problems.

> Rather, as part of their effort to comprehend alien ways of life, anthropologists have been intrigued primarily by the myriad styles of normal drinking behavior that exist when people come under the influence of "demon rum." In keeping with the discipline's holistic approach to data, anthropologists have been at pains to point out how alcohol use (and occasionally, alcohol abuse) is woven into the very fabric of social existence. (p. 2)

Marshall concludes his volume with a summary of the major generalizations to be drawn from the cross-cultural work already done by anthropologists. The first one he lists reads:

> Solitary, addictive, pathological drinking behavior does not occur to any significant extent in small scale, traditional, preindustrial societies; such behavior appears to be a concomitant of complex, modern, industrialized societies. (p. 451)

Thus, for the anthropologist studying traditional society, to concentrate on the dominant concerns of the alcohol field—alcoholism and alcohol problems—would be to miss the major drinking phenomena in those societies. Again, one can argue that that is true in studying alcohol in modern society as well.

Historians, especially those doing social history, may also be prepared to follow out some aspects of Bacon's program. Because of the extraordinary prominence of the public debate about the "liquor problem" in America and other Western nations, there remains a mass of untapped historical materials regarding the place of alcohol. Much of the explicit writing is from a temperance perspective, but that can provide a useful handle on the larger culture, and on questions of social definition. The rebirth of social history in the last 10 years or so has also generated interest in precisely those areas of life most touched by alcohol: family, work, and play. Further, recent historical studies of working-class and ethnic cultures, stimulated by E.P. Thompson and Herbert Gutman, offer other opportunities for studying the place of drinking and alcohol free from some of the assumptions of pathology and disease. It is also worth noting that historical research provides an excellent, efficient, and relatively cheap way of investigating policy questions. The past, including the recent past, provides a wealth of "natural experiments," of changes in laws, sentiment, populations, administrations, enforcement, and community groups.

As has long been recognized, a social science concerned with alcohol in society in all its forms involves a variety of disciplines. The program Bacon outlined, and I have tried to second, is social in the broadest and strongest sense of the word. It may be that at this time anthropologists and historians will lead, with sociologists running to catch up. The others (economists, psychologists, political scientists, geographers, etc.) remain to be heard from.

The eventual fruit of a successful program of sociological, historical, and anthropological research and writing on the place of alcohol in society could include things like descriptions and analyses of drinking styles and cultures, drinking places, songs, rituals, and ceremonies; the symbolism of drink in the larger culture; the relationship of drink to class, gender, age, and so on; as well as studies of the alcohol beverage industry and of government policy and actions of the state. The result would be a parallel conceptual and empirical domain to the present social scientific studies of alcoholics and alcohol problems. The development of such a field would be good, but it was not all I had in mind. There are two additional wishes or hopes I have for a new alcohol social science.

One wish is that through the weight of theory and data the orientation of most work on alcohol would be changed. I hope that new research would not simply exist alongside current ideas, but would challenge them and offer alternative conceptions. When I wrote the previous section, I had in mind specific examples, notably MacAndrew and Edgerton's (1969) book *Drunken Comportment*, which argues strongly against the notion of alcohol as a disinhibitor and superego dissolver. I think it is an extremely important work that has probably made a great difference in steering the last 10 years of sociological, anthropological, and historical work away from the abyss of chemical reductionism. Other examples would be the work of Cahalan, Cisin, and Crossley (1969) and Room (1978), as well as other critiques of the disease notion of alcoholism, the symptoms and states. A recent example would be Gusfield's (1981) analysis of the rhetoric and drama of the drinking and driving issue.

Because I had in mind my favorite examples of what a sociological perspective could bring to the alcohol field, I initially assumed they would be the routine product. That is not the case. All my favorites were written by people knowledgeable about the alcohol field and working from within it. They were arguing against, critiquing, or in some way dismantling orthodoxies of one sort or another. They were explicitly and self-consciously developing alternative sociological explanations and concepts. While some research following the program I have outlined might return to critique the alcohol field, it is not certain that much would. Nor is it certain that researchers would avoid orthodox ideas and frameworks. Clearly such a critique is an important task.

My second wish is that a sociology of alcohol would develop a critical perspective on modern society. In arguing against an emphasis on pathological and problematic drinking—in siding with Bacon against Jellinek—I do not mean to suggest eliminating a concern with social problems. Rather, I mean to open up the question of what are the various relationships between alcohol and other social issues. Because alcohol routinely intersects with so many parts of life, it can be used to get at the unseen or taken for granted sides of everyday life, such as work, family life, play, recreation, rituals and processes of socializing, definitions of masculinity and femininity, assumptions about adolescence, and working-class and middle-class cultures and styles. The point is not only to see other parts of everyday life from a new angle, but to do so critically, with explicitly humanist and political criteria—to raise issues of social justice, freedom, democracy, domination, civil liberties, and oppression. One of the things most worthy of respect in the vigorous 200-year debate about the place of alcohol in modern society—from the early temperance movement on—has been the focus on change and reform. However misdirected it has sometimes been, the discourse about alcohol has usually been organized to talk about problems: something is wrong, usually something rather im portant. I think that is a good thing about the field and I would not want to abandon that.

Alcohol sociology could produce a great many interesting studies of the ways people drink, the laws governing drinking, of the various drinking cultures and subcultures—lots of surveys and ethnographies, histories and phenomenological accounts, all with an unquestioning and uncritical view of the larger social context. The classic concerns of temperance writers and speakers, on the other hand, were serious and critical: family violence, poverty, personal failure, industrial and road accidents, loss of freedom, and human misery. In Europe and America the liquor question has traditionally allowed for discussion of the antisocial consequences of things like routine business practices, monopolistic corporations, advertising, and even the profit motive. I am not suggesting that every piece of work needs to focus on some major problem. Rather, I am urging a conception of the field where it is part of the common normative style to talk frankly about major social problems that are alcohol-related without assuming or requiring that they be alcohol-caused. One example of what this could mean is understanding drunkenness and drinking (including rituals, place, time, environment, and so on) as nonpathological (or even healthy, sane, wise) responses to difficult, oppressive, or impoverished environments. Thus, it may be that, like the family, the drinking culture or place has often been the "haven in a heartless world."

I do not think there is a contradiction between looking at the nonpathological aspect of drinking, as Bacon proposed, and viewing the larger society from a critical perspective. But that combination is by no means inherent in the proposal for an alcohol sociology. The development of such a perspective depends only partly on factors internal to the field; it also depends upon changes in the larger society, in political and intellectual life, and in related academic disciplines. There are some hopeful signs. There are the beginnings of a political economy of alcohol that looks at the industry, the alcohol markets, and the structuring of alcohol consumption by corporations and by the government. The emergence of women's studies has at least opened the possibilities for examining the broad questions of women, drink, the home, saloon, advertising, and so on, beyond the current exclusive focus on women alcoholics. Studies of Indians, blacks, Hispanics, and other Third World groups allow for discussions of colonization, cultural and economic domination, and so on. Research on drink and the work environment allows, theoretically at least, for a variety of questions on work and working conditions, worker control, and management control practices; for example, it should be possible to study employee alcoholism programs in relation to the tradition of scientific management and employer control of the work environment.

The "alcohol problem," as it was first defined in the 19th century and has continued to be discussed in the 20th century, is above all a social and political question. At present, however, the vast majority of research on alcohol, and funding for it, is for the physical sciences: biomedical, chemical, and genetic research. Despite the overwhelming importance of social factors, social science is still the poor cousin of the physical sciences in alcohol research, and the majority of social scientific research is pursuing variations on Jellinek's questions first posed in the early 1940s on alcoholism and problem drinking.

However, we do have an alternative. We can at long last take up Bacon's suggestions—the path not taken—and begin developing a truly social science of alcohol.[2] By now it should be becoming clear that an overweening and exclusive focus on pathology is limiting and self-defeating: as the anthropologists point out, one misses most of what is going on that way. There are many potential areas of study, but they will remain potential unless sociologists and other social scientists take them up. No single researcher and no single research unit can pursue all or even most of the questions available. However, we can begin to define the terrain and contribute to legitimizing—for the alcohol field, funding agencies, governmental health agencies, and the general public—a perspective on alcohol problems that situates them in terms of the larger social forces and issues. The really important contribution of social science is to direct attention and examination to all facets of the place of alcohol in society.

III. Afterword: 1981

The previous sections were written before Ronald Reagan became president in 1980. Since then, his administration sought to drastically cut or totally eliminate federal funding for social science research. In a long article on the cuts, the *Washington Post* (June 19, 1981) reports that at the National Institute of Drug Abuse, a list was compiled of "all the research that used the word 'social' and that became the list for cutting—$9 million in studies, or 20 percent of the agency's budget." One OMB official told of being asked why the Alcohol, Drug Abuse and Mental Health Administration (ADAMHA) couldn't get more funds, and replied: "I guess that's because ADAMHA does a lot of work that's not hard science." One budget officer reported that "we were instructed to reduce science education by half and to reduce support for the social sciences by three quarters."

Although a tiny portion of the federal budget, the social science cuts will have a devastating effect on research. The *Washington Post* reported the following:

> At NSF [National Science Foundation], basic research in three fields—social, behavioral, and economic science—costs little more than half the price of maintaining the Pentagon's military bands. Nevertheless, social science has

> been hacked and the bands remain. [OMB Director David A.] Stockman listed cuts in NSF grants for social, economic, and behavioral research—studies of everything from the gross national product to the origin of man—from $49 million in 1981 to $16 million in 1982.
>
> All new social science grants at the Alcohol, Drug Abuse and Mental Health Administration, between $10 million and $20 million, are targeted as well.
>
> The fields of study being hit include economics, political science, sociology, cognitive psychology, linguistics, anthropology, and social and developmental psychology. In three areas—anthropology, economics, and political science—the NSF is the only U.S. government agency that gives grants for basic research. Basic work in those sciences will be almost wiped off the federal government's books.

All sources seem to agree that the cuts are not primarily for budgetary reasons, but for political reasons. The *Washington Post* points out that "Office of Management and Budget Director David A. Stockman and others have complained for years that social sciences produce little or nothing useful, and that their studies are often used to support liberal social programs." Many social scientists, including conservatives, have protested that the administrations's view of social science is incorrect. Despite these arguments, the administration appears firmly committed to the idea that most social science research is inextricably wedded to liberal politics and that it produces little of worth.

One irony of the cuts is that the most value-neutral, middle-of-the-road, and politically cautious research will be hit the hardest. The type of research advocated in this paper—critical ethnographic and historical studies—has never received much funding and thus will not be severely affected by the cuts. Further, such work can often be done by a single researcher in the field or library. On the other hand, large-scale studies of the general population, or longitudinal research, require large budgets for staff, interviews, mailing, and computer time.

The attempt by social scientists in the United States to make their work politically neutral and noncontroversial has not saved it from the new conservative attack. There will undoubtedly be some attempts to tailor research to the new ideology. But it seems unlikely that federally funded social science will be preserved by further attempts to be middle-of-the-road and neutral. Much social science—and this is especially true for alcohol studies—cannot get any more neutral.

The point is that the *idea* of free and open social science is under attack. Some conservative groups associated with the New Right, notably the Moral Majority, regard social science research as part of "secular humanism," which they believe is undermining the moral character of America. In effect, this implies the elimination of government-sponsored social research. Conservative think tanks and policy institutes will still be able to fund social scientists to find out the things they and their wealthy and corporate sponsors want to know. The point of government-sponsored research is that it can be free from the agendas of private, profit-making corporations. The aim of all science—social as well as physical science—is the unrestricted pursuit of truth about the world in all its richness and complexity. Sadly, that must now be defended.

IV. The View from 1990: Alcohol Studies in a Temperance Culture

In the 1980s, social science research and writing on alcohol were shaped by the rise of neotemperance sentiment and the continued strength of Reaganism. Federal funding for social scientific work on alcohol, drugs, and mental health was not eliminated, but it was channeled in narrower and more problem-oriented directions. In research not funded by federal alcohol grants, the ratio of "problem- and pathology-oriented work" to what we might call "alcohol in society, culture, and history work" is probably higher than it was in 1980; although there is much more of both types now, a greater percentage of all research is problem-oriented.[3]

Contrary to what I seemed to predict, in the 1980s history—not anthropology—led the field in small-scale studies examining the place of alcohol in the larger society. I should have known that: the same issue of the *Drinking and Drug Practices Surveyor* that carried my essay on Bacon also contained another article of mine—an annotated bibliography listing and briefly describing many recent books of good historical work on one alcohol topic or another. There are two simple reasons why historical research has been more plentiful. First, there are many more historians than anthropologists (or for that matter sociologists). Second, because it is easier to gather data for historical work, people in fields such as sociology and literature can more easily do research about alcohol history than they can research a contemporary non-Western culture. There is now an alcohol history organization with a small journal and it holds meetings in conjunction with the American Historical Association. The Alcohol Research Group held an extremely successful conference on the social history of alcohol in 1984 at which nearly 100 papers were presented. There have also been small alcohol history meetings and colloquia. Much of this work is fine history and a real contribution to understanding the place of alcohol in America and other countries.

The questions raised by Bacon's article of most interest to me now center on what have been the obstacles to implementing Bacon's vision and on why the alcohol field developed as it did. In 1943 there were two paths: Jellinek's, which looked at problems and pathologies, and Bacon's, which investigated the place of alcohol in the larger society and culture by examining the drinking patterns of the majority—nonpathological drinking and drunkenness. The proverbial man from Mars might ask: Why did Jellinek's vision and not Bacon's come to dominate social scientific writing on alcohol?

Put that way the question seems a bit absurd to anyone within American culture. One simple answer is that nobody with money would pay very much for the kind of "basic research" Bacon proposed because outside of the social sciences nobody cared much about it. Research on alcohol problems, however, had an audience of concerned politicians, health professionals, and lay persons. Alcohol studies in the 1940s, as a century before in the 1840s, existed because people in America were concerned with alcohol problems. Jellinek's problem-based perspective would always be the dominant one because the alcohol problem orientation had such a sizable constituency in America.[4]

What is not readily apparent is why so many people in America have been so concerned with alcohol problems. A comparative perspective helps us understand the United States better. In the 19th or 20th centuries some European countries also established multidisciplinary fields to study alcohol. Scientific alcohol research and even scientific alcohol journals developed in Finland, Sweden, Norway, and parts of Britain. I term these societies "temperance cultures" because they have had longstanding organized constituencies concerned about drinking as an evil or problem. Alcohol is regarded as a problematic substance in temperance cultures in a way that it is not so regarded in nontemperance cultures. The United States of America was the first temperance culture, the most influential one, and in most respects the most fervent one. All the temperance cultures are Protestant, and all have at some time consumed a significant portion of their alcohol (about 40 percent or more) in the form of distilled spirits. In the 19th and early 20th centuries, the scientific study of alcohol in these temperance cultures reflected the concerns of the temperance movement and regarded alcohol as a toxic and strongly addicting substance. In the mid-20th century, the scientific study of alcohol reflected the concerns of the alcohol movements of that time and focused primarily on alcohol addiction as a problem of some individuals. Since the 19th century, alcohol research has been rooted in and closely associated with the perspectives of popular movements concerned with alcohol problems.

"Well, all right," our man from Mars might say. "That may help explain why Jellinek's approach dominated, but that still does not explain why Bacon's agenda was not addressed better, more completely, and why it did not constitute even an important minority voice in the field. Why didn't the sort of work Bacon proposed ever coalesce into a distinct perspective within the alcohol field?"

Part of the answer, as I suggested in my original essay, was that Bacon did not really develop the perspective in his own work nor did he sponsor much of this work during his long tenure as the director of the Center of Alcohol Studies. Rather, Bacon became "sidetracked," as he put it. In fact, he became heavily involved in promoting Alcoholics Anonymous and the alcoholism movement.

In 1978 Selden Bacon was interviewed for two days by the staff of the Social Research Group. At the end of the second day someone asked Bacon what he thought had been the most important development regarding alcohol in America since the repeal of Prohibition in 1933. A number of things could come to mind as reasonable answers: the increased legitimacy of drinking in America; the movement of most drinking from the saloon and bar room into the home; the more general integration of drinking into so many aspects of daily life; the growth and power of the alcohol beverage industry; the growing body of scientific work on alcohol; the total disappearance of classic temperance forces; the success with which alcohol control was instituted, and the remarkably invisible way it smoothly continues to function. Without a moment's hesitation, Bacon answered "Alcoholics Anonymous."

E.M. Jellinek was the first person at the Yale center in the 1940s to become a promoter of Alcoholics Anonymous and its definition of alcoholism as a disease, but Bacon quickly joined him. Even before Bacon took over as head of the Center of Alcohol Studies, Bacon was a staunch advocate of AA. For years he, AA founder Bill Wilson, and National Council on Alcoholism (NCA) founder Marty Mann constituted what Bacon called a "traveling road show" spreading the word about AA and alcoholism. With his fellow missionaries he advised local and state governments on setting up alcoholism agencies to teach people about the disease of alcoholism and to refer people to the growing organization of self-proclaimed "recovered alcoholics." As head of the Yale and then Rutgers alcohol centers—the only significant university-based alcohol research centers—he helped legitimize AA and its approach. Through the Yale and Rutgers famous summer schools he instructed many social workers, ministers, teachers, and health workers.

For Bacon, for Jellinek's young protégé Mark Keller (who became the long-term editor of the *Quarterly Journal of Studies on Alcohol*), for others at the Yale center, and for many of his generation, Alcoholics Anonymous was a godsend. They had lived through Prohibition and its excesses, including the development of what Robin Room aptly calls "the wet generation" of heavy drinkers rebelling against Victorianism and Prohibition-ism. Bacon and the others abhorred the puritanical moralism of classic temperance and its attempt to force people to give up alcohol because of the problems of a minority of drinkers. They believed in modern science and medicine and wanted to see a modern scientific way of dealing with the problems of alcohol. The charismatic Bill Wilson spoke to Bacon and to others concerned with alcohol issues as effectively as he did to hundreds of thousands of drunks. Wilson offered a near perfect answer to the question of what to do about the real and obvious problem of people who drank too much for their own good: send them to AA.

There were, it is worth remembering, other treatment options possible. European nations, especially the Scandinavian and British countries, also had plenty of binge-drinking, middle-class alcoholics of the type that AA first recruited, and the rest of Europe also had many people who injured their health drinking. Those countries developed a range of options and treatment modes, mostly incorporated into the well-supported public health and state-sponsored medical systems. But in America, with its expensive and privatized health care system, such options did not seem realistic alternatives. Besides, there remained Bill Wilson and the very American-style self-help program of AA.

In the early days, in the 1940s and even for most of the 1950s, AA seemed small, precarious, and precious. AA did seem to challenge orthodoxies—of the century-old temperance movement and also of the completely unresponsive medical establishment. AA was populist, a genuine people's movement providing an alternative to the conventional wisdom and to the failed religious and medical remedies. The rebellious impulse that led Bacon as a young assistant professor to challenge Jellinek's proposals for a "Research Program on Problems of Alcohol" and offer his own radically sociological agenda also led him to embrace AA as his cause and movement.

It is worth tracing some highlights of that development. In the 1940s Alcoholics Anonymous provided the Yale center with a way of conceptualizing drinking problems and a program of action. AA along with the Yale center produced the National Council on Alcoholism, which promoted the ideas even more effectively. In the 1950s, AA members, the Yale center, and NCA convinced state and local governments to form alcoholism agencies and persuaded the American Medical Association (AMA) to proclaim alcoholism a disease. Much recognition and legitimacy followed. In the late 1960s, AA members in Congress, AA members writing letters, the NCA, the AMA, and related friends successfully lobbied to create the first federal alcohol agency: the National Institute on Alcohol Abuse and Alcoholism. In the 1970s, NIAAA began supporting the Employee Assistance Programs that had been independently developed since the 1940s by recovering alcoholics within large corporations.

No one at the beginning, or for many years thereafter, could have imagined the phenomenal legitimacy that AA's ideology would achieve: that undesirable habits and deviant or compulsive behavior of all types would come to be interpreted in AA's language—from workaholics to chocoholics. By the early to mid-1980s AA was not only the dominant treatment for alcoholism in America, it was also the remedy for drug addiction and many other personal problems. From Al-Anon to Gamblers Anonymous to Overeaters Anonymous to Drug Addicts Anonymous to Cocaine Anonymous to Love and Sex Addicts Anonymous to Adult Children of Alcoholics—the 12th step movement had arrived.

In the 1980s, the NCA and other groups lobbied to get warning labels placed on alcohol bottles and cans, the fetal alcohol syndrome was "discovered" and widely warned about, Congress passed legislation using the threat of withholding federal highway funds to force states to raise their drinking age to 21, Mothers Against Drunk Drivers was formed and a new antidrunk-driving sentiment was organized. By the mid-1980s, *Time Magazine* was calling attention to the powerful strand of antidrinking sentiment in America and terming it "neotemperance" or new temperance. Furthermore, all this occurred during a time when alcohol consumption was decreasing.

Despite the failure of constitutional prohibition, temperance culture and ideology never really lost its hold on America. Within American society alcohol remained a powerful symbol of evil, of the loss of self-control and individual responsibility. The other understanding of alcohol—as a social beverage, a pleasure-making substance, a drug that makes one merry—was spread in movies, song, and through advertising; as in the 19th century, that much older and more positive image coexisted alongside the now deeply rooted negative one. Within American culture alcohol became like a wild animal that had been domesticated. But alcohol was only partly tamed: it was never a puppy or kitten. Rather, like a pet wolf, bear, or cougar, alcohol was still thought capable of great and unpredictable damage. That sense of unpredictable danger has for 200 years remained part of the unshakable image of alcohol in American culture. If some individuals seemed to forget the danger, others never did, and the continuing stream of drunkards, recovered or not, provided further testimony to alcohol's profound risks. In America, the public consciousness of and discourse about the dangers and risks of alcohol use have increased since the 1940s due, in part, to AA's rise and enormous success.

As a result of these many years of activity, there is today far more political, medical, governmental, and other organized concern with alcohol as a problem than there was in the 1940s. In short, Bacon's efforts in the alcoholism movement worked against his own antitemperance, antipuritanical research agenda. In ways that Bacon could never have foreseen, his activities promoting the alcoholism movement created an atmosphere that makes an organized program of the kind of sociological work he proposed even more difficult to achieve in 1990.

Finally, there is an additional reason why Bacon's proposed research on nonpathological drinking could not have coalesced into an important minority voice in the alcohol field. Alcohol is not only a drug, beverage, and symbolic object. It is also a commodity. The alcohol field is not neutral in a political-economic sense. The alcohol beverage industry has a great investment in public attitudes toward alcohol and everyone understands that. Describing alcohol nonproblematically always runs the risk of "playing into the hands of the alcohol beverage industry." No serious alcohol researcher could long afford to be accused of that.

Individual researchers will always be able to do work in the spirit of Bacon's proposal—to do research and writing that seeks to locate alcohol in the broad patterns of society, culture, and economy. But as an organized, visible, and *funded* enterprise within the field of alcohol studies in America, Bacon's project, from the moment he proposed it, was probably an impossible dream.[5]

Notes

1. For example, in 1975 Room published a piece that explicitly called for large-scale studies of drinking norms. My repeated mention of drinking norms throughout my essay on Bacon was addressed to the themes and issues of Room's paper.
2. I have tried to do bits of the program outlined here in my own work, some of which is listed in the references.
3. This is, of course, only an informed impression. My sense is also shared by Robin Room (personal communication). However, Room points out that Jellinek's 1942 proposal actually only focused on alcoholism and not on other kinds of problems resulting from drinking. People at the Yale center considered others sorts of alcohol problems to have been part of the temperance movement's domain; both Jellinek and Bacon sought to avoid that association. Many current alcohol research topics—such as drinking by teenagers, drunk driving, the fetal alcohol syndrome, and the range of drinking-related problems—were not actually included in Jellinek's proposal. Since neither Bacon nor Jellinek was embracing the classic temperance concerns, and since Bacon clearly distinguished his proposal from Jellinek's by focusing on the nonpathological, I regard work on all kinds of alcohol problems as fundamentally in the spirit of Jellinek's agenda. Bacon, as my chapter tries to make clear, was proposing something very different.
4. Even the alcohol beverage industry was forced to make its accommodation with the problem orientation. From the late 1940s or early 1950s on, the beverage industry accepted the alcoholism movement's definition of alcohol problems as rooted in the bodies and minds of some drinkers. A problem exists, the industry granted; but the problem, they insisted, lies in the man and not in the bottle.

5. Severely handicapped by the limitations of soft money funding and its uneasy relationship with the University of California, the Alcohol Research Group has over the years helped nurture sociological, historical, and anthropological perspectives on drinking. Many researchers have benefited from contact with that remarkable nexus of intellectual activity. Although there is no organized presence in the alcohol field representing the spirit of Bacon's proposal, a loose but growing network of researchers keeps pursuing topics that Bacon first raised in 1942, and ones he never anticipated. They are mainly affiliated with professional associations in history, anthropology, and sociology. These American scholars have also found strong friends and allies in Europe, especially in Finland, where sociologists form an important voice within alcohol studies.

References

Bacon, S.D. (1943). Sociology and the problems of alcohol: Foundations for a sociologic study of drinking behavior. Introduction by E.M. Jellinek. *Quarterly Journal of Studies on Alcohol, 4*(3), 399–445.

Bacon, S.D. (1976). Concepts. In W. Filstead, J.J. Rossi, & M. Keller (Eds.), *Alcohol problems: New thinking and new directions.* Cambridge, MA: Ballinger Lippincott.

Cahalan, D., Cisin, I., & Crossley, H.M. (1969). *American drinking practices.* Monograph #6. New Brunswick, NJ: Rutgers Center of Alcohol Studies.

Gusfield, J.R. (1981). *The culture of public problems: Drinking-driving and the symbolic order.* Chicago: University of Chicago Press.

Heath, D.B. (1975). A critical review of ethnographic studies of alcohol use. In R.J. Gribbin et al. (Eds.), *Research advances in drug and alcohol problems* (Vol. 2). New York: John Wiley and Sons.

Heath, D.B. (1976). Anthropological perspectives on alcohol: An historical review. In M.W. Everett, J.O. Waddell, & D.B. Heath (Eds.), *Cross cultural approaches to the study of alcohol.* Paris and the Hague: Mouton Publishers.

Heath, D.B., & Cooper, A.M. (1981). *Alcohol use and world cultures: A comprehensive bibliography of anthropological sources.* Toronto: ARF Books.

Jellinek, E.M. (1942). An outline of basic policies for a research program on problems of alcohol. *Quarterly Journal of Studies on Alcohol, 3*(1), 103–124.

Johnson, B.H. (1973). *The alcoholism movement in America: A study in cultural innovation.* Doctoral dissertation, University of Illinois. Ann Arbor, MI: University Microfilms.

Levine, H.G. (1978). The discovery of addiction: Changing conceptions of habitual drunkenness in American history. *Journal of Studies on Alcohol, 39*(1), 143–174.

Levine, H.G. (1980). Temperance and women in 19th century America. In O. Kalant (Ed.), *Research advances in alcohol and drug problems* (Vol. 3). New York: Plenum Press.

Levine, H.G. (1981). The vocabulary of drunkenness. *Journal of Studies on Alcohol, 42*(3), 1038–1051.

Levine, H.G. (1983). The alcohol problem in America: From temperance to Prohibition. *British Journal of Addiction, 79,* 109–119.

Levine, H.G. (1985). The birth of American alcohol control. *Contemporary Drug Problems, 14,* 63–115.

Levine, H.G. (in press). Temperance cultures. In G. Edwards, M. Lader, and C. Drummond (Eds.), *The nature of alcohol and drug related problems.* London: Oxford University Press.

MacAndrew, C., & R.B. Edgerton. (1969). *Drunken comportment: A social explanation.* Chicago: Aldine.

Marshall, M. (Ed.). (1979). *Beliefs, behavior, and alcoholic beverages: A cross cultural survey.* Ann Arbor: University of Michigan Press.

Room, R. (1975). Normative perspectives on alcohol use and problems. *Journal of Drug Issues, 5*(4) 358–368.

Room, R. (1978). *Governing images of alcohol and drug problems: The structure, sources and sequals of conceptualizations of intractable problems.* Doctoral dissertation, University of California, Berkeley. Ann Arbor, MI: University Microfilms.

Room, R. (1979). Priorities in social science research in alcohol. *Journal of Studies on Alcohol* (Suppl. 8, November), 248–268.

CHAPTER 4

Selden D. Bacon as a Teacher, Colleague, Scholar, and Founder of Modern Alcohol Studies

ROBERT STRAUS

Selden Bacon has had a special impact on my life. He has been my teacher, friend, employer, and colleague. Among many superb courses I took at Yale, Selden's was the most demanding and most effective in making students learn to think. He was a master of the Socratic method. Seminars with Selden usually continued far past the scheduled time; I first met my wife, Ruth, one night when Selden's graduate seminar had to adjourn from the library that closed at 10 p.m. to a nearby house where Ruth and one of the other students lived. I was the first graduate student to be recruited by Selden to work in the field of alcohol studies. We collaborated for many years and have maintained both a personal and professional friendship ever since.

Based on more than 40 years of association with Selden Bacon and with the field of alcohol studies, it is my intent to address three topics in this chapter. First, there will be a discussion of some personal experiences as Selden's student and colleague. Second, I will provide some thoughts about the seminal role that Selden has played in the emerging field of alcohol studies and the history of the nation's contemporary response to alcohol problems. Finally, I will include some observations on the changing nature of knowledge about alcohol and the human body, drinking behavior, and alcohol problems, and discuss some implications for research and research training in the future.

Selden Bacon as Teacher and Colleague

In 1945, as a newly married graduate student, I was looking for some part-time work and Selden offered me an opportunity to earn 20 dollars a week by spending 20 hours a week trying to learn something about alcohol use among a population of homeless men. At that time, most of the public image of alcoholism was based on journalistic descriptions of visible populations, such as the men of skid row. Prevailing stereotypes characterized most alcoholics as either derelicts or deranged and assumed that the primary cause of homelessness was alcohol. Selden was skeptical of simplistic explanations for any social phenomena. Alcohol problems quite obviously penetrated many segments of society beyond the homeless and the mentally ill, and it was far too pat to blame the plight of homeless men on drinking alone. I clearly remember Selden telling me that my job was to be that of the bear that "went over the mountain." Without preconceived assumptions, I was to see what I could see.

This opportunity was the beginning of my own career-long involvement in alcohol studies. It provided the basis for my first publication, a 45-page paper on alcohol and the homeless man (Straus, 1946). It was characteristic of Selden that, although he provided the idea, the opportunity, and much guidance for the research, at no time was there any question but that authorship of the resulting publication would be mine. Of special significance for my career was the fact that one of the 203 men who participated in my first alcohol-related study was the subject of a life history (Straus, 1948) and, eventually, of a book based on the rare opportunity to record a human experience prospectively over a 27-year period (Straus, 1974).

Although I chose to write my dissertation in the field of public medical care rather than alcohol studies, I am told that Selden was very forceful in defending it among his faculty colleagues. He also facilitated the offer in 1947 of a faculty position in applied physiology at Yale so that I could become his colleague in alcohol research on a full-time basis.

By the mid-1940s, Yale's Laboratory of Applied Physiology had become the home of the nation's first university-based center for alcohol studies. Beginning in the 1930s with research on the effects of alcohol on the human body, the focus of the laboratory's interests and activities had expanded rapidly under the direction of physiologist and medical historian Howard W. Haggard. Selden Bacon was a faculty member in the Department of Sociology at Yale, specializing in criminology, when he undertook a study for the Connecticut War Council on drunkenness in wartime Connecticut (Bacon & Roth, 1943; Bacon, 1945). This brought him in contact with the alcohol studies group at Yale, and Haggard soon invited Selden to accept a joint appointment in the Laboratory of Applied Physiology. By that time, the functions of the laboratory had been expanded to include the publication of the *Quarterly Journal of Studies on Alcohol,* founded in 1940; a Summer School of Alcohol Studies; and research activities in many disciplines, including anthropology, psychology, economics, law, religion, medicine, psychiatry, history, and education, as well as the various biological sciences. E.M. Jellinek, a biometrician who had joined the laboratory as managing editor of the *Quarterly Journal,* had been designated as director of what was then called the Section of Alcohol Studies. Two experimental "Yale Plan" clinics for treating alcoholics on an outpatient basis, a national voluntary health agency, and a Yale Plan for (detecting and treating alcoholics in) Industry were soon to be launched.

Selden Bacon's first major contribution to the field of alcohol studies was his 1944 monograph entitled, *Sociology and the Problems of Alcohol: Foundations for a Sociologic Study of Drinking Behavior* (Bacon, 1944). This is an absolutely brilliant essay on the nature of social problems and on the need for integration of theory and method in research. It is a conceptual essay in which alcohol, drunkenness, and drinking problems are the vehicles for illustrating a thesis. Yet, 47 years after it was published, this essay is still timely and significant. It remains one of the most important sociological contributions to the field of alcohol studies.

Given the meaningfulness and originality of Bacon's early work in alcohol studies and attributes that have not diminished over the years, one can only wonder what the impact would have been if he had continued to pursue his career primarily as a social scientist. However, this was not to be. Shortly after Bacon joined the Laboratory of Applied Physiology, Howard Haggard was drafted by the university to head a major developmental project and, after completing that assignment, Haggard began a gradual move toward retirement. Jellinek's taste for administration was minimal. Therefore, at first informally, and rather soon officially, Haggard turned to Bacon to provide administrative leadership for the complex collection of research and other activities that were now called the Yale Center of Alcohol Studies. With these responsibilities, Bacon's time for active participation in research was severely diminished. He continued to originate and facilitate numerous research efforts, and he played a major role in conceptualizing, planning, critiquing, and interpreting the meaning of research findings. It was in these capacities that Selden contributed to my own career during the years 1947–1953.

Along with several projects that Selden initiated, facilitated, and entrusted to me to carry out with his occasional support and advice, we had two major collaborative experiences. One was our study of drinking practices and attitudes among nearly 17,000 students in 27 colleges in the United States (Straus & Bacon, 1953). The other was a study of the characteristics of 2,023 male patients seen in nine of the first outpatient clinics for alcoholics established in the United States (Straus & Bacon, 1951). Both studies were conceived as efforts to obtain and interpret some reliable knowledge in a field that was characterized by many strong opinions, stereotypes, and assumptions, but few facts.

The idea for a study of drinking attitudes and practices of college students had been brewing for some time before I came on the scene. It had initially been proposed as a study of Yale students, but this was quickly expanded by Selden to provide for a national perspective and to envision college students as an age and education level segment of the larger American population. The schools selected for inclusion in the study provided a mix of private, public, and sectarian; coeducational, for men or for women only; located in urban and rural settings and in different regions of the country; and having enrollments of varying size. More than a year was devoted to developing and testing the survey instruments and process. Fundamental to our approach was the assumption that students were the experts who would not only give us the answers, but who could help us identify the most significant and relevant questions and how these could be phrased most meaningfully. More than 700 students participated in the development and pretesting of the study, responding to our series of increasingly structured questionnaires and meeting with us in small groups to discuss each question and share their perceptions of how we should approach other students in order to solicit their serious participation. The importance for social science research of involving subjects in developing the questions as well as supplying the answers is one of the most valuable methodological lessons I learned from Selden, and I have tried to pass this on to all the graduate students and fellows who have worked with me in the ensuing 40 years.

My other major collaborative project with Selden was the study we made of characteristics of male patients seen in the first nine outpatient clinics for the treatment of alcoholism established in the United States. The major significance of this research lay in the fact that prior to this time (circa 1950) virtually all the published demographic data on alcoholics had been gathered from captive populations in institutions such as mental hospitals or jails, or from journalistic descriptions of visible populations, such as the men of skid row. Based on such reports there was a prevailing assumption that alcoholics were primarily people whose lives had totally disintegrated or fallen apart; that most were single or no longer married, unemployed or incapable of holding steady jobs, and lacking any residential or community stability. These assumptions were supported by the then prevailing belief within Alcoholics Anonymous that most alcoholics had to go through a progressive disintegration until they "hit bottom" before they would be receptive to help or recovery. Such notions were so widely accepted that the early proposals to establish community treatment resources for alcoholics were justified by the claim that they would help relieve communities of undesirable "drunks" and reduce the cost of maintaining the jails and mental hospitals. It was therefore something of a "find" to discover that the majority of male patients who were attracted to community outpatient clinics for alcoholics were married, living with wives, and they had, relatively speaking, maintained job, residential, and community stability. It was for this reason that the report based on our study was entitled, "Alcoholism and Social Stability." The findings of this study received quite a bit of media attention at the time, and I would like to believe that they contributed to a moderation of the negative stereotypes and stigma of alcoholism.

Selden Bacon as Innovator for the Field

In an introduction to Bacon's 1944 monograph, *Sociology and the Problems of Alcohol,* E.M. Jellinek quite prophetically noted that it might "serve as a kind of prolegomenon to sociological investigations of inebriety." In this monograph and his other major writings, as well as through his teaching and daily interactions with colleagues, Selden actually added a whole new dimension to the study of alcohol issues, and indeed to the study of social problems in general. This was his insistence that the study of pathology begin with the normal; that the study of alcoholism and other alcohol problems begin with the study of drinking. At a time when biological studies on alcohol were, in retrospect, quite simplistic, Selden Bacon established a conceptual framework for the study of social, cultural, environmental, and psychological factors in drinking behavior and alcohol problems that was interactive, comprehensive, continually adaptive to new knowledge, and still timely nearly 50 years later.

Although Selden Bacon's contributions to the alcohol literature are a matter of record and constitute an imposing effort of major impact, I do not believe that many of his other significant roles in the alcohol problems arena are as well known or fully appreciated. Selden Bacon has had a contributing role in the early history of almost every development of significance in the alcohol problems arena in the past 40 years. The following are a few examples.

In the mid-1940s Selden Bacon played a key role in the process that led Connecticut to be the first state in the nation to establish a tax-supported agency to provide treatment for alcoholics and education about alcohol problems. Selden was the first chairman of the Connecticut Commission on Alcoholism and a major force behind its innovative programs, which included the inpatient Blue Hills Hospital. This hospital received patients from and referred patients back to a network of community outpatient clinics throughout the state. The commission used as its models, and assumed responsibility for, the two Yale Plan Clinics. Based on the Connecticut experience, Selden served as a consultant to many other states and communities as the movement to development treatment resources for alcoholics and their families expanded rapidly during the 1950s.

Selden was a founding director of the National Committee for Education on Alcoholism. The committee that was to become the National Council on Alcoholism was founded as a division of the Yale Center of

Alcohol Studies and operated under the sponsorship of the Yale group for several years until it was able to establish its own financial base, identity, and organizational autonomy. Selden played a key role in the expansion of the voluntary health movement for alcoholism in the United States and helped to assist many community groups assess their needs and resources, define goals, and develop responses.

Selden, along with Ralph M. ("Lefty") Henderson, conceived and established the Yale Plan for Industry. They worked with a number of major corporations to help estimate the costs of problem drinking among employees and management, develop ways of identifying the problem drinkers, and make plans for rehabilitation and treatment.

Selden almost individually protected the Yale Center of Alcohol Studies from the essentially hostile environment of its university. Ironically, the very success of the center, which brought it and the university much favorable publicity, was also a distinct threat to its existence. There were many persons of influence among the faculty, alumni, and officers of the university who were not happy to see their university's name so widely associated with such a stigmatized subject as alcoholism. While efforts to force the center's move eventually succeeded (in 1962), Selden quite effectively warded off hostile moves toward the center for more than a decade.

While on the subject of the center at Yale, and subsequently at Rutgers, it is significant to note that Selden helped to pioneer the concept of a multidiscipline, multifunction, university-based center to address the problems of alcohol. He nurtured the concept at Yale and he "sold" the concept to Rutgers University, where the center was to move. He also was instrumental in obtaining the first federally funded center grant in the alcohol field, a grant of almost one million dollars from the National Institute of Mental Health that facilitated the center's move and helped support its early years at Rutgers. Throughout his many years as director of the center at both Yale and Rutgers, Selden maintained a loyalty to and support for colleagues irrespective of their disciplinary leanings or particular scientific orientation. He recognized the importance of documentation and helped obtain support for a specialized library, the abstract archive of alcohol literature, the *Journal of Studies on Alcohol,* and the significant publication program of the center.

Despite his heavy administrative load, Selden maintained his excellence as a scholar and a teacher and continued to provide original, thought-provoking contributions to the alcohol literature and to nurture the careers of successive waves of graduate students, fellows, and junior colleagues.

On the national scene, Selden played a founding role in many of the organizations that have become important components of the alcohol problems arena as it exists today. He was also a member of the Cooperative Commission on the Study of Alcoholism from 1961 to 1966. The commission's *Report to the Nation* (Plaut, 1967) set the stage for the establishment, at President Lyndon Johnson's directive, of the National Advisory Committee on Alcoholism (on which Selden also served) in 1966; that committee's report (1968) included many recommendations that were subsequently incorporated into the legislation that established the National Institute on Alcohol Abuse and Alcoholism (NIAAA). Although, as the field expanded rapidly from this point onward, Selden's degree of participation in key events was relatively less striking, he continued to function in many ways as one of the field's senior scholars.

At the conceptual level, Selden's contributions have continued well past his official retirement (Bacon, 1984, 1987). From my personal perspective, one of his most insightful, original, and prophetic writings was the paper, "Alcoholics Do Not Drink" (Bacon, 1958), in which, based on sociological observations, he anticipated by more than two decades the biological findings that demonstrate that people whom we label as "alcoholics" experience alcohol in ways that are antithetical to the experiences of those whom we think of as "drinkers."

Alcohol Research in Perspective

Persons familiar with the scientific literature on alcohol cannot help but be impressed with the explosion of new knowledge about alcohol and the human body that has been generated in recent years. Supported primarily by research grants from the NIAAA, increasing numbers of gifted investigators from such disciplines as genetics, biochemistry, physiology, and the neurosciences have drastically modified our perceptions about the biology of alcohol and the nature of biologically determined risks for alcoholism and other alcohol-related problems. Simultaneously, and equally significant, methodological advances in the social and behavioral sciences have permitted the generation of new knowledge and the reinterpretation of existing data that have altered our perceptions of factors that influence exposure to alcohol and experiences with drinking. Biologically, our thinking has shifted from a focus on commonalities of response to a recognition of individual variability and heterogeneity. In the social sciences previous notions of cultural and social homogeneity in drinking practices have also given way to new epidemiological insights that reveal much greater diversification than had earlier been recognized. Through modern computer technology, it is also now possible to compare data collected from many different cultures, societies, and periods of time, and to correlate variations and changes in drinking behavior and in the prevalence and nature of drinking problems with such factors as changing beliefs, values, economic conditions, and social controls.

Unfortunately, there has been a glaring lack of familiarity with and appreciation for the nature and significance of biological knowledge on the part of many social and behavioral scientists. The reverse is equally true. Fortunately, there are signs that this condition is changing at least to the extent that an increasing number of scientists working in the field are recognizing the need for knowledge from other sectors that can complement and extend the meaning of their own findings (Institute of Medicine, 1987).

It seems to me that the greatest significance of new knowledge from the biological sciences is in the area of risk determination. We are learning that a number of biological factors appear to influence an individual's relative sensitivity to alcohol, relative risk for developing a dependence on alcohol, and relative vulnerability for developing one or more alcohol-related diseases. We also must recognize that the actual significance for an individual of biologically determined sensitivity or vulnerability depends on whether the individual ever uses alcohol and, for users, on the nature of their cumulative exposure to alcohol or experience with drinking. Exposure to alcohol and experiences with drinking are behavioral phenomenon influenced by factors such as cultural beliefs and values, social customs and pressures, environmental settings, and personality characteristics as they relate to the psychological reinforcement value of alcohol's mood-modifying effects.

Although I believe that there is today a growing recognition of the desirability, even the need, for truly integrative biobehavioral or biopsychosocial conceptual thinking about drinking behavior and alcohol problems, there are unfortunately few individuals with enough depth of understanding in the biological, social, and behavioral sciences to feel adequate to this task. This problem is not unique to alcohol studies. However, because of its important biological and behavioral components, the field of alcohol studies is uniquely suitable for the development of the integration of biological and behavioral theories and the design of research questions and strategies that cross the traditional biological/behavioral barrier. Because we have reached a stage in the generation of knowledge in both the biological and the behavioral sciences where advancements in either will depend on knowledge of the other, there is need for programs to train a cadre of young investigators who will be prepared conceptually, semantically, and methodologically to venture across this barrier. It is from this direction that I am convinced our future advances in alcohol studies will come.

Selden Bacon will probably not agree with me completely on this point. Although, as a program director at both Yale and Rutgers, he supported the efforts of his biological colleagues with loyal tenacity, throughout his career Selden has insisted that the understanding of and effective responses to the problems of alcohol require a social science frame of reference. While respectful of biological research, he has been aware of its limits and skeptical of the importance it has been accorded in relation to other knowledge.

Although most scientists are primarily concerned with the advancement of knowledge, from the perspective of society, the significance of knowledge lies in its applications. Application in the field of alcohol studies means the prevention and treatment of alcohol problems. While our society has received a remarkably good return on its investment in the support of alcohol-related basic research, it has actually made a much larger investment in supporting programs of prevention and treatment. Yet many strategies and programs for prevention and treatment do not appear to reflect recent advances in knowledge. As noted earlier, Selden Bacon's contributions to the alcohol field were not limited to his role as a conceptualizer and scholar. He was also significantly involved as an innovator in the application of knowledge to a variety of strategies and organizational approaches to prevention and treatment. Perhaps it is in his unique role of a scientist who has been directly involved in the application of knowledge that we most need his wisdom today.

A question that remains is whether we have followed the agenda that Bacon proposed in 1944 and, if not, why not. Bacon's suggestions probably had the greatest meaning for and impact on those of his students and colleagues who shared some of the scholarly tradition from which they were derived. I may be the only student of Bacon's who had direct contact with some of the sources of his theoretical roots. I took Keller's last course at Yale, sat in on some of Malinowski's last seminars, and knew Davie, Dollard, and Jellinek. Thus, the systematic agenda that Bacon put forth has always had a familiar ring and I would like to believe that it has influenced most of my work, although as often unconsciously as with purposeful design. The same can probably be said for most of the other graduate students at Yale and later at Rutgers for whom Bacon opened up the vistas of alcohol research. There have also been many students of alcohol from numerous disciplines who were influenced by Bacon's thinking during participation at the Summer Schools of Alcohol Studies that were initiated at Yale in 1943 and have continued at Rutgers since 1962. How much of Bacon's unique way of thinking and perspective was transmitted in those brief courses, it is difficult to ascertain. Clearly, there are serious sociological scholars of alcohol research, such as Robin Room at Berkeley and Harry Levine at CUNY, who are very familiar with Bacon's agenda and have selectively incorporated it into their own.

Because it was published in a specialized journal of alcohol studies with a relatively small readership, Bacon's *Sociology and the Problem of Alcohol* was probably seen by few social scientists outside of the alcohol field.

Beyond these few sociologists and anthropologists, I fear that much of Bacon's agenda has met blind eyes. This is not due to any particular failure on Bacon's part, but simply to the disciplinary provincialism that has characterized most scholarly activities, including alcohol research, in the past decades. Also, society, in its support for research, has favored the biological and other natural sciences over the social and behavioral sciences. This has been true of alcohol studies, as well as in most other areas of inquiry.

Another factor that has limited the impact of Bacon's 1944 agenda was a change that took place within sociology and the other social and behavioral sciences favoring quantitative over qualitative research. Within sociology, there has been a period during which the fruitfulness of ideas has taken a back seat to the worship of numbers; statistical methodologies have been valued over theoretical systems. This is changing somewhat today, and with such change we can anticipate that the value of Bacon's proposals among alcohol scholars will again be enhanced.

As I have noted, I believe that we are now entering a new phase of alcohol research where a mandate for interdisciplinary research is bound to emerge. This will require both methodological and conceptual integration. The theoretical system proposed by Seldon Bacon in *Sociology and the Problems of Alcohol* will lend itself well to this effort. Although oriented to cultural and social phenomena, it is logically expandable to include biological, psychological, and environmental dimensions. Although developed nearly 50 years ago, it has stood the test of time. Although there have been vast changes in the last 50 years in our knowledge about alcohol; in the customs, attitudes, perspectives, values, laws and regulations that govern drinking behavior; in the uses of alcohol and the liabilities of intoxication; and in social provisions for prevention and treatment, Bacon's proposal for the systemic study of drinking behavior and drinking problems could as well help guide research of the 1990s as that of the 1940s.

References

Bacon, S.D. (1944). *Sociology and the problems of alcohol: Foundations for a sociologic study of drinking behavior.* Memoirs of the Section of Studies on Alcohol, Yale University, No. 1. New Haven: Hillhouse Press. Also published in *Quarterly Journal of Studies on Alcohol, 4,* 402–445, 1943.

Bacon, S.D. (1945). *Inebriety, social integration and marriage.* Memoirs of the Section of Studies on Alcohol, Yale University, No. 2. New Haven: Hillhouse Press. Also published in *Quarterly Journal of Studies on Alcohol, 5,* 86–125, 303–339.

Bacon, S.D. (1958). Alcoholics do not drink. *Annals of the American Academy of Political and Social Science, 315,* 55–69.

Bacon, S.D. (1984). Alcohol issues and social science. *Journal of Drug Issues, 15*(1), 7–29,

Bacon, S.D. (1987). Alcohol problem prevention: A "Common Sense" Approach. *Journal of Drug Issues, 17*(4), 369–393.

Bacon, S.E., & Roth, F.L. (1943). *Drunkenness in wartime Connecticut.* Hartford: Connecticut War Council.

Institute of Medicine. (1987). *Causes and consequences of alcohol problems: An agenda for research.* Washington, DC: National Academy Press.

National Advisory Committee on Alcoholism. (1968, December). *Interim report to the secretary of the Department of Health, Education and Welfare.* Washington, DC: Government Printing Office.

Plaut, T.F.A. (1967). *Alcohol problems: A report to the nation by the Cooperative Commission on the Study of Alcoholism.* New York: Oxford University Press.

Straus, R. (1946). Alcohol and the homeless man. *Quarterly Journal of Studies on Alcohol, 7,* 360–404.

Straus, R. (1948). Some sociological concomitants of excessive drinking as revealed in the life history of an itinerant inebriate. *Quarterly Journal of Studies on Alcohol, 9,* 1–52.

Straus, R. (1974). *Escape from custody.* New York: Harper and Row.

Straus, R., & Bacon, S.D. (1951). *Alcoholism and social stability.* New Haven: Hillhouse Press. Also published in *Quarterly Journal of Studies on Alcohol, 12,* 231–260.

Straus, R., & Bacon, S.D. (1953). *Drinking in college.* New Haven: Yale University Press.

CHAPTER 5

The Mutual Relevance of Anthropological and Sociological Perspectives in Alcohol Studies

DWIGHT B. HEATH

The history of alcohol studies is a fascinating topic in itself, with sociological implications that were not lost on many of the principal actors and, on the contrary, were often manipulated by them. Although I do not share the view of some colleagues that we have reached the point where a discipline of "alcohology" is called for (Keller, 1979), there has now been half a century of more or less coherent "alcohol studies" in which perspectives from sociology and anthropology have made important contributions.

In this chapter I will briefly trace a few of those contributions and show how a small cadre of social scientists had a disproportionate impact on the thinking about how and why people drink.[1] Illustrative of this impact is the general acceptance of the proposition that different patterns of belief and behavior concerning drinking result in different kinds and rates of drinking problems among various populations. Similarly, anyone who is planning a prevention project must recognize the need to "target" different populations in different ways, and those engaged in treatment are beginning to accept the need for "culturally appropriate" treatment for various "special populations." Most people assume that it was psychologists dealing with their own patients who focused on relief of anxiety as a prime motivation for drinking; it was, in fact, an anthropologist analyzing older ethnographic reports. Similarly, both the theory about drinking as a means of feeling powerful and that of drinking as an escape from conflict over dependency grew out of data reported by anthropologists who had never thought in those terms. One could even suggest that the elaborate double-blind experiments conducted by psychologists under strictly controlled conditions in the laboratory that confirm the dominant role of "expectancy" in terms of the effects that alcohol may have on mood and action were vividly presaged in MacAndrew and Edgerton's (1969) analysis of ethnographic and historical documents. The role of norms in shaping drinking behavior is now axiomatic, although it was spelled out only gradually since World War II.

The fact that social science perspectives have become integral to the way in which most people think and talk about alcohol is distinctive of the past half-century. This has to do in part with individual authors and concepts, and in part with an intellectual climate in which the boundaries of academic disciplines were not clearly drawn. The present discussion is focused on the United States, where a major portion of the writing on alcohol has been done, and where the links with social science are usually strong.

For many years there has been widespread acceptance of the proposition that the study of alcohol must necessarily be a multidisciplinary, interdisciplinary, or transdisciplinary venture. Alcoholism has been characterized as a "biopsychosocial disease," with explicit acceptance by most researchers that drinking problems, of whatever nature, cannot be understood except as the resultant of a complex interaction of some biological factors with some psychological factors, both of which affect and are affected by some social factors.

In an even broader sense, the cornerstone of interest in sociocultural factors as they relate to alcohol is the simple, yet immensely important, fact that the uniform substance ethanol, when ingested by members of the species *homo sapiens,* has a wide variety of outcomes that tend to be similar within groups and to differ between groups. This is not to say that biochemical and pharmacokinetic processes are irrelevant or more highly variant among populations than within them, but it does emphasize the fact that such processes are often conditioned in large measure by norms, attitudes, values, expectancies, and other things individuals learn about how people and alcohol should interact.

Even the most narrowly focused researcher in whatever field of alcohol studies today tends to proclaim acceptance of the so-called "public health model" of alcoholism, which explicitly recognizes the mutual relatedness of the agent, the host, and the environment. Where we tend to differ is with respect to whether the most meaningful level of analysis of the "host" is, for example, in terms of DNA sequencing, the permeability of cell membranes, personal biography, peer relations, or the internalization of norms through enculturation. Similarly, the "agent" or "vector" can be viewed as simply CH_3CH_2OH, or as a range of different beverages with different meanings, as an economic commodity, a symbol of hospitality, a potentially hazardous substance, or otherwise. And the "environment" can be similarly diverse (Heath, 1989), sometimes emphasizing the physical setting and materiel, sometimes the complex system of legal controls and taxation, or even the whole "intellectual climate" that includes sex roles, age grading, norms about moderation or excess, and the full range of attitudes and values that comprise a society's way of life.

Fermentation is a process that occurs in nature even without human intervention, and it is not surprising that fermented beverages became popular throughout most of the world from early times. Some archaeologists suggest that beer may well have preceded bread as a staple in the human diet (Braidwood et al., 1953; Katz & Voigt, 1987). Climatic limitations account for its absence in the arctic prior to European contact, but we still have no idea why most of North America and the islands of the Pacific lacked indigenous alcohol.

Drinking is usually a social act, important enough to be "explained" by mythology. As Osiris is credited with having introduced beer to the Egyptians, so Dionysus brought wine to the Greeks and Bacchus wine to the Romans, whereas Mayahuel revealed pulque to the ancestors of the Aztecs. Not only is drinking eminently social, it is often hedged about with many rules and rituals that bring it to the attention of those who write the documents that became history, whether as travelers from elsewhere or as lawmakers or chroniclers among their own people. One early document about drinking that has interest for social scientists is the Code of Hammurabi; even 4,200 years ago, the first written codification of law dealt in detail with the regulation of wine sales (Harper, 1904). Furthermore, the Bible contains at least as many passages favorable to drink as unfavorable (O'Brien & Seller, 1982; Seller, 1984).

Alcoholic beverages became important economic commodities early, so that long-range trade played a role in the expansion of many colonial powers; taxes on them often figured prominently in domestic economics. Alcoholic beverages became widely available throughout the world. Occasional attempts at prohibition have been short-lived or ineffective, except in those few instances where they were embedded in a profoundly religious context. Although a small group of social historians are increasingly revealing prescient insights about drinking on the part of various earlier figures, there was little that could be dignified with the label of systematic investigation of alcohol until the second third of the 20th century.

Many accounts of exotic peoples, whether written by explorers, colonial administrators, missionaries, or other travelers, included mention—often deprecating and sensationalized—of native drinking customs. Occasional compilations in the 19th century may have served a vaguely pornographic function; they provide neither sufficient detail nor context to be of much use for understanding (e.g., Dorchester, 1884; Morewood, 1838). Two of the most substantive early accounts of drinking had to do with public drinking establishments: an observational study of behavior in

English pubs (Mass Observation, 1943), and an ambitious effort to ascertain the many functions of American saloons so that functional alternatives could be found (Calkins, 1901). A special irony attaches to those studies inasmuch as recreational drinking has been one of the least studied aspects in subsequent years, and studies in natural settings have been especially scarce in urban Western societies.

The Intellectual and Social Context of Early Sociocultural Studies

The enactment and repeal of nationwide Prohibition in the United States confronted clinicians and scholars with a fascinating set of questions, few of which were addressed at the time. A whole new view of law and order forced itself upon the nation's consciousness, and the rights of both individuals and several states came in for reappraisal. At the same time, the way in which a few long-term heavy drinkers persisted in their habit, even at increasingly evident risk to themselves, piqued the curiosity of a thoughtful physician, Norman Jolliffe. Struck by the degree to which a small population of chronic repeaters so heavily taxed facilities at Bellevue Hospital in uptown Manhattan, he secured a grant from Carnegie Corporation to undertake a thorough review of what was known about alcohol use and abuse. We are fortunate that Mark Keller became his assistant, laying the groundwork for the meticulous and thorough bibliographic documentation that remained a hallmark of alcohol studies until just a few years ago.

From the modest beginning, with E.M. Jellinek as an imaginative coordinator, the program rapidly grew in unforeseen and unforeseeable ways. Having sparked the interest of Howard W. Haggard, the group was invited to Yale University, where they became a section of the newly formed Laboratory of Applied Physiology. The *Quarterly Journal of Studies on Alcohol* rapidly became recognized as a major international and multidisciplinary forum, later to be rescheduled and renamed the *Journal of Studies on Alcohol.* Similarly, the laboratory evolved into an exceptionally active Center of Alcohol Studies, which, among other things, administered the Yale Plan Clinics and the Yale Summer School of Alcohol Studies, in addition to conducting research and publishing the journal and monographs. The center also played key roles in linking the academic community with the treatment community, and linking both of them with public officials in various capacities, as well as with the increasingly vocal constituency of recovering alcoholics. It is an accident of history that the Yale center was born at the same time as Alcoholics Anonymous, and there were interesting links between the organizations for several years, which undoubtedly colored the impact that each had beyond its own circle; that story can better be told by those who were closely involved in it.

For present purposes, it seems important to underscore the sense of openness, experimentation, and intellectual ferment that characterized that group in the 1940s and early 1950s. It is clear that the boundaries among academic disciplines were less sharply drawn then than they tend to be at most universities now—whether because they then had a stronger tradition of liberal education, a different view of scholarship, a less sharply delimited view of professionalism, or for some other reason or combination of reasons.

The work of Shalloo (1941) was cited as combining some psychological insights with clear recognition of sociological variables. Similarly, the psychiatrist Myerson (1940) noted that "motive is not explanation enough," citing "social tradition and social pressure" as partial reasons why Jews and women—both categories with their full share of frustrations, neuroses, and economic and social conflicts—had low rates of alcoholism (at a time when alcoholism was often simplistically "explained" with reference to such frustrations and conflicts).

In some other contexts, I have already provided broad overviews of the literature that deals with alcohol from an anthropological perspective. One article traced the evolution of the topic in chronological order since ancient times (Heath, 1976). At about the same time, I compiled a more encyclopedic review-article that showed how various anthropological and sociological themes and topics, both substantive and theoretical, had been developed with reference to alcohol, up to 1970 (Heath, 1975). A follow-up article, organized in the same manner (Heath, 1986a, 1986b, 1987), complements it and makes the diverse and widely scattered literature of the 1970s more accessible. In this context, I will mention only a few milestones in that rich corpus, and some less important items that illustrate my theme of mutual relevance.

It is gratifying that not only social scientists but administrators, educators, counselors, and alcoholics as well seem comfortable in discussing the relevance of "sociocultural factors" as an integral component of what some call "the public health model"; most anthropologists would be comfortable with what they choose to highlight as relevant—although we might want to add some other aspects that tend rarely to be discussed. Similarly, it is salutary that psychologists, media analysts, and others are venturing to do "cross-cultural studies," even if their usage of that term has more to do with survey instruments and the nation-state than with close and sustained ethnographic study among peoples with strikingly different ways of life (Heath, 1986c).

Until the 1960s, anthropologists had contributed little to the conceptualization and integration of alcohol studies, although drinking behavior was often so striking that it was described in some detail. Part of the reason that the contribution of data was not complemented with interpretation is that few of the anthropologists who went into the field had any intention of studying drinking patterns, much less any sophisticated hypotheses to test or clinical problems to resolve. In many instances, those few anthropologists who paid any attention to alcohol did so for the simple reason that it played important roles in the lives of the people under study. In fact, it is remarkable that the data are so rich on a topic that was at best tangential to their major interests. Without the old-fashioned Boasian view that an ethnographer should learn as much as possible about as many things as possible, we would have few data about drinking and its outcomes in any society. A fieldworker sharply focused on one or a few dependent and independent variables is unlikely to offer much in the way of what I have called "serendipitous by-products of research," many of which constitute the bulk of the historical and ethnographic corpus on the interactions of alcohol and homo sapiens in the vast natural laboratory that is offered by the rich diversity of the human experience.[2]

Although anthropologists have traditionally paid most attention to non-Western peoples, "the sociocultural model" (as it relates to alcohol) is probably best known to many people on the basis of the contrast between Orthodox Jews and Irish Americans. As formulated by Bales (1946), a sociologist who exemplified social psychology in later years, virtually all Jews drink, but few have problems; in contrast, only adult males drink among Irish Americans, but they suffer a high rate of alcoholism. Key differences are both experiential and attitudinal but, in both cases, eminently sociocultural—early as contrasted with late introduction of individuals to the experience of drinking, a familial and social context for Jews as contrasted with the bar for the Irish, a ritual setting as contrasted with a convivial one, and religious meanings as contrasted with secular ones.

Subsequent studies have sharpened our understanding of the drinking behaviors of both populations, affective correlates, varying degrees of commitment to values and norms, and many other factors, but the broad outlines of the contrast between Jews and the Irish remain vivid and widely accepted, even as an acculturative process of gradual convergence may be taking place (cf. Blaine, 1980; Blume, Dropkin, & Sokolow, 1980; Glassner & Berg, 1980; Stivers, 1976).

Jellinek himself had no formal training in the social sciences but brought to the field a familiarity with classics and history, together with a special combination of insight and boldness that helped him to make an indelible mark on alcohol studies during that formative period. Several of his notes on drinking in ancient societies (published posthumously, thanks to the generous editing of Robert Popham) presaged what has become a thriving subfield in history. His broad-brush characterizations of differences in drinking patterns among nations (e.g., Jellinek, 1957) set the tone for much of what continues to be done under the rubric of "cross-cultural studies." The "informant method" that he recommended as an economical and efficient substitute for social surveys as a means of learning about local drinking customs is what an anthropologist might do in a hurried preliminary reconnaissance (Popham, 1976). For that matter, the first definition of alcoholism that grew out of the World Health Organization (WHO) Expert Committee on Mental Health's deliberations under his direction has an anthropological ring in its insistence on cultural relativism, stressing that norms should be judged in relation to the local cultural context: "any form of drinking which in its extent goes beyond the traditional and customary 'dietary' use, or the ordinary compliance with the social drinking customs of the whole community concerned" (WHO, 1951). The very vagueness and relativism of that definition were not appreciated in many circles, however, so the WHO Expert Committee soon produced another definition of alcoholism. Although very different in emphasis, it was similarly anthropological in its insistence that problems can best be identified and evaluated in relation to their cultural context: "Alcoholics are those excessive drinkers whose dependence upon alcohol attained such a degree that it shows a noticeable mental disturbance or an interference with their bodily or mental health, their interpersonal relations, and their smooth social and economic functioning; or who show the prodromal signs of such developments" (WHO, 1952). No social scientist was involved in either committee, but both definitions emphasize characteristics that are socioculturally variable.

The fact that social science played so important a role in the early days of alcohol studies clearly owes something to the individuals and something to the institutions that were involved. The lines between sociology and anthropology were not sharply drawn at Yale, with the tradition of comparative sociology rich in ethnographic detail strongly exemplified in the work and teaching of William Graham Sumner on the one hand, and members of the Institute of Human Relations (including George P. Murdock, John W.M. Whiting, Irvin S. Child, John Dollard, and others) striving valiantly to integrate aspects of psychoanalysis and learning theory with social, structural, and organizational approaches on the other.

Selden Bacon, a sociologist who soon became director of the center, had been a student of Sumner's. The prospectus for social research that he offered early (Bacon, 1943) was ambitiously global in scope, stressing the importance of cultural context, intrasocietal variation as well a normative patterns, normal and salutary drinking as well as the excessive and problematic, together with many other themes. Many people who are not familiar with the field mistakenly assume that the theme is alcohol*ism*, whereas the early researchers were remarkably consistent in attending to alcohol, of which alcoholism is a small part.

Throughout the 1940s and 1950s, a succession of social scientists played a variety of roles in alcohol studies, often behaving in ways that run counter to present-day stereotypes about the division of labor among academic disciplines. For example, graduate students Robert Straus, John Honigmann, and Earl Rubington, together with former high-school principal Raymond McCarthy, conducted research of a type that we would now call "street ethnography" among men we would now call "homeless." Straus and McCarthy (1951) and Rubington (1968) wrote about that experience, with an especially valuable outcome being Straus's (1974) unique longitudinal and phenomenological insight into the life of a chronic problem drinker. Honigmann appears not to have published on that work, although he subsequently wrote prolifically about drinking in a variety of communities in both Austria and the Arctic (Honigmann, 1963, 1964, 1966; Honigmann & Honigmann, 1945, 1965, 1970). The experiment that proved the value of a "drinking diary" written faithfully by trustworthy informants to account for periods of time when observations could not be sustained grew out of the close ties that Phyllis Williams (Williams & Straus, 1950) had developed with friends and neighbors during years of ethnographic work in New Haven's Italian community. Furthermore, a monographic comparison of that community with Italians in Italy again demonstrated that heavy alcohol consumption does not necessarily result in manifest problems (Lolli, Serrianni, Golder, & Luzzatto-Fegiz, 1958), in much the same way that Snyder's (1958) more detailed analysis of Jewish drinking provided more substantive specifics in support of the proposition that children's learning early how to drink, in a supportive family context, with strong symbolic overtones, can serve almost as a form of "immunization" against alcoholism.

Harold Mulford was conducting social surveys on drinking patterns, while Don Cahalan was still engaged in marketing research. Anne Roe, a psychologist, and Jenny Sapir, a social worker, also contributed to the strongly sociocultural tenor of much of the early work on alcohol. Straus collaborated with Bacon in a pioneering study of drinking among college students (Straus & Bacon, 1953); a follow-up by Fillmore, Bacon, and Hyman (1979) provides one of the few longitudinal studies of nonclinical populations.

Donald Horton chose to deal with cultures rather than individuals, and waded into the encyclopedic ethnographic compendium that George P. Murdock had assembled for the U.S. Navy; originally called "the Cross-Cultural Survey," it has evolved into the Human Relations Area Files. Horton won what some might consider an enviable place in the annals of alcohol studies, with one of the most widely quoted lines being his conclusion that "the primary function of alcoholic beverages in all societies is the reduction of anxiety" (Horton, 1943, p. 223). A special irony to be savored by a few social scientists is that Horton appears never to have thought of himself as a student of drinking or of alcohol; his paper was, in essence, an essay in methodology. And, sure enough, the cross-cultural method (now renamed by some practitioners the "holocultural method")—examining the correlation of diverse variables in large samples of societies throughout time and space—is firmly established and has yielded important tests of a vast number of hypotheses about war, marriage, slavery, and many other institutions. Horton went on to teach social studies at a school of education, and never even rejoined the hordes of critics who misinterpreted his method and caviled because some one of the several factors that he had used as an *index* of generalized anxiety might not correlate with drinking in a given society.

The Yale center appears to have been suffused with a heady sense of openness and exploration. Bales's (1946) classic work on Irish and Jewish drinking was developed there, and Ullman (1958), although he only briefly attended the Yale Summer School of Alcohol Studies, went on to make significant contributions that linked drinking and drinking problems with norms and normative behavior. In a sense, he may have "domesticated" drinking problems by bringing them firmly under the umbrella of a general sociological approach. Lemert similarly took special pains to cast alcohol-related issues in a broader context that linked them with other kinds of behavior. Use of the label "deviance" does not seem to have significantly stigmatized his work or diminished his contribution to social science as a means of understanding human behavior, at the same time that he made major contributions to the ethnography of drinking in various parts of the world (Lemert, 1954, 1958, 1962, 1964, 1976).

It seems highly likely that the Yale center had a broad impact quite apart from the research that was conducted there or under its auspices. It was a clearinghouse for information, partly as the home of the *Quarterly Journal of Studies on Alcohol,* and for several years as the seat of the Yale Summer School of Alcohol Studies, which provided a common experience and shared perspectives for virtually all members of the first generation who set major directions in alcohol studies. This is not to imply that there was unanimity or even consensus about what was important and how best to study it, but it does seem important that they all had familiarity with a core of concepts, data, and methods—a sort of academic subcultural tradition, if you will. That degree of intellectual community assured a common vocabulary, familiarity with at least a few historical and current themes and controversies, and a general viewpoint about why alcohol studies mattered in the first place.

We are now removed from that shared subculture by at least a full academic generation. Many of those who are actively engaged in alcohol studies now have no such background in common. There can be no doubt that, in some respects, this distancing has been salutary—even when academics have no intention of enlisting disciples, intellectual inbreeding can channel thinking in a way that is self-limiting. Certainly experimentation with novel methods has strengthened research, and challenges to early assumptions might not have occurred so soon or so effectively if the pattern of such strong continuity had lasted longer. At the same time, there may be a cost inasmuch as contrasting definitions of terms need to be selected and justified, much time and effort may be wasted in "reinventing the wheel," and so forth.

If we take publication of a review article as a fairly reliable sign that a subject has come to be important in the scientific literature, it is striking that both Ullman (1958) and Trice and Pittman (1958) offered very different but thorough and detailed reviews of anthropological and sociological approaches to alcohol studies, both in 1958. An anthology can also be viewed as signaling the "acceptance" of a theme in academia—McCarthy's (1959) volume *Drinking and Intoxication* emphasized "social attitudes and controls," whereas Pittman and Snyder's (1962) *Society, Culture, and Drinking Patterns* unequivocally emphasized the interrelatedness of these three elements. Both volumes were richly sociological and anthropological in their content, reflecting early acceptance of the importance of a sociocultural approach. It is ironic that, in subsequent decades, while both the quality and quantity of social science participation in alcohol studies have increased, the intellectual climate of the times is such that biomedical studies, grossly misunderstood by most laypersons, have eclipsed much of the progress that has been made.

Recent Developments

This is not an appropriate context in which to review sociological and anthropological models and theories about drinking and related problems, although readers who are interested in doing so should see a recent book edited by Chaudron and Wilkinson (1988). What is remarkable in this connection is the large number of relevant models that have emerged from these fields in which the study of alcohol has not played a major role. And the models are exceptional in various respects. Many of the models are in the intermediate range, useful in both practical and conceptual terms—unusual inasmuch as both anthropologists and sociologists tend either to ignore theorizing altogether, or to do it on so macroscopic a scale as to have little immediate applicability. Another remarkable feature of these models is the widespread popularity and acceptance they have enjoyed, even among people whose interests are rarely focused on social or cultural considerations. Also important is the fact that the models to which I refer are primarily inductive, often emerging while data were being analyzed rather than having served to frame the data collection earlier. "All in all, the principal contributions of anthropological and sociological models are impressive, especially in view of the fact that no social scientist has ever appeared to be concerned, in a consistent and systematic way, with delineating such models as the major focus of his or her scholarly efforts" (Heath, 1988c, p. 355).

From about 1960 onward, the pace of work on alcohol by social scientists has accelerated remarkably, with the corpus of published literature roughly doubling every five years. A Tri-Ethnic Research Project dealt with Spanish-Americans, Anglos, and Utes near the University of Colorado, and laid the groundwork for one of the few well articulated psychosocial theories about drinking as it relates to other patterns of behavior (Jessor, Graves, Hanson, & Jessor, 1968; Jessor & Jessor, 1977). The ethnographic literature was again mined systematically using the cross-cultural correlational method that Horton had pioneered, but Field (1962) emphasized aspects of social organization more than anxiety, whereas Bacon, Barry, and Child (in Keller, 1965) focused on aspects of child training, with their conclusion being that heavy drinking and drunkenness tend to be associated with discontinuity between a highly dependent childhood and pressures toward independence in adulthood—"the conflict over dependency model." In 1965, the role of anthropologists in alcohol studies was first highlighted in an article that received distinctive treatment in the journal *Current Anthropology,* which also published comments by several colleagues and a rejoinder by the author all in the same issue (Mandelbaum, 1965). (Some might call it the sincerest form of flattery that an English sociologist published virtually the same text under his own by-line two decades later; others have a different term for such questionable homage.) Popham and Yawney (1966) published a bibliography on culture and alcohol use, and occasional articles began to appear in which drinking patterns of non-Western peoples were systematically related to other aspects, such as cirrhosis (Kunitz, Levy, & Everett, 1969), homicide (Levy & Kunitz, 1971), interethnic relations (Honigmann & Honigmann, 1965), legal change (Heath, 1964, 1971), and so forth.

The First Conference of Research Sociologists on Alcohol was held in Carbondale, Illinois, in 1964, but there appears not to have been another. A few false starts were made toward organizing an interest group, but the Society for the Study of Social Problems (mostly sociologists) was probably the first; the Alcohol and Drug Study Group (mostly anthropologists) began around 1979. The Alcohol Research Group in Berkeley has been a major center for many kinds of social research on alcohol; the Addiction Research Foundation in Toronto and the Finnish Foundation for Social Research on Alcohol are others. During the first few years of its existence, the U.S. National Institute on Alcohol Abuse and Alcoholism (within the

Department of Health, Education and Welfare, subsequently renamed Health and Human Services) strongly favored "the sociocultural model," with explicit endorsement of "responsible drinking," and provided some support for both research (Heath, 1980a) and training (Heath, 1981). In recent years, there has been a strong shift in favor of "the control model," denying that there is such a thing as responsible drinking. Anthropologists have been challenged for "problem deflation" (Room, 1984). There was a timely warning against being "uncritically drawn into a political arena" (Roman, 1984, p. 1), at the same time that engagement in developing employee assistance programs demonstrated the practical relevance of social science perspectives (Trice & Roman, 1978).

The "power theory" (McClelland, Davis, Kalin, & Wanner, 1972) is phrased largely in psychological terms, but derives in large part from ethnographic and folkloric source material. Studies within Western societies have provided insights into family dynamics and the transgenerational transmission of alcohol-related problems (Steinglass, Bennett, Wolin, & Reiss, 1987), as well as about how sex roles and occupations relate to drinking problems (Ablon, 1985; Ames, 1985).

I organized the first international conference on anthropology and alcohol in 1973 (Everett et al., 1976), and first brought research anthropologists together with clinicians for prolonged discussion of the mutual relevance of their work in 1978 (Heath et al., 1981). Marshall (1979) anthologized some recent anthropological papers from around the world, and others did the same for selected regions (Hamer & Steinbring, 1980; Waddell & Everett, 1980). The controversy over "the firewater myth" (Leland, 1976), which had been resolved in favor of cultural differences, has been reopened with a fresh emphasis on possible physiological ("racial") differences (Schaefer, 1981).

A few studies emphasize beneficial aspects of alcohol use (summarized in Heath, 1990), whereas many others focus on problems that are said to be linked with drinking (e.g., Collins, 1981; Klausner & Foulks, 1982). A few social scientists have conscientiously reminded anyone who might listen that it is not very helpful to refer to "drinking problems" or "alcohol-related problems" without any clear indication of what is meant by the word "problem" (e.g., Bacon, 1984, 1985; Heath, 1982; Room, 1976, 1984; Straus, 1976, 1979), but this has had little effect.

Intrasocietal variation is being analyzed in far more detail these days (e.g., Bennett & Ames, 1985), whereas broad statements about "normative patterns" were once adequate (Heath, 1980b). Increasing emphasis on "special populations" is a dominant theme in recent years; these include, for example, youth, women, Hispanics, Asian-Americans, the elderly, blacks, and other categories, apparently selected more on the basis of political than sociological considerations. Some insightful analysis of major themes in American culture as expressed in relation in alcohol can be found in the works of Gusfield (1963, 1981) and Levine (1978). The political economics of alcohol is emerging as a theme of renewed interest, as historians and others look at trade and the dependency theory of economic development (e.g., Sargent, 1979; Singer, 1986).

Implications

Despite increasing rhetoric about the value or necessity of interdisciplinary or multidisciplinary efforts, both in the social sciences in general and in the field of alcohol studies, there appears ironically to have been increasing rather than decreasing compartmentalization during the latter half of this century. An unfortunate by-product of the fact that survey research has taken on a methodological rigor and has won the advantages that go with large-scale quantification is that it is viewed by many as the only relevant enterprise in the social sciences. Qualitative research is appreciated when it is focused on thematic representations in literature, films, or other media, but carries little weight with funding agencies as a means of understanding human behavior in natural settings. Although veteran specialists in both survey and observational techniques agree that such work could most fruitfully be done in a complementary way, dealing with the same populations and the same issues in both quantitative and qualitative terms, there have been few serious efforts at such combined work. A controversy with nonacademic implications (Frankel & Whitehead, 1981) concerns the relative efficacy of legislative controls ("the control model") on the one hand (Schmidt & Popham, 1978) and education ("the sociocultural model") on the other (Parker & Harman, 1978) as means of lessening the damage that results from inappropriate drinking on the part of a small minority of those who drink. This has important policy implications inasmuch as the World Health Organization is encouraging nations around the world to opt for increasingly restrictive legislative controls supposedly because (in their view) education doesn't work (Grant, 1985). Disagreement in this instance has to do not only with the relative validity of the models, but also with large-scale systematic misrepresentation of data in support of the "control model" (for a critique, see Heath, 1988a, 1988b).

There is also some segmentation between those who feel a strong conviction that social science has no meaning unless it is immediately directed to the diminution of human suffering, and those who feel that knowledge can have value even before its practical applications become evident. In spite of these differences, there is still a sense of engagement and enthusiasm among practitioners of alcohol studies that does not always characterize the social scientists I know who are working on other topics.

Part of this has to do with the fact that the subject matter is inherently multidisciplinary. Duster (1983, p. 326) used the image that "alcohol is to social science what dye is to microscopy," opening a window on parts of social systems and relationship among customs that might not otherwise be noticed. Another reason for the continuing excitement is that there exists a genuine "university without walls"—the transdisciplinary collegiality that we have discussed in the early days persists, strengthened by a strongly international participation that contributes significantly to the exchange of data, concepts, and methods.

It is not clear to what extent those early social scientists deliberately overcame boundaries among disciplines, or to what extent those boundaries were less sharply marked than they tend to be now. A specialist in oral history and the sociology of knowledge should address this question more systematically than I can in this limited space. Such a project should be greatly aided by the fact that many of the pioneers to whom we are indebted remain active and productive in the field.

Notes

Portions of this chapter have been read as a paper, with the same title, in a symposium on "Alcohol and Culture," organized by David J. Pittman, at the annual meeting of the Midwest Sociological Society held in Minneapolis, Minn., March 23–26, 1988.

1. Some of the broad comments about trends in the literature prior to 1960 are based on my readings in anthropology and sociology (Heath, 1986a, 1986b, 1987). Discussion of the development of alcohol studies, by contrast, grows out of long discussions with Mark Keller, who played a variety of key roles for half a century. His richly detailed oral history is being tape-recorded for eventual publication. Selden Bacon, Robert Straus, and David Pittman also answered a few of my questions, although none of them has reviewed my comments here.
2. A personal anecdote epitomizes both the openness of the center at Yale and the value of Bacon's (1943) "Foundations for a Sociologic Study of Drinking Behavior." Having just returned from more than a year's fieldwork in the tropical forest of eastern Bolivia, my wife, infant son, and I were noticeably tanned in comparison with most people in New Haven in midwinter. While admiring architectural details of an outlying Yale building, we encountered several men who were engaged in animated conversation, but one (whom I have later come to know and appreciate as Mark Keller) was not so engrossed as to miss the opportunity of asking where we had been. On hearing "Bolivia," another (E.M. Jellinek) promptly asked, "How do they drink there?" We briefly described the Cambas' frequent fiestas, for days at a time, marked by universal drunkenness among adults of both sexes, and no untoward behavior. Mark promptly asked if I would write an article for the journal; as a graduate stu-

dent, I was excited at the prospect but uncertain whether my fieldnotes were adequate. (The focus of my research had been land tenure and social organization, as affected by land reform and revolution.) A quick review of my field notes showed that I had abundant data about drinking, but it was not until I read Bacon's (1943) paper that I knew what key questions should be addressed and what topics must at least be essayed. Fortunately, we had brought back a sample of the beverage, but Leon Greenberg who analyzed it chemically said it was too high a proof for me to have swallowed. Responding to his challenge one Saturday morning, under controlled conditions in his lab, I drank it as I had so often in Bolivia, and he grudgingly added a footnote to the physiology of alcohol, noting that they had never dared press a human guinea pig to do what I had done in ignorance or innocence. The paper I wrote (with invaluable editorial help from Chuck Snyder and Mark) became an instant "classic" (Heath, 1958), partly because it touched on virtually every aspect of drinking and culture that seemed relevant at that stage of thinking in the social sciences, and partly because the Camba clearly drank more, more often, of a beverage with a much higher concentration of alcohol than any other population in the world, and yet there was no evidence of their suffering any deleterious consequences. I cite this often as an illustration of the value of the inductive method in social research; data collected incidentally, with no appreciation of their importance, proved later to have enormous significance in disproving the presumption that problems occur in virtually direct proportion to per capita consumption. Brief restudy in succeeding years showed how Camba drinking changed as their social organization changed, but the unproblematic quality of their binges remained (Heath, 1965, 1971).

References

Ablon J. (1985). Irish-American Catholics in a west coast metropolitan area. In L.A. Bennett & G. Ames (Eds.), *The American experience with alcohol: Contrasting cultural perspectives* (pp. 395–409). New York: Plenum Press.

Ames, G.M. (1985). Middle-class Protestants: Alcohol and the family. In L.A. Bennett and G. Ames (Eds.), *The American experience with alcohol: Contrasting cultural perspectives* (pp. 435–458). New York: Plenum Press.

Bacon, S.D. (1943). Sociology and the problems of alcohol: Foundations for a sociologic study of drinking behavior. *Quarterly Journal of Studies on Alcohol, 4,* 399–445.

Bacon, S.D. (1984). Alcohol issues and social science. *Journal of Drug Issues, 14,* 7–29.

Bacon, S.D. (1985). Conversation with Selden D. Bacon. *British Journal of Addiction, 80,* 115–120.

Bales, R.F. (1946). Cultural differences in rates of alcoholism. *Quarterly Journal of Studies on Alcohol, 6,* 480–499.

Bennett, L.A., & Ames, G.M. (Eds.). (1985). *The American experience with alcohol: Contrasting cultural perspectives.* New York: Plenum Press.

Blaine, A. (Ed.). (1980). *Alcoholism in the Jewish community.* New York: Federation of Jewish Philanthropies.

Blume, S., Dropkin D., & Sokolow, L. (1980). The Jewish alcoholic: A descriptive study. *Alcohol Health and Research World, 4,* 21–26.

Braidwood, R.J., Sauer, J.D., Helbaek, H., Mangelsdorf, P.C., Cutler, H.C., Coon, C.S., Linton, R., Steward, J., & Oppenheim, A.L. (1953). Symposium: Did man once live by beer alone? *American Anthropologist, 55,* 515–526.

Calkins, R. (1901). *Substitutes for the saloon: An investigation for the Committee of Fifty.* Boston: Houghton Mifflin.

Chaudron, C.D., & Wilkinson, D.A. (Eds.). (1988). *Theories on alcoholism.* Toronto: Addiction Research Foundation.

Collins, J.J., Jr. (Ed.). (1981). *Alcohol and crime: Perspectives on the relationship between alcohol consumption and criminal behavior.* New York: Guilford Press.

Dorchester, D. (1884). *The liquor problem in all ages.* New York: Phillips & Hunt.

Duster, T. (1983). Commentary. In R. Room & G. Collins (Eds.), *Alcohol and disinhibition: Nature and meaning of the link.* National Institute on Alcohol Abuse and Alcoholism Monograph 12 (pp. 326–330). Washington, DC: Government Printing Office.

Everett, M.W., Waddell, J.O., & Heath, D.B. (Eds.). (1976). *Cross-cultural approaches to the study of alcohol: An interdisciplinary perspective.* The Hague: Mouton.

Field, P.B. (1962). A new cross-cultural study of drunkenness. In D.J. Pittman & C.R. Synder (Eds.), *Society, culture and drinking patterns* (pp. 48–74). New York: John Wiley and Sons.

Fillmore, K.M., Bacon, S.D., & Hyman, M. (1979). The 27-year longitudinal panel study of drinking by students in college, 1949–1976: Final report. NTIS Report 300–302. Springfield, VA: National Technical Information Service.

Frankel, B.G., & Whitehead, P.C. (1981). Drinking and damage: Theoretical advances and implications for prevention. *Rutgers Center of Alcohol Studies Monograph 14.*

Glassner, B., & Berg, B. (1980). How Jews avoid alcohol problems. *American Sociological Review, 45,* 647–663.

Grant, M. (Ed.). (1985). Alcohol policies. *World Health Organization Regional Publications, European Series 18.* Copenhagen: World Health Organization, Regional Office.

Gusfield, J.R. (1963). *Symbolic crusade: Status politics and the American temperance movement.* Urbana: University of Illinois Press.

Gusfield, J.R. (1981). *The culture of public problems: Drinking-driving and the symbolic order.* Chicago: University of Chicago Press.

Hamer, J., & Steinbring, J. (Eds.). (1980). *Alcohol and native peoples of the north.* Lanham, MD: University Press of America.

Harper, F.F. (1904). *The Code of Hammurabi: King of Babylon.* London: Luzac.

Heath, D.B. (1958). Drinking patterns of the Bolivian Camba. *Quarterly Journal of Studies on Alcohol, 19,* 491–508.

Heath, D.B. (1964). Prohibition and post-repeal drinking patterns among the Navaho. *Quarterly Journal of Studies on Alcohol, 25,* 119–135.

Heath, D.B. (1965). Comments on David Mandelbaum, alcohol and culture. *Current Anthropology, 6,* 289–290.

Heath, D.B. (1971). Peasants, revolution and drinking: Interethnic drinking patterns in two Bolivian communities. *Human Organization, 30,* 179–186.

Heath, D.B. (1975). A critical review of ethnographic studies of alcohol use. In R.J. Gibbins, Y. Israel, H. Kalant, R.E. Popham, W. Schmidt, & R. Smart (Eds.), *Research advances in alcohol and drug problems* (Vol. 2, pp. 1–92). New York: John Wiley and Sons.

Heath, D.B. (1976). Anthropological perspectives on alcohol: An historical review. In M.W. Everett, J.O. Waddell, & D.B. Heath (Eds.), *Cross-cultural approaches to the study of alcohol: An interdisciplinary perspective* (pp. 41–101). The Hague: Mouton.
Heath, D.B. (1980a). Ethnographic approaches in alcohol studies and other policy-related fields. *Practicing Anthroplogy, 3,* 19 et seq.
Heath, D.B. (1980b). A critical review of the sociocultural model of alcohol use. In T.C. Harford, D.A. Parker, & L. Light (Eds.), *Normative approaches to the prevention of alcohol abuse and alcoholism.* (National Institute on Alcohol Abuse and Alcoholism Research Monograph 3, pp. 1–18). Rockville, MD: National Institute on Alcohol Abuse and Alcoholism.
Heath, D.B. (1981). Social science research training on alcohol: Profile of a training grant. *Alcohol Health and Research World, 5*(4), 48–52.
Heath, D.B. (1982). Sociocultural variants in alcoholism. In E.M. Pattison & E. Kaufman (Eds.), *Encyclopedic handbook of alcoholism* (pp. 426–440). New York: Gardner Press.
Heath, D.B. (1986a). Drinking and drunkenness in transcultural perspective, part I. *Transcultural Psychiatric Research Review, 23,* 7–42.
Heath, D.B. (1986b). Drinking and drunkenness in transcultural perspective, part II. *Transcultural Psychiatric Research Review, 23,* 103–126.
Heath, D.B. (1986c). Concluding remarks. In T.F. Babor (Ed.), *Alcohol and culture: Comparative perspectives from Europe and America* (Annals of the New York Academy of Sciences 472, pp. 234–238). New York: Academy of Sciences.
Heath, D.B. (1987). A decade of development in the anthropological study of alcohol use: 1970–1980. In M. Douglas (Ed.), *Constructive drinking: Perspectives on drink from anthropology* (pp. 16–69). Cambridge, England: Cambridge University Press.
Heath, D.B. (1988a). Alcohol control policies and drinking patterns: An international game of politics against science. *Journal of Substance Abuse, 1,* 109–115.
Heath, D.B. (1988b). Quasi-science and public policy: A reply to Robin Room about details and misrepresentations in science. *Journal of Substance Abuse, 1,* 121–125.
Heath, D.B. (1988c). Emerging anthropological theory and models of alcohol abuse and alcoholism. In C.D. Chaudron, & D.A. Wilkinson (Eds.), *Theories on alcoholism* (pp. 353–410). Toronto: Addiction Research Foundation.
Heath, D.B. (1989). Environmental factors in alcohol use and its outcomes. In H.W. Goedde & D.P. Agarwal (Eds.), *Alcoholism: Biomedical and genetic factors* (pp. 312–324). New York: Pergamon Press.
Heath, D.B. (1990). Anthropological and sociocultural perspectives on alcohol as a reinforcer. In W.M. Cox (Ed.), *Why people drink: Parameters of alcohol as a reinforcer* (pp. 263–290). New York: Gardner Press.
Heath, D.B., Waddell, J.O., & Topper, M.D. (Eds.). (1981). Cultural factors in alcohol research and treatment of drinking problems. *Journal of Studies on Alcohol, Suppl. 9.*
Honigmann, J.J. (1963). Dynamics of drinking in an Austrian village. *Ethnology, 2,* 157–169.
Honigmann, J.J. (1964). Survival of a cultural focus. In W.H. Goodenough (Ed.), *Explorations in cultural anthropology: Essays in honor of George P. Murdock* (pp. 227–292). New York: McGraw-Hill.
Honigmann, J.J. (1966). Social disintegration in five northern Canadian communities. *Canadian Review of Sociology and Anthropology, 2,* 199–213.

Honigmann, J.J., & Honigmann, I. (1945). Drinking in an Indian-White community. *Quarterly Journal of Studies on Alcohol, 5,* 575–619.

Honigmann, J.J., & Honigmann, I. (1965). How Baffin Island Eskimos have learned to use alcohol. *Social Forces, 44,* 73–83.

Honigmann, J.J., & Honigmann, I. (1970). *Arctic townsmen: Ethnic backgrounds and modernization.* Ottawa: Canadian Research Centre for Anthropology, Saint Paul University.

Horton, D. (1943). The functions of alcohol in primitive societies: A cross-cultural study. *Quarterly Journal of Studies on Alcohol, 4,* 199–320.

Jellinek, E.M. (1957). The world and its bottle. *World Health, 10*(4), 4–6.

Jessor, R., Graves, T.D., Hanson, R.C., & Jessor, S.L. (1968). *Society, personality and deviant behavior: A study of a tri-ethnic community.* New York: Holt, Rinehart and Winston.

Jessor, R., & Jessor, S.L. (1977). *Problem behavior and psychosocial development: A longitudinal study of youth.* New York: Academic Press.

Katz, S.H., & Voigt, M.M. (1987). Bread and beer: The early use of cereals in the human diet. *Expedition, 28*(2), 23–34.

Keller, M. (Ed.). (1965). A cross-cultural study of drinking. *Quarterly Journal of Studies on Alcohol, Supplement 3.*

Keller, M. (1979). Afterword. In M. Keller (Ed.), Research priorities on alcohol. *Journal of Studies on Alcohol, Supplement 8* (pp. vii-x).

Klausner, S.Z., & Foulks, E.F. (1982). *Eskimo capitalists: Oil, politics, and alcohol.* Totowa, NJ: Littlefield, Adams and Co.

Kunitz, S.J., Levy, J.E., & Everett, M. (1969). Alcoholic cirrhosis among the Navaho. *Quarterly Journal of Studies on Alcohol, 30,* 672–685.

Leland, J. (1976). Firewater myths: North American Indian drinking and alcohol addiction. *Journal of Studies on Alcohol, Monograph 11.*

Lemert, E.M. (1954). Alcohol and the Northwest Coast Indians. *University of California Publications in Culture and Society, 2,* 303–406.

Lemert, E.M. (1958). The use of alcohol in three Salish tribes. *Quarterly Journal of Studies on Alcohol, 19,* 90–107.

Lemert, E.M. (1962). Alcohol use in Polynesia. *Tropical and Geographical Medicine, 14,* 183–191.

Lemert, E.M. (1964). Forms and pathology of drinking in three Polynesian societies. *American Anthropologist, 66,* 361–374.

Lemert, E.M. (1976). Koni, Kona, Kava: Orange beer culture of the Cook Islands. *Journal of Studies on Alcohol, 37,* 565–585.

Levine, H.G. (1978). The discovery of addiction: Changing conceptions of habitual drunkenness in America. *Journal of Studies on Alcohol, 39,* 143–174.

Levy, J.E., & Kunitz, S.J. (1971). Indian reservations, anomie, and social pathologies. *Southwestern Journal of Anthropology, 27,* 97–128.

Lolli, G., Serrianni, E., Golder, G.M., & Luzzatto-Fegiz, P. (1958). *Alcohol in Italian culture: Food and wine in relation to sobriety among Italians and Italian Americans.* Glencoe, IL: Free Press.

MacAndrew, C., & Edgerton, R. (1969). *Drunken comportment: A social explanation.* Chicago: Aldine.

Mandelbaum, D.G. (1965). Alcohol and culture [with comments by V. Erlich, K. Hasan, D. Heath, J. Honigmann, E. Lemert, & W. Madsen]. *Current Anthropology, 6,* 281–294.

Marshall, M. (Ed.). (1979). *Beliefs, behaviors and alcoholic beverages: A cross-cultural survey.* Ann Arbor: University of Michigan Press.

Mass Observation. (1943). *The pub and the people: A work town study.* London: Victor Gollancz.
McCarthy, R.G. (Ed.). (1959). *Drinking and intoxication: Selected readings in social attitudes and controls.* Glencoe, IL: Free Press.
McClelland, D.C., Davis, W.N., Kalin, R., & Wanner, E. (1972). *The drinking man.* New York: Free Press.
Morewood, S. (1838). *A philosophical and statistical history of the invention and customs of ancient and modern nations, in the manufacture and use of inebriating liquors—together with an extensive illustration of the consumption and effects of opium.* Dublin: William Curry, Jun., and William Carson.
Myerson, A. (1940). The social psychology of alcoholism. *Diseases of the Nervous System, 1,* 43–50.
O'Brien, J.M., & Seller, S.C. (1982). Attributes of alcohol in the Old Testament. *Drinking and Drug Practices Surveyor, 18,* 18–24.
Parker, D.A., & Harman, M.S. (1978). The distribution of consumption model of prevention of alcohol problems: A critical assessment. *Journal of Studies on Alcohol, 39,* 377–399.
Pittman, D.J., & Snyder, C.R. (Eds.). (1962). *Society, culture and drinking patterns.* New York: John Wiley and Sons.
Popham, R.E. (Ed.). (1976). Jellinek working papers on drinking patterns and alcohol problems. (Substudy 804–1976). Toronto: Addiction Research Foundation.
Popham, R.E., & Yawney, C.D. (Comps.). (1966). *Culture and alcohol use: A bibliography of anthropological studies.* Toronto: Addiction Research Foundation.
Roman, P.M. (1984). The orientations of sociology toward alcohol and society. *Journals of Drug Issues, 14,* 1–6.
Room, R. (1976). Ambivalence as a sociological explanation: The case of cultural explanations of alcohol problems. *American Sociological Review, 41,* 1047–1065.
Room, R. (1984). Alcohol and ethnography: A case of "problem deflation"? [with comments by M. Agar, J. Beckett, L. Bennett, S. Casswell, D. Heath, J. Leland, J. Levy, W. Madsen, M. Marshall, J. Moscalewicz, J.C. Negrete, M. Rodin, L. Sackett, M. Sargent, D. Strug, & J. Waddell]. *Current Anthropology, 25,* 169–191.
Rubington, E. (1968). The bottle gang. *Quarterly Journal of Studies on Alcohol, 29,* 943–955.
Sargent, M. (1979). *Drinking and alcoholism in Australia: A power relations theory.* Melbourne: Longman Cheshire.
Schaefer, J.M. (1981). Firewater myths revisited: Review of findings and some new directions. In D.B. Heath, J.O. Waddell, & M.D. Topper (Eds.), Cultural factors in alcohol research and treatment of drinking problems, *Journal of Studies on Alcohol Supplement 9* (pp. 99–117).
Schmidt, W., & Popham, R.E. (1978). The single distribution theory of alcohol consumption: A rejoinder to the critique of Parker and Harman. *Journal of Studies on Alcohol, 39,* 400–419.
Seller, S. (1984). Attributes of alcohol in the New Testament. *Drinking and Drug Practices Surveyor, 19,* 18–22.
Shalloo, J.P. (1941). Some cultural factors in the etiology of alcoholism. *Quarterly Journal of Studies on Alcohol, 2,* 464–478.
Singer, M. (1986). Toward a political-economy of alcoholism: The missing link in the anthropology of drinking. *Social Science and Medicine, 23,* 113–130.
Snyder, C.R. (1958). *Alcohol and the Jews: A cultural study of drinking and society.* Glencoe, IL: Free Press.

Steinglass, P., Bennett, L.A., Wolin, S.J., & Reiss, D. (1987). *The alcoholic family.* New York: Basic Books.

Stivers, R.A. (1976). *A hair of the dog: Irish drinking and American stereotype.* University Park: Pennsylvania State University Press.

Straus, R. (1974). *Escape from custody.* New York: Harper and Row.

Straus, R. (1976). Problem drinking in the perspective of social change, 1940–1973. In W.J. Filstead, J.J. Rossi, & M. Keller (Eds.), *Alcohol and alcohol problems: New thinking and new directions* (pp. 29–56). Cambridge, MA: Ballinger.

Straus, R. (1979). The challenge of reconceptualization. In M. Keller (Ed.), Research priorities on alcohol, *Journal of Studies on Alcohol Supplement 8* (pp. 279–288).

Straus, R., & Bacon, S.D. (1953). *Drinking in college.* New Haven, CT: Yale University Press.

Straus, R., & McCarthy, R.G. (1951). Nonaddictive pathological drinking patterns of homeless men. *Quarterly Journal of Studies on Alcohol, 12,* 601–611.

Trice, H.M., & Pittman, D.J. (1958). Social organization and alcoholism: A review of significant research since 1940. *Social Problems, 5,* 294–307.

Trice, H.M., & Roman, P.M. (1978). *Spirits and demons at work: Alcohol and other drugs on the job* (2nd ed.). Ithaca: New York State School of Industrial and Labor Relations.

Ullman, A.D. (1958). Sociocultural backgrounds of alcoholism. *Annals of the American Academy of Political and Social Sciences, 315,* 48–54.

Waddell, J.O., & Everett, M.W. (Eds.). (1980). *Drinking behavior among southwestern Indians: An anthropological perspective.* Tucson: University of Arizona Press.

Williams, P., & Strauss, R. (1950). Drinking patterns of Italians in New Haven. *Quarterly Journal of Studies on Alcohol, 11,* 51–91, 250–308, 452–483, 586–629.

World Health Organization, Expert Committee on Mental Health. (1951). Report of the first session of the alcoholism subcommittee. *WHO Technical Report Series 42.* Geneva: World Health Organization.

World Health Organization, Expert Committee on Mental Health (1952). Second report of the alcoholism subcommittee. *WHO Technical Report Series 48.* Geneva: World Health Organization.

CHAPTER 6

Beyond the "Exotic and the Pathologic": Alcohol Problems, Norm Qualities, and Sociological Theories of Deviance

JAMES D. ORCUTT

When Selden Bacon examined the state of alcohol research in 1943, he found its vision seriously impaired. In one of the most colorful passages of "Sociology and the Problems of Alcohol" (1943), he highlighted the limitations of a focus on "pathologic" drinking that had "dominated all the studies in this field":

> This exotic fraction of drinking behavior has attracted all the attention, just as the comet or shooting star elicits more comment than do the millions of "ordinary" stars. In the average citizen this imbalance of interest is not blameworthy; in the scientifically trained student, however, it is blameworthy. It is perhaps true that most sciences had their genesis in the observation and analysis of the immediately painful and the extraordinary, but that stage should by now have been passed. The entire field of social science may be freely criticized on this score; in many instances it may still be found gazing in starry-eyed wonder at the occasional volcanoes, emeralds, and icebergs ... when it has a gigantic earthcrust as its field of inquiry. (p. 409)

In subsequent sections of this programmatic essay, Bacon set about the task of providing a clearer view of the broad range of drinking practices and sociological problems that had been obscured by the myopic "study of the exotic and the pathologic." First, he outlined a conceptual framework that stressed the fundamental significance of cultural rules, sanctions, and sanctioning agents in providing the motivation, "determining the incidence and defining the limits of socially accepted drinking behavior." Then, he mapped out a rich empirical terrain waiting to be explored: the "stratification" or differentiation of drinking patterns in complex society; historical and cultural variations in moral reactions to alcohol problems; the process of socialization into both "normal and abnormal" drinking behavior. Finally, he recommended a number of methodological techniques that the sociological observer might find valuable, including the judicious use of quantitative results from questionnaires "for particular rather than for general purposes, as a subsidiary rather than as a basic method, and toward the close rather than at the beginning of the collection of data." In his concluding remarks, Bacon (1943) was optimistic about the theoretical benefits that would flow from the vigorous program of empirical exercise he prescribed for the field of alcohol studies:

> With a sufficient body of attested data which are amenable to comparison and synthesis, theoretical formulations can be evolved which will be of far greater utility than those arising from the speculative postulates of the armchair philosopher. (p. 445)

Just a few years later, Bacon apparently set aside his reservations about the questionnaire as a "subsidiary method" of data collection to collaborate with Robert Straus on the massive College Drinking Survey (Straus & Bacon, 1953). However, in more important respects, *Drinking in College* was a faithful and fruitful example of the broad-scale approach to drinking behavior that Bacon proposed a decade earlier. Aside from the original contribution of Straus and Bacon's own findings to sociological knowledge about "normal" drinking practices among "ordinary" college students, other researchers at Yale University drew upon these data for insights into the peculiar features of alcohol use within specific religious/ethnic groups. Most notably, in *Alcohol and the Jews* (1958/1978), Charles Snyder used qualitative evidence from interviews with adult Jews and quantitative results from Straus and Bacon's student survey to draw his influential portrait of the distinctive pattern of moderate consumption and sobriety among Jewish drinkers. At the same time, Jerome Skolnick (1958), following an intriguing lead from Straus and Bacon about Mormon drinking problems, mined their data for additional evidence of relatively frequent intoxication and "social complications" among student drinkers from abstinent Protestant backgrounds.

Throughout the late 1950s and the 1960s, the empirical literature in the field of alcohol studies grew well beyond its earlier limits as a host of survey researchers (e.g., Cahalan, Cisin, & Crossley, 1969; Knupfer & Room, 1964; Mulford, 1964) answered Bacon's call for description of "the drinking behavior of... representative sample[s] of the drinking population" (1943). Subsequent increases in federal support for sociological research (Wiener, 1981) and attention to its policy implications (Moore & Gerstein, 1981) seemed to bear out Bacon's earnest hope that social science would eventually play a more active part in the "solution" and "control" of alcohol problems.

However, I will argue here that theoretical developments in the field of alcohol studies have not lived up to Bacon's optimistic expectations. Although sociologists have steadily increased their distance from the "disease model" of alcoholism and other individualistic explanations of alcohol problems (Room, 1983), there are few signs of progress toward more adequate sociological formulations. Perhaps the most prominent and systematic effort at a theoretical "comparison and synthesis" of a body of "attested data" on drinking patterns is Mizruchi and Perrucci's (1962, 1970) typological analysis of norm qualities and deviant drinking behavior. In addition, this framework has inspired a relatively active line of research on the normative contexts of drinking behavior and alcohol problems (e.g., Abu-Laban & Larsen, 1968; Krohn, Akers, Radosevich, & Lanza-Kaduce, 1982; Larsen & Abu-Laban, 1968; Linsky, Colby, & Straus, 1986; Orcutt, 1978; Shore, 1985; Vaughn, 1983). I will approach the norm qualities tradition in much the same way that Bacon assessed the emerging field of alcohol studies in 1943. That is, I begin by showing how Mizruchi and Perrucci's framework and the research it has generated are still impaired by a narrow focus on "the exotic and the pathologic." Ironically, an "exotic fraction of drinking behavior"—Jewish sobriety versus the alleged excesses of traditionally abstinent Protestants—attracts most of the attention in Mizruchi and Perrucci's analysis. Moreover, despite broad advances in research on interpersonal aspects of alcohol use and social reactions to deviant drinking, the norm qualities tradition continues to labor under a reductionist vision of alcohol problems as "pathologic" products of personal "idiosyncracy" and social "disorganization."

The main objective of my critical analysis of the norm qualities tradition is to identify some crucial points at which alcohol studies would benefit from the guidance of theoretical work in a closely related area—the sociology of deviance. Many of the key concepts and analytical problems that Bacon outlined in 1943 also attracted the interest of Merton (1938, 1957), Sutherland (1939, 1947/1956), Lemert (1951), and other theorists who laid the foundations for the modern field of deviance. Even as Bacon was calling for an expanded empirical focus on normal as well as abnormal drinking behavior, Merton (1938) and Sutherland (1939) had already taken initial steps beyond the "exotic and pathologic" and toward more general theories of deviant behavior as a sociologically "normal" phenomenon. In later sections of this chapter, I will indicate how insights from the theoretical perspectives that evolved from that early work—the normative perspective on deviant behavior and the relativistic perspective on social reactions to deviance (Orcutt, 1983)—might be useful in transcending some persisting limitations in the norm quality tradition and the field of alcohol studies more generally.

The Norm Qualities Tradition

Mizruchi and Perrucci's Framework

Mizruchi and Perrucci's initial statement of their norm qualities framework, which appeared in 1962 in the *American Sociological Review* (*ASR*), focused mainly on a distinction between the "prescriptive" and "proscriptive" content of cultural norms toward alcohol use. A later version (Mizruchi & Perrucci, 1967, 1970) included additional discussion of a third "general aspect" of normative systems which they termed "permissiveness." In certain respects, Mizruchi and Perrucci's framework bears a family resemblance to earlier attempts to typify and contrast distinctive cultural orientations toward drinking and intoxication (e.g., Bacon, 1957; Bales, 1946; Pittman, 1967; Ullman, 1958). Invariably, these studies used Jewish drinking practices as a concrete referent for one "ideal" type against which less benign cultural patterns were compared or judged. Accordingly, Mizruchi and Perrucci's (1970) characterization of "prescriptive" norms as "an elaborate system of explicit directives as to what, when, where, with whom, how much, and why one is expected to consume alcoholic beverages" draws heavily on Snyder's (1958/1978) portrait of Jewish drinking culture. Furthermore, in characterizing the defining mandate of "proscriptive" norms—"do not drink"—Mizruchi and Perrucci parallel earlier descriptions of the abstinent cultures of Mormons, ascetic Protestants, and Islamic societies (e.g., Bacon, 1957; Bales, 1946; Pittman, 1967).

However, in other ways, Mizruchi and Perrucci's formulation departs from most of the descriptive typologies and ameliorative proposals that parade under the misleading banner of the "sociocultural model" of alcohol problems (Frankel & Whitehead, 1981; Harford, Parker, & Light, 1980). First, rather than locating their work within the applied realm of alcohol studies, Mizruchi and Perrucci (1962) begin their *ASR* article by avowing a basic interest in the sociological "problem of order" and linking their analysis of "the functional significance of certain aspects of norms in social systems" to Durkheim's (1893/1933) concern with forms of social solidarity. In the service of their theoretical objectives, they pitch their "hypothetical scheme" at a higher level of analytical abstraction than other typologies of drinking cultures. Of course, the reified "norm qualities" retain little of the concrete richness and depth of the case studies of particular religious/ethnic groups from which these conceptual dimensions were induced. However, by freeing this abstract framework from its original empirical moorings, Mizruchi and Perrucci are able not only to explore its theoretical implications for social systems analysis but also to suggest how it could be readily applied to a variety of new research problems, such as sexual behavior or interpersonal aggression (1962).

Second, Mizruchi and Perrucci's framework appears to steer clear of the psychodynamic imagery imbedded in the notion of cultural "ambivalence" toward alcohol (Room, 1976). In fact, the norm qualities approach clearly eschews the "de-emphasis of the content of norms" that undergirds other explanations of excessive drinking as a irrational product of ambivalence or normative conflict (Room, 1976). Instead, Mizruchi and Perrucci see drinking behavior as patterned where normative directives are present and as highly variable or "idiosyncratic" where norms are absent—a line of argument that Room (1976) at least finds viable as an explanation of problematic drinking in the "limited social milieux of particular individuals or small aggregates." Yet, in their revised statement, Mizruchi and Perrucci (1970) suggest that the cultural condition of "permissiveness" and a corresponding lack of effective limits or directives for drinking behavior can "persist over very long periods of time" in complex societies. They explicitly link this macrosociological conception of permissiveness to Merton's (1957) analysis of the strain toward anomie in American society—"since permissiveness and anomie represent aspects of the same phenomenon" (Mizruchi & Perrucci, 1970). Thus, unlike most theorists concerned with drinking cultures, Mizruchi and Perrucci take some preliminary steps toward an integration of the study of alcohol problems with relevant theoretical work in the sociology of deviance.

Finally, in addition to focusing on drinking behavior, Mizruchi and Perrucci (1962) briefly sketch out a unique typology of "group reactions to system strain" produced by "alcohol pathology." They note that their treatment of normative, collective reactions to alcohol problems complements Merton (1957) and Parsons's (1951) theoretical analyses of deviant, individual reactions to strain. More significantly, Mizruchi and Perrucci's illustrations of how ethnic, religious, and professional interest groups attempt to reshape moral boundaries and construct new rules or cultural patterns anticipated some major advances in relativistic work on drug- and alcohol-related deviance in subsequent years (e.g, Becker, 1963; Gusfield, 1963, 1967).

Resolving the Alcohologist's Paradox

While Mizruchi and Perrucci's framework charts a more theoretically oriented course for the study of alcohol problems than have other typologies of drinking cultures, the "prescriptive" and "proscriptive" ideal types still reflect a fascination with "exotic" or paradoxical cases that is endemic to the field of alcohol studies. Clearly, the field's longstanding interest in Jewish drinkers is not simply inspired by admiration of their sobriety; rather, the deeper significance of this case seems to lie in the "truth" that "Jews... are among those having *extremely low rates of alcoholism accompanied by relatively high frequency of drinking*. The evidence is sufficiently clear to cause us to abandon frequency of drinking as a cause of alcoholism" (Ullman, 1958, emphasis added).

On the other hand, the case of backsliding Mormons is rarely cited merely to remind us that some nonconformity among abstinent groups is to be expected; instead, it provides a more instructive lesson in moral irony: "Total abstinence teaching... , in some people, *inadvertently encourages the behavior it most deplores*" (Skolnick, 1958, emphasis added).

Even if Mizruchi and Perrucci show little interest in the antiprohibitionist policy implications that advocates of the "sociocultural model" draw from these cases (see Room, 1976), the "appeal" of the norm qualities framework ultimately rests on its elegant resolution of the alcohologist's paradox of sober drinkers and drunken abstainers. Mizruchi and Perrucci's pivotal distinction between the orderly prescriptive world of Jewish drinking and the stark proscriptive environment of abstinent groups—where individuals drink only at their own risk—offers a plausible and thoroughly cultural account for these "exotic" cases.

Unfortunately, in the process of constructing this clean distinction "from a study of the data on cultural factors in drinking behavior and further induction into an appropriate theoretical model," Mizruchi and Perrucci (1970) gloss over some crucial features of Jewish drinking culture. Snyder (1964) has already registered his concern about the lack of attention to group solidarity among Jews and their social integration into the religious community in Mizruchi and Perrucci's framework, "which focuses exclusively on the quality of the drinking norms." Moreover, Mizruchi and Perrucci's narrow rendition of Jewish drinking norms as "prescriptive" is difficult to reconcile with Snyder's (1958/1978) evidence on negative attitudes toward intoxication and strong reactions to drunkenness, especially among Orthodox Jews. As Snyder points out, "many of the negative ideas and sentiments associated with drunkenness and the drunkard [in Orthodox Judaism] are not unlike those of the ascetic Protestant sects." If anything, the "data on cultural factors" in Jewish drinking reveal a rich mixture of prescriptive *and proscriptive* elements (cf. Frankel & Whitehead, 1981). Yet, through "further induction into an appropriate theoretical model," Mizruchi and Perrucci amputate the latter themes to sharpen the conceptual contrast between Jewish prescription and ascetic Protestant proscription.

Mizruchi and Perrucci perform a more complex "inductive" operation in piecing together the proscriptive type from various analyses of survey data on drinking behavior. They begin by noting that Snyder (1958/1978) and Skolnick's (1958) reanalyses of Straus and Bacon's (1953) data found relatively high rates of intoxication and "social complications" for drinkers "among ascetic Protestant and Mormon groups . . . [a]s contrasted with . . . Jewish students" (Mizruchi & Perrucci, 1970). Of course, these familiar results only show the paradoxical consequences of drinking in abstinent cultures, and shed little light on the normative conditions that promote "extremities of deviant reactions for these groups" (Mizruchi & Perrucci, 1970). However, Mizruchi and Perrucci find the missing inductive link for their account in Mulford and Miller's (1959, 1960a, 1960b) study of drinking and "definitions of alcohol" among adult Iowans. In the following description of Mulford and Miller's cumulative scale of various social and personal effects that individuals associate with the use of alcohol, Mizruchi and Perrucci (1970) highlight its relevance to their own interest in the normative orientation of drinking behavior:

> Differentiating between drinking behavior that is directed by normative systems and that which involves idiosyncratic decisions regarding alcohol consumption, [Mulford and Miller] developed a scale of "personal-effects definitions," which makes a distinction between relatively normative and non-normative drinking behavior. (p. 396)

Mizruchi and Perrucci go on to note that Mulford and Miller's results make it "clear that the focus on 'personal-effects' on the part of the drinker *as contrasted with a more normative orientation* is associated with problem drinking" (1970, emphasis added). As further evidence of the "convergence" between the Iowa research and their own conceptual interpretation of ascetic Protestant and Mormon drinking, Mizruchi and Perrucci (1970) quote a lengthy passage in which Mulford and Miller suggest, in part, that heavy consumption by personal-effects drinkers may be "a reflection of the *relative absence of social norms* in the situations" where they do much of their drinking (from Mulford & Miller 1960a; Mizruchi and Perrucci's emphasis). When these "data" from Mulford and Miller's work are grafted onto the body of evidence from Straus and Bacon's survey, Mizruchi and Perrucci (1970) have all the pieces they need to construct their proscriptive type: the "extreme reactions" of Mormons et al. are a reflection of non-normative, idiosyncratic drinking in a cultural wasteland where there is "an almost complete absence of directives."

Regrettably, Mizruchi and Perrucci's monotonic image of "non-normative drinking" is a poor reproduction of Mulford and Miller's more variegated, symbolic interactionist treatment of definitions of alcohol. In Mulford and Miller's view (1960a), all "drinking behavior, including alcoholism" occurs within a complex "framework of social rules, i.e., prescriptions and proscriptions" in which individuals encounter "a variety of definitions" through interaction "with representatives of first one subculture and then another." Most importantly, the hard distinction between "normative and non-normative drinking behavior" is *nowhere* to be found in Mulford and Miller's (1959, 1960a, 1960b) relatively brief comments on their conceptions and measures of "drinking for interpersonal or social effects" and "drinking to induce direct personal effects." Furthermore, Mizruchi and Perrucci's implication of contrast or discontinuity between drinkers with "a more normative orientation" versus problem drinkers who "focus on 'personal effects' " fails to heed Mulford and Miller's clear words of caution against such polarized interpretations of their cumulative scale of definitions of alcohol: "That is, the personal-effects drinkers agree to items in the upper part of the scale *in addition to accepting the social-effects statements lower on the scale*" (1960a, emphasis added). Thus, if anything, personal-effects drinkers attach a *more comprehensive set of meanings* to alcohol than do those individuals who restrict their definitions to social effects. Finally, Mizruchi and Perrucci's strong, cultural emphasis on "the *relative absence of social norms*" tends to direct attention away from Mulford and Miller's (1960a) original focus on the limited social milieux of certain drinking situations, such as parties or public places, where personal-effects drinkers are less likely to encounter the restrictions of "intimate group norms."

In sum, when Mizruchi and Perrucci's typifications of prescriptive and proscriptive norm qualities are held against the very empirical sources from which they were presumably drawn, this conceptual framework does not fare well according to their own criterion of "goodness of fit" with sociological studies of drinking behavior (1970). At best, their one-dimensional caricature of Jewish drinking culture bears only a partial resemblance to the multifaceted social world described in Snyder's work. In adapting the notion of "personal-effects drinking" to their account of abstinent cultures, Mizruchi and Perrucci literally force a "meaningless" interpretation onto Mulford and Miller's empirical and conceptual analysis of definitions of alcohol in a richly textured symbolic environment. Thus, Mizruchi and Perrucci can only achieve a clear conceptual separation between the prescriptive and proscriptive ideal types by selectively severing or distorting their overlapping connections to the empirical patterns to which they purportedly correspond. Although they are proffered as authentic reproductions of the empirical world, these pristine conceptual creatures—the contrasting norm qualities—are more rarefied, more "exotic" than anything found in the literature on drinking cultures.

Regulation, Constraint, and Permissiveness

As I noted earlier, Mizruchi and Perrucci (1962) initially targeted their work on the sociological problem of order rather than on the study of alcohol problems. Therefore, even if they fail to provide an adequate account of the empirical characteristics of drinking cultures, it is important to consider their answer to the broader question of how qualitative characteristics of the "norms themselves provide an inherent potential for system maintenance and system mal-integration." Assessing their framework in terms of its contribution to functional analysis, I would suggest that Mizruchi and Perrucci have succeeded in distilling two distinct, Durkheimian functions—*regulation* and *constraint*—from the prescriptive and proscriptive content of norms toward alcohol use. In their pure form, prescriptive norms maintain order through behavioral regulation, that is, directing members of a group or society "to act in a particular way [and] spelling out the forms of behavior to which ... members must conform" (Mizruchi & Perrucci, 1962). Conversely, proscriptive norms maintain or impose order on behavior by constraining "participants in the social structure to avoid, abstain, desist, and reject all forms" of nonconforming activity.

This reading of Mizruchi and Perrucci's framework as an analysis of distinct functions of normative integration sheds additional light on the important implication that the two norm qualities inherently "predispose group members to different kinds of deviant reactions." In a well-integrated social order based purely on the function of normative regulation, behavior will tend to vary around prescriptive guidelines only as a matter of degree—that is, conformity and nonconformity stand on a continuum of acts that are *more or less regulated* within this normative order. However, in the case of a social order integrated strictly through the function of normative constraint, any behavior that falls outside of the proscriptive norms is by definition *unconstrained* or dis-orderly. These contrasting images of deviation remain beneath the surface of Mizruchi and Perrucci's (1962) formal hypotheses that "pathology will be low" where norms are prescriptive and "pathology will be high" where norms are proscriptive. Nonetheless, they are clearly apparent in an empirical characterization of "gradual and predictable patterns of deviant behavior" among Jewish drinkers and, more significantly, in the presumption that "non-normative" drinking accounts for the "extremities of deviant reactions" and "pathology" among Mormons and ascetic Protestants. The conclusion that personal-effects drinking, intoxication, or social complications among abstinent groups are manifestations of inherently unconstrained, idiosyncratic behavior follows directly from the underlying logic of Mizruchi and Perrucci's functional analysis—if not from the alcohol research they cite as supporting evidence.

However, Mizruchi and Perrucci (1970) explore a much broader and more complex problem of cultural order in their extended treatment of "permissiveness and anomie." Here, they move beyond the relatively well-integrated preserves of the Jewish and Mormon subcultures to consider a normative orientation that they see as more typical of "drinking in a good many American social contexts." Permissiveness takes the form of "individual determination of limits" and is manifested in the patently anomic behavior of "patternless drinking." How do Mizruchi and Perrucci arrive at this remarkable characterization of post-Prohibition drinking cultures and practices in the United States and other industrial societies? They abandon their claim to an inductive warrant for the construct of permissiveness by pointing out that "the evidence with respect to this pattern or lack of pattern is scanty." Instead, this cultural typification appears to represent a *logical synthesis* of the functions of constraint and regulation or, rather, of the negation of these conditions of social order. That is, given historical or cultural circumstances where "the rules [both] limiting and directing man's desires" have lost their functional significance, the norm qualities framework leads almost automatically to the deduction that behavior will be unconstrained, poorly regulated, and essentially "patternless" at the level of the normative system.

While Mizruchi and Perrucci liken their analysis of permissiveness to Merton's (1957) theory of anomie and deviant behavior, their treatment of cultural disintegration and patternless drinking bears a much stronger resemblance to more primitive conceptions of social disorganization and personal pathology (Matza, 1969; Mills, 1942; Rubington & Weinberg, 1981). For instance, Mizruchi and Perrucci's (1970) heroic effort to explain the massive disruption of normative order entailed in their view of permissiveness would mesh well with any early textbook account of "social disorganization and inebriety":

> Our [approach] suggests that the source of the weakening of social controls on drinking is a reflection of general societal transformation related to industrialization and urbanization.... Given the kind of rapid transformation which characterizes the responses of subsystems to increase in population, change in type of production and modification of family functions,... there is a simultaneous change in group attitudes toward the normative order. (p. 261)

As Matza (1969) has shown, classic conceptions of social disorganization and pathology failed to comprehend the mundane realities of cultural diversity and stable group relations documented in ethnographic research on urban deviance. By the same token, Room (1976) notes that "any explanation of alcoholism in America which focuses on an absence of norms fits rather awkwardly with obtrusive social facts." Whereas Mizruchi and Perrucci (1970) suggest that an association between permissiveness and deviant drinking is illustrated by high rates of arrest for drunkenness in Finland, Room (1976) indicates that it is difficult to conceive of such a society as having abandoned social controls and normative constraints on drinking. Furthermore, in citing Allardt's (1957) investigation of permissive (i.e., positive) attitudes toward drinking in Finland, Mizruchi and Perrucci (1970) fail to consider a striking contradiction between their individualistic inference that "permissiveness . . . allows the person rather than the group to determine the range of appropriate conduct" and Allardt's (1957) own assessment of the sociological implications of his research:

> The . . . most important single finding of this study is the "discovery" of the effect of small drinking groups on actual drinking. . . . [D]rinking groups are intervening factors between the general, culturally determined, social norms and people.

However, the "obtrusive social facts" of group-mediated drinking patterns and subcultural variation in definitions of alcohol (cf. Mulford & Miller, 1960a) cannot be readily accommodated in Mizruchi and Perrucci's analytical model of a loosely integrated, mass society "with only two levels, the individual and the whole system" (Room, 1967).

Deviant Drinking as Pathology

Somewhat surprisingly, Mizruchi and Perrucci's (1970) discussion of permissiveness avoids the medical imagery of "drinking pathology" that was liberally employed in their initial analysis (1962) and focuses, instead, on "deviant drinking behavior." Perhaps this shift in terminology was intended to align their arguments with Merton's (1957) more general approach to the cultural sources of norm-violating behavior in U.S. society. Yet, their treatment of permissiveness depicts conditions under which the sociological concept of "norm-violation" is particularly problematic. If normative standards are vague or absent and if they have no patterned or reliable relationship to variations in drinking behavior, then neither members of the permissive society *nor sociologists* can rely on cultural rules to draw the line between "normal" and "deviant" drinking. This absence of normative criteria leaves it up to individual drinkers "to determine the range of appropriate conduct" (1970); but how would a scientific analyst determine which forms of individual conduct constitute "deviant drinking behavior"?

Mizruchi and Perrucci's answer to this definitional question remains as they originally phrased it in their earlier paper: "pathological reactions . . . represent deviations which are threats to the personal well-being of group members, e.g., problems of drinking or psychosis" (1970). Here, Mizruchi and Perrucci escape the ambiguity of a normatively grounded definition by conceptualizing "deviance" as *objectively threatening or harmful forms of personal pathology*—as a property that ultimately resides in the factual order of individual aberration or defect rather than in the normative order of shared meaning. Only by separating their concept of deviant drinking from normative criteria can Mizruchi and Perrucci sensibly argue that an erosion or disappearance of proscriptive standards during periods of permissiveness will be associated with "a wider range of deviance" (1970). That is, the less the extent to which drinking behavior violates moral or legal norms, the looser the constraints on "man's desires" and the greater the range and rate of pathological forms of "deviance."

Mizruchi and Perrucci's conception of deviant drinking as inherently pathological behavior diverges even more sharply from relativistic conceptions of "deviance" as a social reaction or label that audiences apply to actors and/or behavior. As I indicated earlier, their 1962 paper briefly outlined a typology of "group reactions" that reflected "various efforts to cope with . . . system strain" produced by "alcohol pathology." However, two relativistic analyses of deviance published that same year conceptualized the relationship between individual behavior and social reactions quite differently. First, contrary to Mizruchi and Perrucci's assumption that group reactions are essentially by-products of individual pathology, Kitsuse (1962) contended that "forms of behavior *per se* do not activate the processes of societal reaction which sociologically differentiate deviants from nondeviants." Moreover, as Erikson (1962) subsequently argued, the attribution of "pathology" and other deviant qualities to individuals or behavior is fundamentally contingent on the definitions and reactions of social audiences:

> Deviance is not a property *inherent in* certain forms of behavior, it is a property *conferred upon* these forms by the audiences which directly or indirectly witness them. Sociologically, then, the critical variable in the study of deviance is the social *audience* rather than the individual *person*.

Sociologists in the field of alcohol studies soon demonstrated the utility of this relativistic perspective in micro-level analyses of the labeling of individuals as "alcoholics" (Roman & Trice, 1968; Trice & Roman, 1970) and in historical work on the symbolic politics that have transformed public definitions of alcohol problems as "sin" or "sickness" (Gusfield, 1963, 1967). Yet, even as these studies were gaining new insights into the social construction of the "disease model" and its implications for labeling processes, Mizruchi and Perrucci's (1970) revised analysis stood pat on the position that "drinking pathology" is both objectively given and causally prior to "group reactions" (see Woolgar & Pawluch, 1985). To the extent that reactions vary in intensity over time or differ from group to group, these patterns are direct "consequences of the... differential strain" induced by variations in the rate or nature of alcohol pathology (1970). Accordingly, Mizruchi and Perrucci's objectivist approach discounts any symbolic functions (Gusfield, 1967) or "spurious" properties (Lemert, 1951) of group reactions. From their viewpoint, high rates of arrest for drunkenness in Finland and the abstinence movement in the United States reflect little more than the instrumental efforts of "governmental agencies... and religious groups [to] cope with problems of alcohol" (1970).

In the end, Mizruchi and Perrucci's norm qualities framework offers few productive avenues for sociological analyses of "group reactions" or for cultural explanations of the "factual order" of drinking behavior in complex societies. Both of these phenomena are more deeply rooted in the psychological substratum of individual pathology. As Bacon (1943) noted long ago, there is no scarcity of physiological and psychiatric theories that might explain how "man's desires" or various personal idiosyncracies drive individuals to drink in ways that threaten their physical or mental well-being. However, sociology would appear to have little to contribute to the systematic study of behavior—whether normal or symptomatic—which is essentially "patternless" at the group and cultural levels of analysis. Thus, in moving from reified conceptions of "exotic" drinking cultures to a synthetic view of the "pathologic" character of drinking problems in North American society, Mizruchi and Perrucci leave the field of alcohol studies pretty much as Bacon found it in 1943.

Deviance Theory and Norm Qualities Research

Mizruchi and Perrucci's (1962) overriding theoretical concern with the functional significance of the content of drinking norms led them away from other avenues that Bacon (1943) originally recommended as promising directions for sociological inquiry into alcohol problems. For instance, Bacon stressed the need for research on socialization—"the teaching and learning of drinking behavior"—and included sanctions and sanctioning agents as key elements in his framework for the sociological study of drinking. Although Mizruchi and Perrucci granted that social order and, by implication, the integration of drinking cultures also depend on whether a "normative system is effectively transmitted (socialization) or collectively controlled (sanctions)," they noted that the contribution of these interpersonal mechanisms to "the process of system maintenance... does not concern us here." As we have seen, this inattention to the *social* bases of individual motives and group reactions placed serious constraints on their interpretation of evidence on specific drinking cultures and their comprehension of the patterning of alcohol problems in complex societies. Rather than letting matters rest with this critical indictment, I now turn to some alternative theoretical strategies for reopening these important lines of inquiry and broadening the sociological foundations of research in the norm qualities tradition.

Differential Association and Socialization into Deviant Drinking

Shortly after Bacon's (1943) essay appeared, Edwin Sutherland (1947/1956) presented his final statement of differential association theory—a formulation that still stands as the most systematic and influential sociological explanation of criminality and other forms of norm-violating behavior (Akers, 1985; Gaylord & Galliher, 1988; Orcutt, 1987). Sutherland's approach to deviant socialization is squarely in line with Bacon's (1943) principle that "drinking behavior is subject to the same mode of analysis as any other form of behavior, whether it be table manners . . . or earning a living." Most importantly, in emphasizing the intimate connection between individual motivation and *normal processes of group interaction*, differential association theory focuses attention on a crucial social link in the development of various drinking practices and motives that is missing in Mizruchi and Perrucci's reductionistic account of "drinking pathology."

There is certainly ample evidence to support a theory that holds that the "principal part of the learning of [deviant drinking behavior] occurs within intimate personal groups" (Sutherland, 1947/1956). Consider the conclusions of two relatively recent and quite thorough reviews of the huge "body of attested data" on adolescent drinking and drug use:

> All of the research finds that the best predictors of the adolescent's abstinence or use, the frequency or pattern of use, and attitudes toward use are the patterns of use and attitudes of family and peers. This generalization would appear to be one of the very few (perhaps the only) unequivocal findings from teenage drug and drinking studies. (Radosevich, Lanza-Kaduce, Akers, & Krohn, 1980)

> The most consistent and reproducible finding in drug [and alcohol] research is the strong relationship between an individual's drug behavior and the concurrent drug use of his friends, either as perceived by the adolescent or as reported by the friends. (Kandel, 1980)

While other reviewers are a bit more reserved than Radosevich et al. (cf. Thompson & Wilsnack, 1984) or place somewhat greater weight on family influences than Kandel (cf. Barnes, 1977), it is clear that Sutherland's basic model of primary group socialization fares extremely well against Mizruchi and Perrucci's criterion of "goodness of fit" with studies of adolescent drinking and drug use.

Evidence from a series of investigations based on Mizruchi and Perrucci's (1962) initial article provides a more direct indication of the relevance of differential association theory to research questions and conceptual limitations of the norm qualities tradition. In a well-known study conducted in 1966, Larsen and Abu-Laban (1968) carried the reified contrast between proscriptive and prescriptive norms one step further: they operationalized Mizruchi and Perrucci's ideal types as forced-choice, single-item measures that could be readily administered in surveys of general populations. A questionnaire that was distributed to a sample of 440 Canadian adults asked them to indicate what each of five different groups (parents, religious group, immediate family, close friends, and coworkers) "made known to you [about] rules concerning drinking of alcoholic beverages." In addition to providing response options for proscriptive norms ("one should not drink") and prescriptive norms ("definite guidelines concerning acceptable drinking behavior"), Larsen and Abu-Laban also gave respondents a third option that allowed for the obvious possibility that certain reference groups, such as coworkers, might have "made known to me either few guidelines concerning drinking or none at all." Rather than treating this residual category as the homely offspring of a forced-choice design (i.e., "neither of the above"), Larsen and Abu-Laban adopted it as a measure of "nonscriptive norms." They stressed that this oxymoron "should not be equated with 'permissive norms,' although the two terms have in common the notion of individual determination of limits."

Notwithstanding Larsen and Abu-Laban's proprietary claim to the contrary, there appears to be little difference between their conceptual rendition of nonscriptive norms and Mizruchi and Perrucci's (1967, 1970) revised discussion of the anomic condition of permissiveness. Even though this revision (Mizruchi & Perrucci, 1967) appeared shortly after Larsen and Abu-Laban collected their data, they merely cite it in passing at the beginning of their research report (Larsen & Abu-Laban, 1968) and, then, completely ignore Mizruchi and Perrucci's arguments about the adverse consequences of patternless drinking in permissive cultures. In fact, they only identify Mizruchi and Perrucci as sponsors of the more limited proposition that "there is a relatively high degree of deviant drinking behavior associated with *proscriptive* norms" (Larsen & Abu-Laban, 1968, emphasis added). These moves eventually put Larsen and Abu-Laban in the position to highlight their own, independent "discovery" of the problematic implications of nonscriptive norms:

> Our most significant finding... is that among persons who drink, those who have received few if any, guidelines (nonscriptions)... are more likely to become heavy drinkers than those who are given specific directives.... This finding may appear to support Mizruchi and Perrucci's proposition.... However, it is important to note the difference in the social context to which their gener-

alization and ours applies. Their hypothesis was intended to refer to drinkers from an abstinent background; our finding refers to drinkers from a nonscriptive environment.

Not only are there good reasons to question Larsen and Abu-Laban's claims about the unique theoretical significance of this finding, but there are also compelling statistical reasons to wonder whether this and other results of their study have any significance at all. Larsen and Abu-Laban's (1968) analyses of cross-tabular relationships between a categorized quantity-frequency index of alcohol consumption and the measures of norm qualities were based on data from 180 usable questionnaires mailed back by respondents—41 percent of those originally distributed. Aside from the potential problems of self-selection bias and correlated error in an alcohol survey with such a poor return rate, virtually all of the percentage differences shown in Larsen and Abu-Laban's comparisons of drinking patterns are small, based on relatively few cases, and well within the range of random variation. There is absolutely no way to draw meaningful conclusions about the modest differences between "prescriptive" and "nonscriptive" drinkers who responded to this survey. One, and only one, pattern is crystal clear in these data: In marked contrast to respondents from all other normative contexts, a majority of individuals exposed to proscriptive norms in their current primary group relations (i.e., family, friends, or coworkers) abstained from alcohol use. Of those who specified immediate family or close friends as sources of proscriptive norms, no fewer than 90 percent were either abstainers or very infrequent drinkers (once a month or less). Thus, if Larsen and Abu-Laban's study demonstrates anything, it suggests that differential association with primary group definitions unfavorable to alcohol use effectively regulates abstinent behavior and constrains deviant drinking among adults.

Krohn and his associates (1982) added some new twists to this line of inquiry in an investigation of norm qualities and adolescent drinking and marijuana use. After noting considerable conceptual ambiguity in earlier discussions of ambiguous "normative climates," they introduced separate conceptions and measures of *permissive* ("unconditional approval of use") and *ascriptive* norms ("don't care one way or the other, don't think much about it"). These response options and two other categories for *prescriptive* ("sometimes approve, sometimes disapprove, depending on circumstances") and *proscriptive* norms ("total disapproval") were employed in questionnaire items designed to tap respondents' own attitudes or normative orientations and the perceived norms of socially significant adult (e.g., parents), peer (e.g., close friends), and religious reference groups. In their analysis of self-reported frequency of alcohol and marijuana use and "abuse" (a checklist of problems during or soon after use) among a large sample of 7th- to 12th-grade students, Krohn et al. not only treated the four norm quality responses as nominal categories but, alternatively, as an ordinal scale of *permissiveness* (1 = proscriptive, 2 = ascriptive, 3 = prescriptive, and 4 = permissive). Of course, this representation of norm qualities as different points on a single gradient is quite far removed from Mizruchi and Perrucci's (1962) discrete, ideal types of drinking culture. However, Krohn et al. (1982) were also interested in some "obvious affinities" between the norm qualities tradition and differential association theory, and had found this ordinal scale useful as a measure of Sutherland's concept of a ratio of favorable/unfavorable definitions in an earlier analysis of these data (Akers, Krohn, Lanza-Kaduce, & Radosevich, 1979).

As in Larsen and Abu-Laban's (1968) results, one group of respondents stood apart from other adolescents in the Krohn et al. survey: those individuals who identified their close friends as a proscriptive reference group reported markedly low levels of use and abuse of alcohol. Whereas a majority of respondents in every other norm quality/reference group context drank at least occasionally, only 15.9 percent of the teenagers whose peers totally disapproved of alcohol use were themselves drinkers. Furthermore, "abuse" of alcohol was almost nonexistent (1.4 percent) among the relatively small number of drinkers with proscriptive friends, as compared to at least 10 percent "abusers" among drinkers in all other normative contexts. The contrast between these respondents and those who associated with permissive peers was particularly striking. Nearly four of five teenagers in the latter group (79.5 percent) were drinkers, and a quarter of these drinkers (25.3 percent) were classified as "abusers." Finally, respondents who perceived their close friends as being either mixed (prescriptive) or indifferent (ascriptive) toward alcohol use fell between these polar types. In both of these cases, slightly more than half of the respondents reported that they drank at least occasionally (57.2 percent and 53.0 percent respectively) and, among these drinkers, approximately 10 percent scored as alcohol "abusers."

Here and elsewhere in their analysis, Krohn et al. (1982) found little evidence of the unconstrained drinking that Mizruchi and Perrucci (1962) attributed to proscriptive contexts or that Larsen and Abu-Laban (1968) claimed to see among nonscriptive/ascriptive drinkers who presumably received few normative guidelines. However, results like those above are more interpretable if the norm quality measures are translated as different "states" in the process of primary group association with favorable or unfavorable definitions of teenage drinking. Thus, an excess of definitions unfavorable to alcohol use (proscriptive norms) from one's closest friends virtually amounts to a sufficient condition for abstinent behavior or, at most, highly constrained drinking among adolescents as well as among the adults surveyed by Larsen and Abu-Laban. At the other extreme, an excess of favorable definitions from permissive peers yields the highest rate of teenage drinking. And, where peer group definitions remain in the more or less "balanced" states of mixed approval/disapproval or general indifference toward alcohol use, the theory of differential association "predicts" what Krohn et al. observed: an equiprobable outcome—close to a 50-50 split in respondents' decisions to drink or not to drink (Orcutt, 1987).

A less contrived and more convincing demonstration of the relevance of Sutherland's theory comes from multivariate analyses in which Krohn and his associates (Akers et al., 1979; Krohn et al., 1982) employed their ordinal scale of permissiveness. Krohn et al. (1982) showed that the permissiveness of peers, as compared to the normative influences of adults or religious groups, is by far the best predictor of a teenager's own normative orientation toward drinking. In turn, the latter variable—which reflects respondents' personal definitions or motives toward the act of drinking—mediated the effects of reference group norm qualities on alcohol use and abuse. However, when Akers et al. (1979) analyzed these data, they added another variable that was an even more powerful predictor of teenage drinking: differential association with close friends who use alcohol. In regression equations including measures of norm qualities and numerous other predictors, personal definitions and differential peer association with patterns of drinking together accounted for nearly all the explained variance in alcohol use.

Findings like these are hardly new, of course, but sociologists have often overlooked such "ordinary" patterns of group association and social motivation while searching for more "exotic" cultural variations in drinking behavior. Even Straus and Bacon (1953) were not immune to this tendency. In *Drinking in College*, they devoted only one table and a single sentence to one of the strongest predictors of alcohol use among all the items included in their survey:

> Students who reported that the majority of their close friends drink were for the most part drinkers themselves, while those whose close friends mostly abstain were with little exception, abstainers too.

In retrospect, it is interesting to speculate what might have become of the norm qualities tradition and the mythic image of Mormons as "drunken abstainers" if Straus and Bacon had worked this simple measure of differential association—extent of drinking by close friends—into their analyses of religious affiliation and drinking behavior.

Other classic works, such as Snyder's (1958/1978) case study of Jewish drinking practices, abundantly illustrate how individuals learn distinctive techniques and definitions of alcohol use through frequent, sustained, and intense association with these patterns of behavior and meaning in primary group contexts. However, unlike the norm qualities tradition—for which Jewish sobriety is the shibboleth of prescriptive culture—differential association theory also draws support from Snyder's brief account of episodes of "unruly" drinking and frequent intoxication reported by respondents who were military veterans. As Glassner and Berg (1980) point out after replicating Snyder's finding, many Jews "radically increased their consumption and instances of intoxication... during college and military service [when] *most of their friends were non-Jews*" (emphasis added). Only Orthodox Jews, "who seem to stay together during these experiences," retained a reasonable resemblance to the stereotype of sober drinkers. No matter whether these servicemen and collegians continued to "drink like Jews" or started to "drink like Gentiles," the process of differential association is reflected in the fact that they all tended to drink like their companions in intimate personal groups.

Most important, perhaps, the research of Krohn and his associates (Akers et al., 1979; Krohn et al., 1982) suggests that Sutherland's group-centered theory of deviant socialization can fill some yawning gaps in Mizruchi and Perrucci's (1970) analysis of patternless drinking in the permissive society. These studies reveal that the normative and factual orders of teenage drinking are complex but hardly anomic or "patternless." Across the full range of drinking practices—abstinence, recreational drinking, and "abuse"—Krohn et al. (1982) find that a teenager's own taste or distaste for alcohol corresponds closely to the favorable or unfavorable definitions of his or her peers. Against individualistic conceptions of idiosyncratic motives and unconstrained behavior, Akers et al. (1979) provide support for a view of teenage drinking as a thoroughly social act that is constructed and given meaning through a process of differential association in primary groups. In place of Mizruchi and Perrucci's bleak, two-level model of cultural disintegration and personal pathology, Sutherland's theory offers a coherent and truly sociological account of how "drinking groups are intervening factors between the general, culturally determined, social norms and people" (Allardt, 1957).

Social Reactions to Disorderly Drinking

Given the place and time in which he wrote his remarks on sanctions and sanctioning agents—at Yale, less than a decade after the repeal of Prohibition (Beauchamp, 1980)—Bacon (1943) was understandably skeptical about the effectiveness of "negative sanctions of overt character... in relation to the drinking habits of... large groups." A subsequent generation of sociocultural researchers shared Bacon's jaundiced view of proscriptive measures and his concern with the question of "why these techniques... have shown the poorest results for the control of drinking, particularly excessive drinking" (cf. Aaron & Musto, 1981). Of course, Mizruchi and Perrucci's (1962, 1970) analyses of disorderly drinking in abstinent cultures and postprohibitionist societies sidestepped this question by focusing on an absence of regulatory norms rather than on any adverse consequences of negative sanctions per se. Instead, their objectivist interpretation of group reactions merely treated punitive sanctions—for example, certain "social complications" or high arrest rates—as indicators or by-products of drinking pathology and system strain. However, as illustrated in their discussion of relatively high rates of arrest for drunkenness in Finland (1970), this position implies that the prevalence of heavy drinking or "pathology" in different populations will tend to be directly related to aggregate levels of negative sanctioning.

Linsky, Colby, and Straus (1986) recently examined this and other epidemiological implications of the norm quality framework in an analysis of data for the 50 states of the United States. When these researchers correlated state rates of alcohol consumption and death from liver cirrhosis with rates of arrest for alcohol-related offenses (driving while intoxicated [DWI], and other alcohol-related arrests), they found either no relationship or modest inverse relationships between the indicators of heavy drinking and the measures of negative sanctioning. Thus, contrary to Mizruchi and Perrucci's (1962) notion of "group reactions to system strain," Linsky et al. observed that "DWI arrests and arrests for other alcohol-related offenses do not arise as a response or reaction to heavy drinking behavior."

A more theoretically central aim of the Linsky et al. (1986) study was to relate these measures of alcohol problems to a state-level index of norm content based on (1) membership in fundamentalist churches, (2) population living in legally "dry" areas, (3) rate of on-premise liquor outlets, and (4) degree of restriction on alcohol sales. Linsky and his associates found strong inverse relationships between this composite Proscriptive Norm Index and their indicators of heavy drinking for the 50 states. That is, the most proscriptive states generally had the lowest rates of alcohol consumption and cirrhosis mortality. Consistent with the individual-level studies I reviewed above and with Sutherland's (1947/1956) macro-level extension of his theory, these results suggest that the epidemiology of heavy drinking is an expression of the differential social and political organization of states for or against alcohol use.

Linsky et al. (1986) also found that the Proscriptive Norm Index was directly related to rates of arrest for DWI and other alcohol-related offenses—indicators that they (as would Mizruchi and Perrucci) initially viewed as measures of "disruptive alcohol-related behavior." However, they recognized that interpretation of "the somewhat paradoxical finding of... high DWI and other alcohol-related arrest rates... within proscriptive or dry communities" hinged on a familiar issue in deviance research: Are arrest rates a valid measure of variations in "disruptive," law-violating *behavior*? Or, as relativistic analysts argue, is it more meaningful to treat arrest statistics as measures of differential *social control*—as a complex product of "both the true incidence of the behavior in question and the society's reaction to that behavior" (Linsky et al., 1986; see also Hindelang, 1978; Kitsuse & Cicourel, 1963).

To shed additional light on this issue, Linsky et al. (1986) examined survey data that provided sample estimates of the actual level (i.e., prevalence "over the last month") of *self-reported DWI behavior* among adults in 24 states. Eight of the states that ranked as most "proscriptive" on the Proscriptive Norm Index showed the *lowest* mean proportion of adults who reported having driven while intoxicated; yet, these same states had by far the *highest* average rate of DWI arrest based on official statistics. The eight states ranked as most "permissive" normatively were marked by the *highest* average level of self-reported DWI behavior but the *lowest* average official rate of DWI arrest. The average ratio of official DWI arrests to self-reported DWI behavior was more than two-and-a-half times greater in "proscriptive" states than in "permissive" states. As Linsky et al. (1986) point out in their concluding arguments, these intriguing patterns suggest that

> strong normative proscriptions regarding alcohol seemingly produce results opposite to their intent, not so much by increasing disruptive comportment, but rather by increasing the [social reaction] against such behavior once it occurs.... [S]ocieties that fear alcohol soon encounter problems with disruptive alcoholics.

In sum, Linsky and his associates (1986) build a convincing case against the position that law enforcement and other group reactions are subsidiary, instrumental responses to the "factual order" of alcohol pathology. The disruptive behavior of individuals simply does not explain the social facts of DWI arrest rates. While inferences from state-level data about social reactions to a "fear" of alcohol in "dry communities" are certainly open to some question, Linsky et al. leave little doubt about the need for a sociological theory of sanctioning processes to account adequately for the production and distribution of alcohol-related problems in the United States.

The relativistic logic that Linsky et al. (1986) employed in their study of legal sanctions can be extended to earlier evidence on informal sanctions that figured prominently in the development of the norm qualities tradition. One of the major inductive supports for Mizruchi and Perrucci's (1962) typification of "the extremities of deviant reactions" in proscriptive cultures was Skolnick's (1958) reanalysis of Straus and Bacon's (1953) data on "social complications associated with drinking" among students from certain abstinent religious backgrounds. Although Mizruchi and Perrucci refer to a "marked increase [in social complications] for the ascetic Protestant groups," Skolnick actually reported that one such group—Methodist drinkers—did not differ significantly from the more "permissive" Episcopalians. The only notable difference among Protestent drinkers involved "non-affiliates of abstinence background" (NAAB), who were slightly less than twice as likely as Episcopalians to report "damaged friendships" as a social complication of alcohol use.

Now, the same interpretative issue that Linsky et al. (1986) confronted in their analysis of DWI arrest rates arises here: Are damaged friendships and other forms of social complication (loss of a job, arrest) primarily manifestations of the respondent's *injurious drinking behavior* (Straus & Bacon, 1953)? Or do such items more immediately reflect the workings and outcomes of social reaction processes, for example, exclusionary reactions to NAAB drinkers by their former, abstinent friends (Orcutt, 1973)? The latter interpretation, of course, implies that it is not unconstrained drinking but relatively harsh sanctions applied to drinkers in general that account for the arguably "marked" association of an abstinent background with reports of social complications. In line with this argument, Straus and Bacon reasoned that the high incidence of social complications they found among Mormons "can be explained in part in terms of... drinking sanctions [for] it is readily conceivable that students who drink even in moderation would risk losing the respect of their Mormon friends."

However, Cahalan and Room (1974) have provided the most thorough empirical examination of this issue. Taking advantage of the cluster sampling methodology employed in several national surveys of drinking practices, these researchers were able to articulate data aggregated at the level of neighborhood sampling units—more closely approximating "dry [or wet] communities" than Linsky et al. (1986)—with individual-level data on drinking behavior and problematic consequences. First, Cahalan and Room found that a cluster-level measure of temperance sentiment (the attitudinal response that there is "nothing good" about drinking) was by far the strongest correlate of cluster scores on an index of "tangible consequences" (consisting mostly of problems with relatives, friends, neighbors, job, or police). The prevalence of heavy drinking was essentially unrelated to tangible consequences across clusters. Again, it appears that communities that "fear alcohol" do indeed encounter or, perhaps, create a disproportionate share of drinking problems.

Next, Cahalan and Room (1974) classified clusters as "wet," "medium," or "dry" according to their average levels of alcohol consumption/abstinence and examined the drinking behavior and social consequences for individual respondents within these clusters. One particularly striking result emerged when they held individual consumption relatively constant by focusing only on "heaviest intake" drinkers. Whereas roughly one-quarter of these heavy drinkers in wet or medium clusters reported moderate or severe social consequences, nearly two-thirds of comparable drinkers in dry clusters experienced either moderate or severe social consequences. Clearly, something about drinkers in dry communities—or about the community's reaction to them—sets them apart as unusually troubled or troublesome individuals.

Yet, after a detailed examination of numerous background and personality measures available for these respondents, Cahalan and Room (1974) found very little systematic support for the possibility that psychological defects or other distinctive traits lead certain individuals to drink heavily in spite of community disapproval and adverse social consequences. Instead, their evidence favored the view that "the specially strong association of tangible consequences with heavy drinking in dry areas is more a matter of the strong local reactions to heavy drinking behavior than of strong inherent maladjustments of the heavy drinker."

These findings, joined with those of Linsky et al. (1986), add new significance to Mizruchi and Perrucci's (1962) characterization of the "extremities of deviant reactions" in proscriptive cultures. If drinking is a more complicated, consequential, and problematic act in dry than in wet communities, the primary source of these troubles appears to be relatively extreme *social* reactions to deviant drinkers—arrest, disrupted relationships, and disrepute among those who see "nothing good" about drinking. The "attested data" seem to show that the deviant drinker in proscriptive contexts is not distinguished by any peculiar signs of unconstrained or pathological behavior but by "social processes external to the individual expelling him from 'normality' " (Cahalan & Room, 1974). Although data on arrests and social complications are of dubious value as epidemiological measures of drinking pathology, the most recent research in the norm qualities tradition indicates that a relativistic reading of these patterns of sanctioning can yield important sociological insights into the moral/political anatomy of alcohol problems and the social constitution of disorderly drinking.

Conclusion

It would appear that Mizruchi and Perrucci (1962) did not reach far enough into Durkheim's work when they anchored their approach to the problem of order on *The Division of Labor in Society* (1893/1933). The relativistic framework that Linsky et al. (1986) and Cahalan and Room (1975) found useful for understanding social reactions to disorderly drinking rests instead on Durkheim's brilliant analysis of "the normal and the pathological" in *The Rules of Sociological Method* (1895/1966). With characteristic irony, Durkheim chose the phenomenon of crime, "whose pathological character appears incontestable," to illustrate his rules for determining the normality of social facts. He observed, first, that crime is a ubiquitous fact of collective life: it has been present "in all societies of all types" and even its increasing rate in modern societies "indisputably [displays] all the symptoms of normality." Second, and more importantly, Durkheim contended that "crime is normal because a society exempt from it is utterly impossible." He employed a classic mental experiment to advance this argument:

> Imagine a society of saints, a perfect cloister of exemplary individuals. Crimes, properly so called, will there be unknown; but faults which appear venial to the layman will create there the same scandal that the ordinary offense does in ordinary consciousnesses. If, then, this society has the power to judge and punish, it will define these acts as criminal and will treat them as such. (1895/1966)

And so might Durkheim see as "indisputably normal" the social facts of drinking behavior and alcohol problems in dry communities, among Mormons, or for other latter-day embodiments of a proscriptive "society of saints." When drinking is an act that offends "very strong collective sentiments," heavy consumption (Larsen & Abu-Laban, 1968), "abusive" drinking (Krohn et al., 1982), or driving while intoxicated (Linsky et al., 1986) will tend to be "more rare" than in permissive social circles. And, yet, the moral alchemy of this intense collective sentiment that there is "nothing good" about drinking (Cahalan & Room, 1974) transforms the "venial faults" and "inevitable divergences" of some family members, friends, or neighbors into alcohol problems:

> What confers this [problematic] character upon them is not the intrinsic quality of a given act but that definition which the collective conscience lends them. If the collective conscience is stronger, if it has enough authority practically to suppress these divergences, it will also be more sensitive, more exact-

> ing, . . . reacting against the slightest deviations with the energy it otherwise displays only against more considerable infractions. (Durkheim, 1895/1966)

Thus, I presume that Durkheim would treat Mizruchi and Perrucci's notion of the "pathological character" of alcohol problems with the same sense of sociological irony as he did the 19th-century views of criminological positivists. Judged against his rules, the "factual order" of deviant drinking is a decidedly "normal" product of social reactions.

However, Durkheim was mainly concerned with social facts at the level of collectivities, and he only briefly adumbrated an explanation of individual deviation from "collective types"—the "inevitable divergences" found even in a "society of saints" (1895/1966). He asserted that absolute conformity to the collective type is "utterly impossible; for the immediate physical milieu in which each of us is placed, the hereditary antecedents, and the social influences vary from one individual to the next, and consequently diversify consciousness." Here, Durkheim might find a congenial colleague in Sutherland, who, in his own way, did much to advance the study of crime as a "normal" fact of group life. Sutherland's (1947/1956) account of deviant motivation as a product of variations in the primary group milieu, antecedent socialization, and ongoing influences of differential association puts sociological flesh on Durkheim's skeletal model of the diversification of individual "consciousness." More to the point of this chapter, neither Durkheim nor Sutherland completely begged the question of individual deviation by reducing it to "idiosyncracy" or "pathology."

In this modest effort to demonstrate the bearing of deviance theory on research in the field of alcohol studies, I have limited most of my attention to the empirical materials from which Mizruchi and Perrucci (1962, 1970) constructed their hypothetical "collective types" of normative order. Also, in reframing these materials around the problems of socialization and sanctioning of deviant drinking, I have focused on just a small part of Bacon's (1943) grand blueprint for the sociological study of drinking behavior. My main objective has been to suggest that the project of transcending "the exotic and the pathologic"—which is largely unfinished after half a century—will progress more rapidly if powerful theoretical tools from the sociology of deviance are put to work on the rich resources of alcohol studies.

Acknowledgments

I wish to thank Harold Mulford and Robert Perrucci for giving me their reactions to an incomplete draft. I am especially grateful to Richard Wilsnack for his insightful reading of my critique and his thoughtful suggestions for moving beyond it.

References

Aaron, P. & Musto, D. (1981). Temperance and Prohibition in America: A historical overview. In M. H. Moore & D. R. Gerstein (Eds.), *Alcohol and public policy: Beyond the shadow of Prohibition* (pp. 127–81). Washington, DC: National Academy Press.

Abu-Laban, B., & Larsen, D.E. (1968). The qualities and sources of norms and definitions of alcohol. *Sociology and Social Research, 53*, 34–43.

Akers, R.L. (1985). *Deviant behavior: A social learning approach* (3rd ed.). Belmont, CA: Wadsworth.

Akers, R.L., Krohn, M.D., Lanza-Kaduce, L., & Radosevich, M. (1979). Social learning and deviant behavior: A specific test of a general theory. *American Sociological Review, 44*, 636–655.

Allardt, E. (1957). Drinking norms and drinking habits. In E. Allardt, T. Markkanen, & M. Takala (Eds.), *Drinking and drinkers* (pp. 7–109). Helsinki: Finnish Foundation for Alcohol Studies.

Bacon, S.D. (1943). Sociology and the problems of alcohol: Foundations for a sociologic study of drinking behavior. *Quarterly Journal of Studies on Alcohol, 4*, 402–445.

Bacon, S.D. (1957). Social settings conducive to alcoholism: A sociological approach to a medical problem. *Journal of the American Medical Association, 164*, 177–181.

Bales, R.F. (1946). Cultural differences in rates of alcoholism. *Quarterly Journal of Studies on Alcohol, 6*, 480–499.

Barnes, G.M. (1977). The development of adolescent drinking behavior: An evaluative review of the impact of the socialization process within the family. *Adolescence, 12*, 571–591.

Beauchamp, D.E. (1980). *Beyond alcoholism: Alcohol and public policy*. Philadelphia: Temple University Press.

Becker, H.S. (1963). *Outsiders: Studies in the sociology of deviance*. New York: Free Press.

Cahalan, D., Cisin, I.H., & Crossley, H.M. (1969). *American drinking practices*. New Brunswick, NJ: Center of Alcohol Studies.

Cahalan, D., & Room, R. (1974). *Problem drinking among American men*. New Brunswick, NJ: Center of Alcohol Studies.

Durkheim, E. (1933). *The division of labor in society*. (George Simpson, Trans.). New York: Free Press. (Original work published 1893)

Durkheim, E. (1966). *The rules of sociological method*. (Sarah A. Solovay & John H. Mueller, Trans.; George E.G. Catlin, Ed.). New York: Free Press. (Original work published 1895)

Erikson, K.T. (1962). Notes on the sociology of deviance. *Social Problems, 9*, 307–314.

Frankel, B.G., & Whitehead, P.C. (1981). *Drinking and damage: Theoretical advances and implications for prevention*. New Brunswick, NJ: Center of Alcohol Studies.

Gaylord, M.S., & Galliher, J.F. (1988). *The criminology of Edwin Sutherland*. New Brunswick, NJ: Transaction Books.

Glassner, B., & Berg, B. (1980). How Jews avoid alcohol problems. *American Sociological Review, 45*, 647–664.

Gusfield, J.R. (1963). *Symbolic crusade: Status politics and the American Temperance movement.* Urbana: University of Illinois Press.
Gusfield, J.R. (1967). Moral passage: The symbolic process in public designations of deviance. *Social Problems, 15*, 175–188.
Harford, T.C., Parker, D.A., & Light, L. (Eds.). (1980). *Normative approaches to the prevention of alcohol abuse and alcoholism.* Rockville, MD: National Institute on Alcohol Abuse and Alcoholism.
Hindelang, M.J. (1978). Race and involvement in common law personal crimes. *American Sociological Review, 43*, 93–109.
Kandel, D.B. (1980). Drug and drinking behavior among youth. In A. Inkeles, N.J. Smelser, & R.H. Turner (Eds.), *Annual Review of Sociology*, Vol. 6 (pp. 235–285). Palo Alto, CA: Annual Reviews.
Kitsuse, J.I. (1962). Societal reaction to deviant behavior. *Social Problems, 9*, 247–256.
Kitsuse, J.I., & Cicourel, A.V. (1963). A note on the use of official statistics. *Social Problems, 11*, 131–139.
Knupfer, G., & Room, R. (1964). Age, sex, and social class as factors in amount of drinking in a metropolitan community. *Social Problems, 12*, 224-240.
Krohn, M.D., Akers, R.L., Radosevich, M.J., & Lanza-Kaduce, L. (1982). Norm qualities and adolescent drinking and drug behavior: The effects of norm quality and reference group on using and abusing alcohol and marijuana. *Journal of Drug Issues, 12*, 343–359.
Larsen, D.E., & Abu-Laban, B. (1968). Norm qualities and deviant drinking behavior. *Social Problems, 15*, 441–450.
Lemert, E.M. (1951). *Social pathology*. New York: McGraw-Hill.
Linsky, A.S., Colby, J.P., & Straus, M.A. (1986). Drinking norms and alcohol-related problems in the United States. *Journal of Studies on Alcohol, 47*, 384–393.
Matza, D. (1969). *Becoming deviant.* Englewood Cliffs, NJ: Prentice-Hall.
Merton, R.K. (1938). Social structure and anomie. *American Sociological Review, 3*, 672–682.
Merton, R.K. (1957). *Social theory and social structure* (rev. ed.). New York: Free Press.
Mills, C.W. (1942). The professional ideology of social pathologists. *American Journal of Sociology, 49*, 165–180.
Mizruchi, E.H., & Perrucci, R. (1962). Norm qualities and differential effects of deviant behavior: An exploratory analysis. *American Sociological Review, 27*, 391–399.
Mizruchi, E.H., & Perrucci, R. (1967). Norm qualities and deviant behavior. In E. H. Mizruchi (Ed.), *The substance of sociology* (pp. 259–270). New York: Appleton-Century-Crofts.
Mizruchi, E.H., & Perrucci, R. (1970). Prescription, proscription and permissiveness: Aspects of norms and deviant drinking behavior. In G.L. Maddox (Ed.), *The domesticated drug: Drinking among collegians* (pp. 234–253). New Haven: College & University Press.
Moore, M.H. & Gerstein, D.R. (Eds.). (1981). *Alcohol and public policy: Beyond the shadow of Prohibition.* Washington, DC: National Academy Press.
Mulford, H.A. (1964). Drinking and deviant drinking, U.S.A., 1963. *Quarterly Journal of Studies on Alcohol, 25*, 634–650.
Mulford, H.A., & Miller, D.E. (1959). Drinking behavior related to definitions of alcohol: A report of research in progress. *American Sociological Review, 24*, 385–389.

Mulford, H.A., & Miller, D.E. (1960a). Drinking in Iowa. III. A scale of definitions of alcohol related to drinking behavior. *Quarterly Journal of Studies on Alcohol, 21*, 267–278.

Mulford, H.A., & Miller, D.E. (1960b). Drinking in Iowa. V. Drinking and alcoholic drinking. *Quarterly Journal of Studies on Alcohol, 21*, 483–499.

Orcutt, J.D. (1973). Societal reaction and the response to deviation in small groups. *Social Forces, 52*, 259–267.

Orcutt, J.D. (1978). Normative definitions of intoxicated states: A test of several sociological theories. *Social Problems, 25*, 385–396.

Orcutt, J.D. (1983). *Analyzing deviance*. Homewood, IL: Dorsey Press.

Orcutt, J.D. (1987). Differential association and marijuana use: A closer look at Sutherland (with a little help from Becker). *Criminology, 25*, 341–358.

Parsons, T. (1951). *The social system*. New York: Free Press.

Pittman, D.J. (1967). International overview: Social and cultural factors in drinking patterns, pathological and nonpathological. In D.J. Pittman (Eds.), *Alcoholism* (pp. 3–20). New York: Harper & Row.

Radosevich, M., Lanza-Kaduce, L., Akers, R.L., & Krohn, M.D. (1980). The sociology of adolescent drug and drinking behavior: A review of the state of the field: Part II. *Deviant Behavior, 1*, 145–169.

Roman, P.M., & Trice, H.M. (1968). The sick role, labelling theory and the deviant drinker. *International Journal of Social Psychiatry, 12*, 245–251.

Room, R. (1976). Ambivalence as a sociological explanation: The case of cultural explanations of alcohol problems. *American Sociological Review, 41*, 1047–1065.

Room, R. (1983). Sociological aspects of the disease concept of alcoholism. In R. Smart, F. Glaser, Y. Israel, H. Galant, R. Popham, & W. Schmidt (Eds.), *Research advances in alcohol and drug problems* (Vol. 7, pp. 47–91). New York and London: Plenum Press.

Rubington, E., & Weinberg, M.S. (1981). *The study of social problems: Five perspectives* (3rd ed.). New York: Oxford.

Shore, E.R. (1985). Norms regarding drinking behavior in the business environment. *Journal of Social Psychology, 125*, 735–741.

Skolnick, J.H. (1958). Religious affiliation and drinking behavior. *Quarterly Journal of Studies on Alcohol, 19*, 452–470.

Snyder, C.R. (1964). Inebriety, alcoholism, and anomie. In M.B. Clinard (Eds.), *Anomie and deviant behavior* (pp. 189–212). New York: Free Press.

Snyder, C.R. (1978). *Alcohol and the Jews*. Carbondale, IL: Southern Illinois University Press. (Original work published 1958)

Straus, R., & Bacon, S.D. (1953). *Drinking in college*. New Haven: Yale University Press.

Sutherland, E.H. (1939). *Principles of criminology* (3rd ed.). Philadelphia: Lippincott.

Sutherland, E.H. (1956). A statement of the theory. In A. Cohen, A. Lindesmith, & K. Schuessler (Eds.), *The Sutherland papers* (pp. 7–12). Bloomington: Indiana University Press. (Original work published 1947)

Thompson, K.M., & Wilsnack, R.W. (1984). Drinking and drinking problems among female adolescents: Patterns and influences. In S.C. Wilsnack & L.J. Beckman (Eds.), *Alcohol problems in women: Antecedents, consequences, and intervention* (pp. 37–65). New York: Guilford Press.

Trice, H.M., & Roman, P.M. (1970). Delabeling, relabeling, and Alcoholics Anonymous. *Social Problems, 17*, 538–546.

Ullman, A.D. (1958). Sociocultural backgrounds of alcoholism. *Annals of the American Academy of Political and Social Science, 315*, 48–54.
Vaughn, S.M. (1983). The normative structures of college students and patterns of drinking behavior. *Sociological Focus, 16*, 181–193.
Wiener, C. (1981). *The politics of alcoholism: Building an arena around a social problem*. New Brunswick, NJ: Transaction Books.
Woolgar, S., & Pawluch, D. (1985). Ontological gerrymandering: The anatomy of social problems explanations. *Social Problems, 32*, 214–217.

CHAPTER 7

A 1990s View of Research on Youthful Alcohol Use: Building from the Work of Selden D. Bacon

FLORENCE KELLNER ANDREWS

I last saw Selden Bacon in the summer of 1976, when I defended my doctoral dissertation. Since that time, I put down new roots in Canada and have become influenced by a different society, different colleagues, and altered circumstances. These influences, although appreciable and constantly present, are tempered by my experiences with Selden Bacon and the very vital atmosphere of the Rutgers Center of Alcohol Studies, where I did my graduate work. The influence of Selden Bacon, who was overwhelmingly generous with his time, unflagging in his demands for as reasonable a product as a very green graduate student could produce, and unfettered in his criticism of faulty thinking and sophomoric writing, remains with me and continues to benefit my teaching and my research. When I ask a student, "So what?" about a piece of work that lacks statements concerning theoretical or practical implications, or "How do you know?" about a piece missing statements of evidence, it is Bacon who is doing the talking.

Back then, such help was not always easy to take. I was in awe of this man whose spoken prose was elegant, forceful, and of a quality that could be transcribed with no editing whatsoever. His impressive command of language made one of his first comments to me about a written product, "Is English your native language?" more of a motivation to communicate effectively than the offense that it could have been. Every week or so, I sat in his office with a note pad, attempting to record some fraction of his comments on my most recent ten pages. These sessions became a source of amusement for others on the center staff, and I shall always appreciate the concern of Lucille Hynda, the center administrator. Lucille kept her eye on Bacon's door, asking people nearby, "Is he still talking?" When I came out, she was ready to receive and to humor a dazed 22-year-old, who was convinced that she would never become an academic: "If he didn't think you were worth it, would he have just spent two hours with you?" Someday, I just might be "worth it"; at the very least, I continue to work in the alcohol field in research and in teaching.

I am located in Ottawa, Canada's capital. The combination of graduate training in alcohol studies and being close to federal agencies has provided a number of good research opportunities. My first full-time employment here was as a research assistant for the Royal Commission of Inquirey into the Non-Medical Use of Drugs. Work for this agency entailed two reports: one, on the involvement of drugs with crime; the other, on the patterns of use and abuse of alcohol in Canada. One of the consequences of this early work experience was to familiarize me with data on alcohol use from Canadian federal sources. In the past 10 years, I have collaborated with a number of government researchers analyzing national data on alcohol. Presently, I am involved in a large project that concerns changes in Canadian drinking patterns over the past 10 years, with special emphasis on youth, the elderly, and the concurrent recreational use of alcohol and other drugs.

In this chapter, I shall consider developments in research on alcohol use by young people since the publication of *Drinking in College*. I shall also assess some of the important changes in social influences upon youthful drinking since mid-century.

In 1953, Robert Straus and Selden Bacon's *Drinking in College* was published. This work stands as a most farsighted study in its scope, its presentation of the dimensions of alcohol use, and its methods. The research sample included nearly 16,000 students from 27 colleges. The colleges were selected to represent geographic regions, funding sources (public or private endowments), religious affiliations, and race and gender.

This study has not been replicated on a national scale in the four decades since it was carried out. Instead, most recent studies on collegiate drinking draw samples from a single institution of higher education. At the very best, the research covers one region of the country. For example, a study by Wechsler and McFadden (1979) sampled students from 34 New England colleges and Temple's (1986) research included a large sample from two California universities.

Although present statistical and computational technology facilitate more complex analysis than was common in social science in the 1950s, the conceptualization of alcohol use and the identification of aspects of use laid out by Bacon (1943) in "Sociology and the Problems of Alcohol" and employed in *Drinking in College* remain a standard for formulating the most relevant questions about alcohol use. Particularly important are the quantity-frequency measures, frequency of intoxication, and alcohol-associated problems. These features of conventional drinking and problem drinking are still used to assess contemporary patterns and problems.

Perhaps the best example of the early methodological sophistication of this research was the recognition of longitudinal research in the alcohol field and the potential of the large sample as a source for follow-up research. Straus and Bacon had the foresight to ask their respondents to give permission for later contact, and more than three-quarters of them gave consent. Fillmore's (1974, 1975) research, which followed up 200 subjects from this sample a generation later, demonstrated the utility of longitudinal work. Associations between early alcohol use and later use were documented, especially where alcohol problems were concerned. Early planning of longitudinal research anticipated some of the more recent work on noncollege, general population samples. The RAND study (Polich, Armor, & Braiker, 1981), research from the American Drinking Practices sample (Cahalan, 1970; Cahalan & Room, 1972), and, more recently, the Ecological Catchment Area studies (e.g., see Maddox, Robins, & Rosenberg, 1986) are yielding finely grained longitudinal information about the people, circumstances, and fluctuations in alcohol use and problem use.

A recent reading of *Drinking in College* reminded me that what are now basic assumptions about alcohol use had to be argued and empirically supported in early writing. Prominent among these assumptions are the ubiquity of alcohol use, as well as its normalcy: most people use alcohol in ways that are not problematic. Although these ideas are now taken for granted by alcohol researchers, they remain unsettled in society at large. The recent health movement has introduced an ideology of restriction in food and recreational substance intake that may eventually translate into decreased intake of alcohol and tobacco. At the very least, ambivalence about alcohol use should be heightened.

Regarding ambivalence, the Straus and Bacon study demonstrated an inverse relationship between the proportion of drinkers and the incidence of intoxication in ethnic and religious groups. These results suggested that those groups with well-established regulations regarding the use of alcohol (indicated by high proportions of users) produce low rates of problem drinking. Those groups with prohibitive attitudes about drinking provide no regulation once drinking begins. Further research on this problem has produced inconsistent results, and the issue of the effect of norms and attitudes on drinking behavior and on problem drinking remains a subject of some debate (e.g., see Frankel & Whitehead, 1981; Peele, 1987a, 1987b; Room, 1987).

The following discussion highlights stability and change in the conditions affecting alcohol use among young people since the publication of *Drinking in College*. It also identifies the most crucial directions that have been followed by investigators of alcohol issues among youth, and makes some proposals regarding future directions.

Youth and Alcohol: Broadening the Research Base

Describing the socioeconomic variability of the sample, the authors of *Drinking in College* state that the college population comes from "nearly every part of American society." Indeed, in their careful selection for diversity of financial support, sectarian affiliation, and region, substantial variability in cultural origin and economic level was attained. Variation in the college population is even greater at the present time. In 1950, less than 30 percent of the American population between the ages of 18 and 24 were enrolled in colleges (U.S. Bureau of the Census, 1968). As of 1986, more than 50 percent of the same population attended college (U.S. Bureau of Census, 1989). The increased universality of higher education also increased the degree to which the college population reflects the general population of the same age.

There is an interesting anomaly in recent surveys of national drinking practices: their sampling units are usually "households" such that persons residing in all types of institutional settings are not sampled. Thus, people living in college residences are omitted from the surveys, thereby excluding a critical sector of the late teens/young adult cohort (e.g., see Norton & Colliver, 1988). This is true in Canada (Andrews & Rootman, 1984) as well as in the United States. Studies using national household samples, while they have a broad population base, most likely misrepresent this particular age cohort in serious but unknown ways, because half of them are in college, and many of these are residing in nonhousehold settings.

The campus scene itself has changed since 1950, with an increase in alcohol outlets on campus. *Drinking in College* reported a negligible amount of on-campus and sorority/fraternity house drinking. Students tended to drink in private homes or in public, licensed establishments. At present, an increase in on-campus outlets of alcoholic beverages, and a decrease in *in loco parentis* policies at the colleges themselves certainly provide more opportunities for collegians to use alcohol.

Although place of residence was a questionnaire item in Straus and Bacon's study, it was not included in their analysis. I am assuming that their sample was probably made up of almost all on-campus residents. With the increased access of institutions of higher education in large population centers, more students live off-campus and some of these remain at home with parents while attending college. Place and type of residence surely have implications for the social control of drinking, as well as the social control of other behaviors. Recent findings that collegiate drinking is associated with a combination of familial drinking norms and family cohesion (Wiggins & Wiggins, 1987) and that participation in all-male, age-homogeneous drinking contexts is associated with heavier drinking

(Orford, Waller, & Peto, 1974) indicate that residence, with its related restrictions (or lack of them), is an extremely important variable in differential drinking practices.

To summarize, although increased universality of higher education has made collegians more representative of the general population of the late teens/early twenties age group than was the case in 1950, alcohol and substance use studies of this age group have not taken advantage of this opportunity to represent an important sector of 18- to 24-year-olds on a national scale. With few exceptions (e.g., Blane and Hewett, 1977; Hanson, 1971; Wechsler and McFadden, 1979), there has been little attempt to assess national trends in alcohol use and abuse in the college population. National population surveys will miss collegians who are on-campus residents, because sampling frames exclude residents in institutional settings.

A valid assessment of trends in college drinking would require a national sampling of college entrants, with the sampling frame including adequate representation of the more traditional four-year college of the type studied by Straus and Bacon (1953), as well as the newer state and community colleges. Such sampling would represent a cross-section of those people who are most likely to assume the more lucrative, more influential jobs in the future. On the other hand, national studies of youth and their life styles (including drinking practices) should include college residences in the sampling frames. Such inclusion would allow for the assessing of these possible influences: type of living arrangement and their associated controls; the possible effects of social class differences upon youthful alcohol use; and, finally, the influence of the collegiate environment itself. The convenience for academics of doing local studies of the alcohol use of mainly full-time students in academies where they teach has made us lose sight of some important issues of representativeness.

Several fundamental questions should be addressed. To what extent does the four-year, full-time (or two-year, or part-time) student status represent a peculiar time of life? For the student, responsibilities for self and toward others are in transition and are, perhaps, more relaxed than during the pre- and postcollege periods when family and/or work requirements tend to have greater importance and exert more regulation upon the lives of individuals. How does the alcohol use of collegians compare with that of noncollegians of the same age?

As is evident in ***Drinking in College*** and in Bacon's early prescription for sociological investigations of alcohol issues (Bacon, 1944), he intended alcohol studies to be integrated into and informed by the larger social context. Efforts to answer these questions would force us to address the issues of sociocultural context and the connection of drinking with leisure, work, custom, and ceremony—issues that originally influenced the direction of Bacon's work.

Precollege Drinking

Straus and Bacon reported that postchildhood drinking, including intoxication, tended to take place before starting college. These findings have been echoed in subsequent studies involving college students, indicating that the level of alcohol use and the frequency of intoxication in high school is strongly associated in the college context (Gliksman, 1988; Humphrey & Friedman, 1986; Samson, Maxwell, & Doyle, 1989; Wechsler & McFadden, 1979; Wechsler & Rohman, 1981). The earlier the age of first alcohol use, the greater the level of alcohol intake and the frequency of drunkenness.

Because parental drinking behaviors, attitudes, and control continue to have an impact upon youthful drinking, with peer group influence increasing with age (Barnes & Welte, 1986; Harford & Spiegler, 1983; Wiggens & Wiggens, 1987), it is advantageous to view college drinking as at least partly a continuation of patterns of alcohol use set during high school. The study of first-year university entrants by Louis Gliksman (1988) is exemplary in this regard. Rather than relying upon retrospective reports, Gliksman surveyed his sample during the transition from high school to university by administering a questionnaire about drinking behaviors to students before entry into the freshman year of an Ontario university. The same students were queried eight months later, when they were nearing the end of their first year. The increase in levels of alcohol use, as well as problems associated with this use provide a good analysis of the impact of college entry upon high school graduates, as far as alcohol is concerned. Moreover, this study provides solid information concerning the areas at which educational and preventive measures may be directed. Because the age of initiation to alcohol use seems to be declining in North America (Whitehead, 1982), students come to college with a longer history of drinking and intoxication than was the case at midcentury. To properly assess the impact of both their precollege and college experience upon drinking, research incorporating methods similar to the Gliksman design is in order.

The Use of Alcohol and Other Drugs Among Young People

A relatively new phenomenon of the past generation is the widespread recreational use of drugs other than alcohol and tobacco. Surveys of school and college populations have documented considerable prevalence of illicit substance use, the most common of these being cannabis

(Barnes & Welte, 1986; Brown & Skiffington, 1987; Coombs, Fawzy, & Gerber, 1986; Murray, Perry, O'Connell, & Schmid, 1986; Penning & Barnes, 1982; Swadi, 1988; Welte & Barnes, 1987). Sadava and Wayda (1982) have observed that studies of nonmedical drug use are most valid when there is a recognition that many people use more than one substance concurrently. In their review of literature on concurrent use, these authors documented strong associations between heavy alcohol use and cigarette smoking, heavy alcohol use and marijuana use, and marijuana use and the use of illicit drugs other than marijuana. Other work has shown similar associations (Haberman, 1969; Simpson & Sells, 1974).

Studies on youthful drinking and drug use have documented a positive association between illicit drug use and the use of distilled spirits (Wechsler & Thum, 1973); between the frequency of alcohol use and the use of illicit drugs (Prendergast & Schaefer, 1976; Wechsler & McFadden, 1976; Wechsler & Rohman, 1981); and between the frequency of alcohol intoxication and the use of uncommon drugs (Wechsler, 1976). Recently, the "stepping stone" hypothesis has been revived and supported concerning a progression in the use of alcohol, tobacco, cannabis, and illicit substances other than cannabis (Jessor, 1979; Kandel, 1978; Kandel & Andrews, 1987; O'Donnell & Clayton, 1982; White, Johnson, & Garrison, 1983).

The use of marijuana has been singled out by a number of investigators as being implicative because it is associated with other illicit drug use and with heavier use of licit substances. Cannabis has also been associated with unconventional or deviant behaviors, such as early sexual experience, protest activities, and delinquency (Jessor & Jessor, 1977; White et al., 1983). Although there is evidence that marijuana may be considered a "boundary" drug between the use of licit and illicit substances (e.g., Jessor, 1979; O'Donnell & Clayton, 1982; White, Johnson, & Horwitz, 1986), for the general population for whom more exotic or narcotic drugs are either not as available or desirable, the importance of the use of cannabis seems to lie in the area of its association with tobacco use and with the heavier use of alcoholic beverages. Thus, marijuana use can be seen to signal involvement with licit drugs that are potentially dangerous.

I suggest, then, that the "stepping stone" hypothesis merits further elaboration and testing, especially with reference to the most commonly used drugs: alcohol, tobacco, and cannabis. While a chronological progression from alcohol through cannabis has been well established, the feedback from, or perhaps the interaction between, cannabis use and alcohol use has not been investigated. The youth culture ideology of the 1960s and early 1970s claimed cannabis to be a viable replacement for alcohol—the latter being the drug of the older generation and war; the former, of peace and the younger generation. What seems to have happened is the widespread use of both of these two drugs. What may be the real public health danger of marijuana use is not in its boundary-crossing effects resulting in involvement with other more scarce drugs, but rather in its influence upon the increased use of the licit and readily available alcohol and tobacco.

During the last four decades, *Drinking in College* and other surveys have documented a reliable association between motivation for drinking and the volume of alcohol used. Heavier drinkers and those with alcohol-associated problems are more likely to use alcohol for internal effects than to comply with social custom. Bacon's "Alcoholics Do Not Drink" (1958) still provides a most cogent insight into the phenomenon of personal-effects usage, as well as addiction. It is reasonable to claim that habitual and heavy cannabis users "do not smoke." Whether or not the use takes place in a social situation, the result is a psychophysical one that encourages participants' attention to their internal states, rather than to the external, social situation.

While marijuana use alone—even heavy use—has never been shown to have addictive properties or the devastating behavioral and physical effects possible with alcohol abuse, one of its real dangers may lie in its effects upon alcohol use itself. If youthful initiation into the use of alcohol is quickly followed by initiation to cannabis use, personal effects learning and motivation may be easily transferred to the use of alcohol, if they were not there to begin with.

At present, these notions are speculative, for it is mainly correlations and some data concerning the chronology of use of various drugs that are available. What is needed are some short-term longitudinal studies of young cohorts, using panels frequent enough to pinpoint acquisition and the associated motivation and effects. If this is done over the course of about a decade, detailed knowledge of the interaction effects on a social-psychological level could be gained. At the very least, substance use among the young could be better understood if queries are directed to assessing more than one drug. We know that a sizable proportion of young people are using more than one substance (Norton & Colliver, 1988) and that multiple substance use is associated with heavier alcohol use. What remains to be done is to specify the nature of these associations and their links to the cultural context in which they take place.

Conclusions

Rereading *Drinking in College* and reviewing some recent investigations of youthful alcohol use, I have been surprised in a number of ways. The endurance of the early work by Straus and Bacon remains evident, even in those recent reports that do not refer to the early study. Concepts and measures that differentiate between drinking patterns and their contexts that were employed in *Drinking in College* have become working assumptions in later research. To mention a few examples, the identification of "warning signs" of alcohol addiction and the construction of social complications scales have been refined and qualified over the years. Nevertheless, the difference between alcohol-related psychophysical damage and disturbing behavior associated with drinking remains a critical analytic distinction. Findings on gender differences in attitudes and practices of young drinkers—especially the fact that women are more rule-bound than men—deserve further examination today; another topic deserving attention is the relationship between parental drinking practices and parental advice and the attitudes and drinking behaviors of their offspring. The issues just mentioned are only a sampling of the useful conceptual and empirical contributions found in *Drinking in College* and in Bacon's other work.

More than anyone else, Bacon developed the groundwork for the social-scientific study of alcohol use. For social scientists, his contributions contain a very important lesson: that the links between drinking customs and surrounding social arrangements be well documented. It is only through such documentation that understanding any custom is possible. Moreover, Bacon's recommendations concerning the study of drinking may serve as a model for research on other complex social phenomena (e.g., sexual behavior, sports, and religious practices). Alcohol use and phenomena such as these have their bright as well as their dark sides—where ambivalence and emotionality generate powerful extreme opinions that at once become divorced from reality and generate some unhealthy realities. Bacon's refusal to sacrifice validity to fad and ideology has been translated into an uncompromising mode of investigation that eschews simple questions and prevents simple answers.

Acknowledgments

I am grateful for the assistance of Jane M. Franklin and Maggie Sullivan, who gathered much of the bibliography for this chapter and who brought important recent studies to my attention. This chapter was supported in part by Health and Welfare, Canada (NHRDP Grant 6606-3948-DA).

References

Andrews, F.K., & Rootman, I. (1984). *The relationship of norms and beliefs to the use of alcohol, tobacco, and cannabis among young Canadians.* Paper presented at the Society for the Study of Social Problems, San Antonio.

Bacon, S.D. (1943). Sociology and the problems of alcohol: Foundations for sociologic study of drinking behavior. *Quarterly Journal of Studies on Alcohol, 3*, 402-445.

Bacon, S.D. (1958). Alcoholics do not drink. *Annals of the Academy of Political and Social Science, 315*, 55-69.

Barnes, G.M., & Welte, J.W. (1986). Patterns and predictors of alcohol use among 7-12th grade students in New York State. *Journal of Studies on Alcohol, 47*, 53-61.

Blane, H.T., & Hewett, L.E. (1977). *Alcohol and youth: An analysis of the literature.* Final report, NIAAA contract ADM 281-75-0026. Pittsburgh: University of Pittsburgh.

Brown, P.M., & Skiffington, E.W. (1987). Patterns of marijuana and alcohol use attitudes for Pennsylvania 11th graders. *International Journal of the Addictions, 22*, 567-573.

Cahalan, D. (1970). *Problem drinkers: A national survey.* San Francisco: Jossey-Bass.

Cahalan, D., & Room, R. (1972). Problem drinking among American men aged 21 to 59. *American Journal of Public Health, 62*, 1473-1482.

Chafetz, M.E. (1967). Alcoholism prevention and reality. *Quarterly Journal of Studies on Alcohol, 28*, 345-348.

Coombs, R.H., Fawzy, F.I., & Gerber, B.E. (1986). Patterns of cigarette, alcohol, and other drug use among children and adolescents: A longitudinal study. *International Journal of the Addictions, 21*, 897-913.

Fillmore, K. (1974). Drinking and problem drinking in early childhood and middle age. *Quarterly Journal of Studies on Alcohol, 35*, 819-840.

Fillmore, K. (1975). Relationships between specific drinking problems in early adulthood and middle age. *Journal of Studies on Alcohol, 36*, 882-907.

Frankel, B.G., & Whitehead, P.C. (1981). Drinking and damage: Theoretical advances and implication for prevention (Monograph 14). New Brunswick, NJ: Center of Alcohol Studies.

Gliksman, L. (1988). Consequences of alcohol use: Behavior changes and problems during first year of university. *International Journal of the Addictions, 23*, 1281-1295.

Haberman, P.W. (1969). Drinking and other self-indulgences: Complements or counter-attractions? *International Journal of the Addictions, 4*, 157-167.

Hanson, D.J. (1977). Trends in drinking attitudes and behaviors among college students. *Journal of Alcohol and Drug Education, 22*, 17-22.

Harford, T.C., & Grant, B.F. (1987). Psychosocial factors in adolescent drinking contexts. *Journal of Studies on Alcohol, 48*, 551-557.

Harford, T.C., & Speigler, D.L. (1983). Developmental trends in adolescent drinking. *Journal of Studies on Alcohol, 44*, 181-188.

Humphrey, J.A., & Friedman, J. (1986). The onset of drinking and intoxication among university students. *Journal of Studies on Alcohol, 47*, 455-458.

Jessor, R.L. (1979). Marijuana: A review of recent psychosocial research. In R.L. Dupont, A. Goldstein, & J. O'Donnell (Eds.), *Handbook on drug abuse* (pp. 337-355.) Washington, DC: Government Printing Office.

Jessor, R.L., & Jessor, S. (1977). *Problem behavior and social development: A longitudinal study of youth.* New York: Academic Press.

Jessor, R.L., & Jessor, S. (1978). Theory testing in longitudinal research on marijuana use. In D. B. Kandel (Ed.), *Longitudinal research on drug use: Empirical findings and methodological issues.* Washington, DC: Hemisphere, Halsted-Wiley.

Kandel, D.B. (1978). Convergences on prospective longitudinal surveys of drug use in normal populations. In D.B. Kandel (Ed.), *Longitudinal research on drug use: Empirical findings and methodological issues.* Washington, DC: Hemisphere, Halsted-Wiley.

Kandel, D.B., & Andrews, K. (1987). Processes of adolescent socialization by parents and peers. *International Journal of the Addictions, 22,* 319-342.

Linsky, A.S., Colby, J.P., Jr., & Straus, M.A. (1986). Drinking norms and alcohol-related problems in the United States. *Journal of Studies on Alcohol, 47,* 384-393.

Maddox, G., Robins, L.N., & Rosenberg, N. (Eds.). (1986). *The nature and extent of alcohol problems among the elderly.* New York: Springer Publishing.

McCarty, D., Morrisen, S., & Mills, K.C. (1976). Attitudes, beliefs, and alcohol use: An analysis of relationships. *Journal of Studies on Alcohol, 37,* 328-341.

Murray, D.M., Perry, C.L., O'Connell, C., & Schmid, L. (1987). Seventh-grade cigarette, alcohol, and marijuana use: Distribution in a North Central U.S. metropolitan population. *International Journal of the Addictions, 22,* 357-376.

Norton, R., & Colliver, J. (1988). Prevalence and patterns of combined alcohol and marijuana use. *Journal of Studies on Alcohol, 49,* 378-380.

O'Donnell, J.A., & Clayton, R.R. (1982). The stepping stone hypotheses—marijuana, heroin, causality. *Chemical Dependencies: Behavioral and Biomedical Issues, 4,* 229-241.

Orford, J., Waller, S., & Peto, J. (1974). Drinking behavior and attitudes and their correlates among university students in England. *Quarterly Journal of Studies on Alcohol, 35,* 1316-1374.

Peele, S. (1987a). The limitations of control-of-supply models for explaining and prevention alcoholism and drug addiction. *Journal of Studies on Alcohol, 48,* 61-77.

Peele, S. (1987b). What does addiction have to do with level of consumption? A response to R. Room. *Journal of Studies on Alcohol, 48,* 84-88.

Penning, M., & Barnes, G.E. (1982). Adolescent marijuana use: A review. *International Journal of the Addictions, 17,* 749-791.

Polich, J.M., Armor, D.J., & Braiker, H.B. (1981). *The course of alcoholism: Four years of treatment.* New York: John Wiley and Son.

Prendergast, T.J., & Schaefer, E.S. (1976). Correlates of drinking and drunkeness among high-school students. *Quarterly Journal of Studies on Alcohol, 35,* 232-242.

Room, R. (1975). Normative perspectives on alcohol use and problems. *Journal of Drug Issues, 5,* 358-368.

Room, R. (1987). Alcohol control, addiction and processes of change: Comment on "the limitations of control-of-supply models for explaining and preventing alcoholism and drug addiction." *Journal of Studies on Alcohol, 48,* 78-83.

Sadava, S.W., & Wayda, C. (1982). *Concurrent multiple drug use: A critical review.* Paper presented at the 43rd convention of the Canadian Psychological Association, Montreal, P.Q.

Samson, H.H., Maxwell, C.O., & Doyle, T.F. (1989). The relation of initial alcohol experiences to current consumption in a college population. *Journal of Studies on Alcohol, 50,* 254-260.

Simpson, D.D., & Sells, S.B. (1974). Patterns of multiple drug abuse: 1969–1971. *International Journal of the Addictions, 9,* 301-314.

Skolnick, J.H. (1958). Religious affiliation and drinking behavior. *Quarterly Journal of Studies on Alcohol, 19,* 452-470.

Straus, R., & Bacon, S.D. (1953). *Drinking in college*. New Haven: Yale University Press.

Swadi, H. (1988). Drug and substance use among 3,333 London adolescents. *British Journal of Addictions, 83*, 935-942.

Temple, M. (1986). Trends in collegiate drinking in California, 1979–1984. *Journal of Studies on Alcohol, 43*, 274-282.

U.S. Bureau of the Census. (1968). *Statistical Abstract of the United States, 1968* (89th ed.). Washington, DC.

U.S. Bureau of the Census. (1989). *Statistical Abstract of the United States, 1989* (109th ed.). Washington, DC.

Wechsler, H. (1976). Alcohol intoxication and drug use among teenagers. *Journal of Studies on Alcohol, 37*, 1672-1677.

Wechsler, H., & McFadden, M. (1976). Sex differences in adolescent alcohol and drug use. *Journal of Studies on Alcohol, 37*, 1291-1301.

Wechsler, H., & McFadden, M. (1979). Drinking among college students in New England: Extent, social correlates and consequences of alcohol use. *Journal of Studies on Alcohol, 40*, 969-996.

Wechsler, H., & Rohman, M. (1981). Extensive users of alcohol among college students. *Journal of Studies on Alcohol, 42*, 149-155.

Wechsler, H., & Thum, D. (1973). Teenage drinking, drug use and social correlates. *Quarterly Journal of Studies on Alcohol, 34*, 1220-1227.

Welte, J.W., & Barnes, G.M. (1987). Alcohol use among adolescent minority groups. *Journal of Studies on Alcohol, 48*, 329-335.

White, H.R. (1987). Longitudinal stability and dimensional structure of problem drinking in adolescence. *Journal of Studies on Alcohol, 48*, 541-550.

White, H.R., Johnson, V., & Garrison, C.G. (1983). *The drug-crime nexus among adolescents and their peers*. Paper presented at the annual meeting of the American Sociological Society, Detroit.

White, H.R., Johnson, V., & Horwitz, A. (1986). An application of three deviance theories to adolescent substance use. *International Journal of the Addictions, 21*, 347-366.

Whitehead, P. (1982). *Young drinkers in Canada: A review of recent epidemiology*. Unpublished manuscript, Health Promotion Directorate, Health and Welfare, Canada.

Wiggins, J.A., & Wiggins, B.B. (1987). Drinking at a southern university: Its description and correlates. *Journal of Studies on Alcohol, 48*, 319-324.

CHAPTER 8

Science, Social Movements, and Cynicism: Appreciating the Political Context of Sociological Research in Alcohol Studies

ARMAND L. MAUSS

Nearly half a century ago, Selden Bacon (1943) laid out what he called "the foundations" for a sociological study of drinking behavior. He cogently argued in this pioneering essay that alcohol abuse or "inebriety" should not be studied in isolation from the larger context of the sociology of alcohol use in the culture more generally. For the most part, this sensible instruction, like many others in Bacon's essay, has been honored only in the breach. Only a few enlightened anthropologists and sociologists (e.g., Critchlow, 1986; Gusfield, 1981; Heath, 1987; McAndrew & Edgerton, 1969; Pittman & Snyder, 1962; Room, 1975, 1976) have tried to direct our attention to the more general issue of the part alcohol plays in our own and other cultures.

The research agenda that Bacon called for, however, has been largely ignored, as the prodigious research output in the field of alcohol studies (like that of other drug studies) has continued to focus on the *problem* of alcohol, that is, upon the problematic concomitants of alcohol use. In more recent years, and no doubt in some frustration, Bacon (1987) has assumed a kind of "fall-back" position, in which he urges upon our subdiscipline at least the "common sense" of applying truly scientific precision at all levels of our research on the "problem," whether in definition, measurement, or analysis.

The "Social Problems" Heritage and Its Champions

In retrospect, it seems clear enough why Bacon's sensible approach has proved too idealistic for the arena in which we have been working. At least two conditions have been crucial in determining the nature of the research done on substance use and abuse in North America: (1) the pragmatic, social problem-solving intellectual heritage of American sociology, best exemplified by the Chicago School (Kurtz, 1984; Lewis & Smith, 1980); and (2) the moralistic and meliorative tradition in American politics.

Until recently, nearly all American sociologists (and probably other social scientists) have taken for granted the Chicago School assumption (contra Durkheim, 1950) that "social problems" are analogous to medical problems. For example, just as a cancer is an objectively identifiable and undesirable pathology in the human body, so too are social problems (e.g., poverty, prostitution, alcohol addiction) objectively discernible and treatable: cancers on the body politic, as it were. In line with the conventional sociological faith of the times, Bacon himself began his essay with a section entitled "The Scientific Approach to Social Problems" (1943, p. 402), in which he made use of an analogy between inebriety and syphilis (p. 404). From that time to this, furthermore, social scientists have generally accepted the definitions of politicians as to what constitute the most serious "social problems" of the day, especially with the growing dependence upon federal research funding.

Though Bacon was taking a fully legitimate scientific position in his 1943 treatise, this position was doomed from the start by the two historic conditions mentioned above. In a strictly scientific sense, it may be true that the only rational way to understand substance use ("excessive" or not) is within a given sociocultural context, as Bacon insisted. Yet it is not within that larger context that the use of alcohol and other substances becomes a *social problem;* it remains only a *kind of behavior.* Such behavior becomes problematic only to the extent that it offends the moral sensibilities or challenges the interests of politically powerful segments of the society (Mauss, 1975, 1989; Schneider, 1985).

Thus, while Bacon (1943) was scientifically correct in urging upon us a broader context for studying what he called the "alcohol problems," he seems not to have recognized that the most appropriate context was actually the *political* one. He certainly did recognize (1943, pp. 413–415, 419–420) the parts played by "chartering" and "sanctioning" as determinants of drinking behavior, but not, apparently, that those very factors are but arenas for political struggle. Yet he could not have been politically naive, considering his own personal struggles with competing interests from his earliest days at Yale (Blocker, 1989, chap. 5).

By the time of his 1987 call for at least a "common sense" approach to the study of alcohol problems, Bacon had become much more concerned about the variety of interest groups and social movements involved in the field of alcohol studies. Yet he still seemed unaware that the conflicts and ambiguities he so deplored among these groups in theory, definition, measurement, and policy prescriptions actually serve a variety of political interests and are thus unlikely ever to be resolved at the scientific level.

Alcoholics Anonymous (AA) has become a "successful" recovery program on the basis of a theory about an unidentified physiological factor producing in some drinkers a disease that can never be totally overcome: these are the "alcoholics," and complete abstinence is their only hope for a normal life. Could anyone expect AA to accept any newly discovered "scientific facts" to the contrary? Could anyone expect that the large and growing corps of career alcoholism treatment counselors who are themselves AA members would condone any treatment program built on a contradictory premise, however well demonstrated "scientifically"?

How about the medical community, in which thousands of careers have been built by large federal grants for research and treatment on the "disease" of alcoholism? Could we expect much tolerance (not to say enthusiasm) for a new "scientific" discovery that seriously undermines the "disease" definition? Certainly the countervailing evidence and cogent arguments in the recent book by Herbert Fingarette (1987) have scarcely been given any notice. Even the "common sense" that Bacon (1987) now urges on us is unlikely to motivate research on alternatives to the "disease" definition, as long as it works so well for the confluence of interests represented by AA, the National Council on Alcoholism (NCA), the medical community, and middle-class alcoholics and their families (for whom the disease concept legitimates a relatively benign and exculpatory definition of the condition).

Even at the level of measurement and data, where Bacon (1987, p. 386) deplores, among other things, such a wide range of estimates for the proportion of highway fatalities caused by alcohol use (17 to 50 percent), it is not easy to see whose interests would be served by the discovery of a "definite" or "fully reliable" empirical figure. As things stand, the beverage industries can embrace the lower figure (and quibble even about how much of *that* should be attributed instead to driver age and inexperience), while the zealots of Mothers Against Drunk Driving (MADD) can continue to use the higher figure for leverage in getting passage of increasingly restrictive and punitive drunk-driving legislation.

Then there is the morass of prevention programs, which Bacon (1987) rightly includes among the efforts made to date that so badly need formal, scientific evaluation in the field of alcohol studies. Most states of the union now mandate alcohol and drug abuse prevention programs in their public schools. Conscientious educators, both within public school systems and in private businesses, have built careers from constructing and marketing a plethora of school-based prevention programs. The nearly unanimous verdict in the professional literature (e.g., Moskowitz, 1989; Tobler, 1986) that none of these programs has had any impact on student alcohol or drug use has not slowed the growth in the marketing of such programs. Nor has it dampened the confidence placed in them by the "czar" of our current national "War on Drugs." In such a political arena, "common sense," alas, will play no more part in finding solutions to our alcohol and drug "problems" than will the broader social science approach that Bacon urged on us half a century ago.

Entrepreneurship and the Social Construction of the Problem

That sociological research in alcohol studies has largely failed to attend to Bacon's recommendations, whether his early or more recent ones, is thus not attributable to weaknesses or fallacies in the recommendations themselves. It is attributable, rather, to the highly politicized context in which the research has been sponsored and carried out. One has only to recall the controversy over "controlled drinking" occasioned by the Rand Report (Armor, Polich, & Stambul, 1976), or over the "responsible drinking" issue in NIAAA (National Institute on Alcohol Abuse and Alcoholism) under Ernest Noble (1978) a little later. These are but two examples of conflicts that have regularly occurred among "alcohologists" in recent years over the very definition of the phenomenon to be studied. The Congress and federal funding agencies, of course, have usually provided the battlegrounds where the contending interest groups have struggled over how and where the funds for research, prevention, and treatment are to be allocated.

Much of the contention has focused on the question of just what "the problem" is. The various interest groups attempting to define the problem for the nation can be considered "entrepreneurs" of different kinds.

Market Entrepreneurs

First are the "market entrepreneurs" of the kind familiar in any branch of commerce. These are the *suppliers* of the substances, whether in business legally (as is the case usually for alcohol, tobacco, and caffeine) or illegally (as in the case of opiates, cocaine, and the like).

For these entrepreneurs, the "alcohol problem" is simply that certain individuals drink irresponsibily or for the wrong reasons; the alcohol (or other substance) is itself a source of harmless pleasure when used responsibly and ought not to be regarded as a problem per se.

Moral Entrepreneurs

Then there are the "moral entrepreneurs" (as Becker has called them), who promote a variety of *normative positions* on the substance in question. These positions range along a continuum between total prohibition, at the one extreme, and total freedom or libertarianism, at the other. Intermediate moral positions include prohibition for minors, but not for adults; prohibition for expectant mothers, but not for others; prohibition for drivers, but not for others; prohibition in the workplace, the restaurant, or the airplane (depending on the substance), but not elsewhere; and so on. Whatever the stance of the moral entrepreneur, it is based primarily upon some sort of ethical premises (even where health concerns are cited).

For all such moral entrepreneurs, the "alcohol problem" is found in our *rules* (and/or laws) governing the use of the substance. Strict libertarian moralists don't want *any* rules, while prohibition moralists want rules that outlaw even the possession of the substance.

Salvation Entrepreneurs

The success of the first two categories—market and moral—and, indeed, the struggle between the two, have ironically generated growth industries for other kinds of entrepreneurship. One of these might be called "salvation entrepreneurs" because of the close analogy they exhibit to religious sects. Whether the product in question is religion or some other "program," the salvation entrepreneur has the gospel that will save our society from the scourge of whatever the target substance is.

One sector of the salvation market is the *prevention* sector, which includes those involved in either public or private organizations that are marketing prevention packages. The most pervasive example is to be found in school-based prevention programs like those of CORK (*Cork Communicator,* 1981) or "Here's Looking at You" (Mooney, Roberts, Fitzmahan & Gregory, 1979—now privately produced) or CASPAR (DiCicco, 1978; Hubbard, 1978—created by curriculum experts in a school system). (See also Thompson, Daugherty & Carver, 1984; Williams & Vejnoska, 1981.) All these programs carry the implicit promise of significantly reducing the resort to substance use by school children (i.e., "saving our kids"), and the market has proved almost limitless. Whether the focus is upon schools or upon the larger society, though, prevention entrepreneurs define the "alcohol problem" (or "drug problem") as one of *inadequate psychological inoculation,* a problem, that is, of inadequate information, understanding, self-esteem, skills in decision making, peer-pressure resistance, or some other psychosocial inadequacy that this or that program can ameliorate or eliminate.

A second sector of the "salvation" market is the *treatment* sector, consisting of the variety of public and private treatment and rehabilitation programs intended to retrieve those who have fallen into dependency on alcohol or other substances. Some of these are explicitly religious in their orientation, of course, such as the Salvation Army, the Union Gospel Missions, and a variety of religious missions commonly found in our inner cities (Bibby & Mauss, 1974; Mauss, 1982; Rooney, 1980). Others are private organizations (some for profit, some not), such as AA, the Schick program or the Betty Ford Center, while still others are established by public social welfare agencies.

Treatment entrepreneurs define the "alcohol problem" as a physiological, spiritual, and/or psychological flaw in the user that can be overcome only (or at least best) by the prescribed treatment modality. A variety of modalities can be seen in this bewildering array of agencies and programs, but all offer the explicit hope of salvation from ruinous addiction to alcohol and/or other substances. (This market, including AA, has proved viable enough to have generated many imitators for such newly discovered "addictions" as overeating, gambling, and sexual excesses.)

Government Entrepreneurs

The final general category of entrepreneurs is "government entrepreneurs," a market sector consisting of all those who have careers devoted wholly or partly to applying government policy at the federal, state, or local levels relating to substance use or abuse. Here I would include funding agencies like NIAAA and the National Institute on Drug Abuse (NIDA), professional researchers largely dependent upon such funding (especially at universities), agencies administering funds and programs for prevention or treatment, and various law enforcement agencies.

For these entrepreneurs, the "alcohol problem" is whatever current legislation and public policy *say* that it is, which may vary from one decade to the next (or even from year to year). Depending upon exigencies of policy or funding, government agencies may be allies to any or all of the other kinds of entrepreneurs. However, it is important to remember also that government agencies and their personnel always have agendas of their own and thus must be regarded as entrepreneurs in their own rights, as well as allies.

Social Control and Social Movements

It is obvious that none of these different kinds of entrepreneurs (including government-funded academics) can be expected to have much of a stake in the kind of general sociocultural research on alcohol that Bacon once called for. As has been often observed (and deplored) since Bacon's early essay (e.g., Roman, 1984), research in the alcohol field has been preoccupied with deviant behavior. I would put it somewhat differently, and, I think, more precisely, by stating that the field has actually been preoccupied with *social control,* where alcohol and other substances are concerned. That is the focus shared by all the kinds of entrepreneurs I have just discussed.

Whether looking to law enforcement, to the medical profession, to schools, to churches, to universities, or to treatment programs, all such entrepreneurs are actually dealing in one way or another with drinking and drug use as behavior to be controlled. Even the beverage industries have promoted the idea that the control of "excessive" or "irresponsible" drinking is "the problem." Most parents, teachers, employers, and politicians take essentially the same position. That is why Bacon's agenda has been so seldom addressed: It is simply in no one's interest to see alcohol use as anything but a subject for social control and the derivative public policy.

We have had enough good studies in the history of the social control of alcohol use to be aware of the several social movements, from the very beginning of our national history, that have advocated various control measures (e.g., Blocker, 1989; Blumberg, 1978, 1980; Gusfield, 1963; Levine, 1984; Rumbarger, 1989). The present century, of course, began in the throes of a powerful prohibition movement, which eventually achieved constitutional status lasting until the 1930s. During Prohibition, we placed chief reliance on law enforcement agencies for social control, though a great many Protestant denominations also proved important as social control agencies. After the repeal of Prohibition, the social control pendulum began to swing away from the hard-nosed punitive approach. The new look in social control placed greater reliance on a combination of voluntary, educational, and medical approaches represented by such organizations as AA, the National Committee for Education on Alcoholism (NCEA, later the NCA), and the founding of the Yale Center of Alcohol Studies, of which Selden Bacon eventually became head (Blocker, 1989).

This immediate post-Prohibition period was notable also for the general acceptance of the "disease" definition of alcoholism, thanks largely to the entrepreneurship of E.M. Jellinek. Both the drinking and the production of alcoholic beverages were slow to recover from the Prohibition policy, partly because of the depression of the 1930s, so that even as late as 1960 American levels of alcohol consumption remained fairly low (Blocker, 1989). Given the widespread discreditation of Prohibition, and the relatively low levels of alcohol use, the social movement comprised of the likes of Yale, AA, and NCA met little opposition in gaining general acceptance of a basically medical definition of alcohol abuse, and thus of a reliance primarily on the medical profession for social control (except in the case of inner-city derelicts).

Medicine, education, and voluntary (nonprofessional) rehabilitation programs constituted the nation's general approach to social control up to the "Age of Aquarius" (the 1960s), when suddenly the already modest level of concern over alcohol abuse was displaced almost entirely by the outbreak of a "drug use epidemic." It seemed that our youth had abandoned alcohol for more exotic and dangerous drugs, such as marijuana, LSD, and various barbiturates and amphetamines (Musto, 1988). It was not reassuring to the nation that this new resort to illicit and dangerous drugs was especially associated with a youth counterculture that seemed to reject all of the society's sacred myths and values. The use of law enforcement against that counterculture more generally included proscription of its favorite drugs, but resources did not permit much social control activity against alcohol use (legal or otherwise), despite some sentiment for resorting once again to more legal controls over alcohol use (Blocker, 1989).

The current (and rather ambiguous) phase of our nation's control policies over various substances began about 1970 (Wiener, 1981). A widespread perception of a whole generation hooked on drugs, both on the campuses and in the military, made political action seem imperative. Accordingly, the National Institute on Mental Health (NIMH), which had been the main federal agency promoting research in the various "social problems," was divided three ways. The NIAAA was given responsibility for funding various alcohol-related work, NIDA was given similar jurisdiction over all the other drugs, and NIMH kept the rest of the "social problems."

With the new government commitment to scientific research in these two problem areas, social control through law enforcement began to recede somewhat in favor of experimentation with different (noncriminal) modes of social control. Marijuana use was decriminalized in many jurisdictions. The legal drinking age was dropped by a year or more in several states. Most states decriminalized public drunkenness, thereby freeing police to concentrate on "real crime." With the Vietnam War winding down, both the counterculture and its drugs seemed to become less visible. Meanwhile, a new emphasis began on education as a way of preventing alcohol and drug abuse. By the late 1970s, the pendulum had swung quite far away from the legal coercion upon which social control had relied earlier in the century.

Since then, during the 1980s, we have seen the pendulum start to swing back again toward reliance on law enforcement (Blocker, 1989). This development is most conspicuous, of course, in the great national "drug war" that has recently been launched, but it can be seen as well in a much greater effort to control alcohol use through law enforcement. Partly under the influence of social movement organizations like MADD (Reinarman, 1988), a national crackdown on drunk driving is underway. Nearly all states that had reduced the age for legal drinking have now raised the age back up to 21. Warning labels are starting to appear on containers of alcoholic beverages. The "disease" definition of alcoholism is under increasing attack by research professionals, especially sociologists. We thus seem to be entering a new cycle in social control policy where substance use is concerned.

Social Movements and Funding Cycles

Each new swing of the pendulum has been associated with a more or less powerful social movement that has promoted the change in social control policy. There is no need to recount further the histories of all these social movements, which are, in any case, fairly well known (Blocker, 1989). The main point to emphasize is that each new movement has depended for its success upon its ability to mobilize various kinds of resources (Zald & McCarthy, 1987). These resources take many forms, of which funding is only the most obvious. Also included in the requisite resources would be legitimation and/or sponsorship by government agencies; alliances among various social movement organizations and among the different kinds of entrepreneurs mentioned earlier; promotion by the mass media; and many other things. Sad commentary though it may be, the history of research in the field of alcohol use is much more readily understandable in the context of the comings and goings of social movements than by reference to the thoughtful scientific agenda proposed half a century ago by Selden Bacon.

Let me offer some concrete illustrations, first at the national level and then from my personal research experience.

The National Scene

When NIAAA and NIDA were formed out of NIMH, NIAAA started with a disadvantage. In 1970, the national campaign against illicit drugs was far more prominent than the rather subdued counterpart for alcohol use. To be sure, NIAAA was started with a lot of support in Congress, due in large part to the personal efforts of recovering alcoholic Senator Harold Hughes and the lobbying of the NCA. Yet it was the "great drug scare" of the 1960s that had been the most influential in the breakup of NIMH, and in 1970 it was the illicit drugs that received most of the national attention. As the 1970s progressed, however, bringing an end to the Vietnam War and campus unrest, it became increasingly difficult to convince Congress to fund NIDA as generously as it had done in the beginning. By the mid-1970s, drug use was peaking or actually starting to decline among the youth (Johnston, O'Malley, & Bachman, 1989). One seldom heard any more of "bad trips," and enforcement against marijuana became lax, even where the substance had not been decriminalized.

It was at this juncture that NIDA was able to capitalize on a feature that NIAAA did not have: the list of dangerous and illicit drugs under NIDA jurisdiction was almost endless, whereas NIAAA had only *one* drug—alcohol—with which to worry the nation. As marijuana and LSD began to fade somewhat as vexing drugs, NIDA was able to focus national attention on various barbiturates and amphetamines, then on heroin specifically, then on cocaine, then on tobacco, etc. (though not necessarily in that order). Tobacco too could yield more than one horror, for after smoking we discovered the dangers of "smokeless" tobacco (snuffing and chewing). Cocaine, too, is back again; it wasn't so serious when it was favored merely as a weekend recreational habit for "yuppies," but now that it is pervasive and conspicuous in our inner cities (in the form of "crack"), it appears that no cost is too great to get rid of it. One looks in vain for correlations between changes in the actual *use* of the various drugs and the changes in emphasis that they have received from year to year.

Meanwhile, back at NIAAA, how did the agency entrepreneurs hold their own against competition with NIDA for tight funds, and thereby serve the various other alcohol entrepreneurs that make up the NIAAA constituency? They couldn't switch emphasis to another drug, as NIDA could. Instead, they began to rotate their emphasis to various alcohol-using *populations.* A review of several years' back issues of the NIAAA *Information and Feature Service Newsletter* reveals a very astute strategy: whereas NIAAA had begun with a general focus on drinking problems, implicitly referring to adult male problems, it then began to focus its attention (and its requests for proposals) on the subpopulation of public inebriates. Then the emphasis switched to female drinkers, then to youth, then to mothers (fetal alcohol syndrome), then to blacks, Hispanics, Native Americans, and other ethnic groups, then to the elderly, then to homosexuals, AIDS victims, and so on (again, not necessarily in that order). Now we are back to the youth again.

Note that, as in the case of NIDA, it is difficult to find any correlations between these changing NIAAA emphases and any changes in alcohol use patterns in the various subpopulations. Indeed, it would be naive in the extreme to believe that these changing emphases in either agency were driven by changing epidemiologies, changing theoretical frameworks, or any other scientific considerations. Such changes can, however, be understood easily as a quest for continued and/or increased resources by agency entrepreneurs, who are responding partly to their own agendas and partly to the resource needs of the various moral, salvation, and research entrepreneurs with which they are allied.

This strategy for mobilizing resources proved so successful at NIAAA that eventually it was embraced by NIDA, as well. NIAAA is thus likely to continue at some disadvantage in the struggle for federal funds, especially in light of the latest "drug war" that has been launched. There is no implication intended here that there is anything unethical about such a strategy. Any social movement organization can be expected to engage in a constant quest for resources and allies to promote its programs. That quest, however, will not necessarily promote the cause of science.

Personal Research Experience

What occurs at the national level is often reflected locally as well. In the early 1970s, when there was a special national focus on "public inebriates," I received a small Washington state grant to study and compare the alcohol rehabilitation efforts of various downtown religious missions with those of secular social agencies in a major northwest city. A number of studies by my students and myself issued from this project. Out of this work, we eventually formed a relationship with one of the missions that was willing to join us in installing and evaluating a novel rehabilitation program for "skid row" inebriates. Funding for the first year came from the sponsoring religious denomination and from a one-year United Way grant. We had a unique theory to try, and we evaluated program outcomes for more than a year, maintaining post-treatment contact with a fairly large proportion of the clients, especially considering their transient skid-row way of life. We wrote an extensive project report that included a systematic evaluation of program outcomes indicating that our rehabilitation theory had great promise (Fagan & Mauss, 1986).

When the time came to fund the project for a second year, the mission and its sponsoring denomination were still willing to provide part of the resources, and we researchers were willing to continue donating our time and expertise. When we went back to the United Way for refunding, however, we got a rude lesson in the politics of social movements and resource mobilization. It seemed that a new national focus had emerged on alcohol-related problems among *women*, and United Way, like other funding agencies, had discovered that women were beginning to appear on skid row (sometimes with children). Accordingly, funding would now have to be divided between the main traditional programs for male inebriates and the initiation of some new programs for females. Our mission was thus told that funding would be forthcoming for any new program that they wanted to start for *women*, but not for continuation of "new or experimental" programs for men.

I took a copy of our evaluation report to the appropriate United Way board, who agreed that it was an impressive report about a promising program. However, their funds for male programs were all spoken for by the more "traditional" programs offered by the Salvation Army and others. When I asked about evaluations of such "traditional" programs, which might permit a comparison of "success rates" with our program, it was conceded that none of the traditional programs had ever yielded an evaluation. Nevertheless, they would still get the money. Longstanding alliances built around the exchange of resources can hardly be expected to yield to scientific research evidence! The long and short of it was that the mission replaced our program with a new one for women, which at least kept their United Way money coming.

In the mid-1970s, we initiated another evaluation research project in the same major northwest city. In line with the new Uniform Alcoholism and Intoxication Treatment Act, the state legislature had decided to decriminalize public inebriation. Thereafter, the police and their "drunk tanks" were no longer to be responsible for collecting the incapacitated public inebriates. Instead, the latter were to be rounded up in vans operated by civilian social service employees and taken to overnight "detox" centers. The next day, their condition was to be assessed and, once restored to a modicum of health and hygiene, they were to be referred to a regular treatment program. Social control responsibility (and the concomitant requisite funding) was thus transferred from law enforcement agencies to a newly created social welfare apparatus of special detox centers and treatment programs for public inebriates. The idea was, of course, to replace the classical "revolving door" with humane and voluntary treatment.

We were able to obtain arrest records for the appropriate census tracts indicating the nature and magnitude of police activity for two years *before* and two years *after* the effective date of decriminalizaton. We were also able to get access to the records for the first two years of the new civilian detox pickup and treatment program operating in the corresponding census tracts. From all these records, we were able to determine whether, in fact, the new decriminalization law was having its intended effect: whether police handling of public inebriates had indeed been replaced by detox and treatment regimens at the hands of social work professionals. What we learned was that while arrest rates for public drunkenness went down to zero (such no longer being a legal arrest category), arrest rates for certain other "typical" skid row offenses (e.g., urinating in public) drastically increased. After all, the police still had to respond to complaints from downtown merchants and otherwise occupy their peacekeeping shifts in the skid row areas.

Meanwhile, the records of the new detox program showed that almost none of the men brought in for short-term detoxification ever appeared at the follow-up treatment agencies, despite appointments and commitments to do so. In other words, public inebriates were now circulating in and out of public detox *even faster* than they had earlier circulated in and out of jail (where at least they were kept off the streets with 30-day sentences). Indeed, when we added together the arrest rates for the residual skid row offenses and the rates of pickup by the detox vans, we found that public inebriates were circulating through the entire system at about *four times* the earlier rate of circulation through the courts and jails—though, of course, they were doing so on a much more humane and comfortable basis.

We reported our findings to the appropriate authorities, and we published an article about our work in *Social Problems* (Fagan & Mauss, 1978). A year later, we investigated the feasibility of a follow-up study of the same general kind. We were not well received this time. We learned that the findings from our previous work (which had received some media coverage) had seriously jeopardized the funding *both* of the police department *and* of the new detox program. One innovation, however, had occurred in the intervening year: admission to detox now took place in two stages. The first stage was "medical admission" to a ward in the main public hospital. From there, only the minority of "more promising" detox patients were transferred to a second stage of (longer term) detox at the civilian detox center. Only those in the *second stage,* furthermore, were counted in the processing statistics, which, of course, began to look a lot better. We were not able to gain access to the more "confidential medical records" of admissions to the *first* (hospital) stage of detox.

Once again, we saw that even the "common sense" called for by Bacon (1987) could not be given precedence over the need for resources by agencies spawned in the new movement to redefine and decriminalize the condition called "public inebriation." The agencies had been given mandates deriving ultimately from social movement entrepreneurship rather than from rational, scientific theories or data. Jobs and careers were on the line. The city government did not know what else to do with public inebriates, who themselves were enjoying more creature comforts than ever before. No interested party stood to benefit by further investigation or evaluation of the new program.

A third illustration showing how social movement imperatives frustrate scientific investigation can be seen in the renewed campaign against youthful drinking, which began a decade or so ago. While available evidence gave no indication of a recent increase in juvenile drinking, or of a special concern about it on the parts of parents or educators, a new campaign was launched mainly by NIAAA itself to attract national attention and funds to the problem of drinking by minors (Chauncey, 1980). Millions of dollars were poured into research, development, and evaluation of school-based prevention programs built on the "psychosocial inoculation" premise. Starting in the late 1970s, my colleagues and I became involved in testing one program in particular. Generous NIAAA funding was made available both for the curriculum development team (based in a northwest public school district) and for our research team. We conducted a comprehensive, longitudinal evaluation of this program over a three-year period ending in the early 1980s.

At that point, we issued a project report giving our initial three-year findings. We tried to be as upbeat and optimistic as possible, pointing out the modest and spotty evidence of program impact on certain psychosocial traits like knowledge and self-esteem. On *one* measure (out of many) at *one* grade level, we were even able to point to evidence of impact on actual drinking behavior. All in all, though, it was clear from any but the most superficial reading of our report that the prevention program had had very little impact on anything except student knowledge and information. In particular, the evidence of impact on actual drinking *behavior* was almost nonexistent.

It soon became apparent that we had made a serious mistake in allowing any optimistic observations, however few, to intrude into our report; both the curriculum creators and NIAAA took our report to be a vindication of the program! Accordingly, federal funds were made available for its adoption in many new "demonstration sites" around the nation, as well as for the development of revised versions of it that would take into account the new drug scare. The school specialists who had first developed the curriculum, meanwhile, left the school system and went into the commercial production and marketing of this now "highly successful" prevention program. In its later versions, it has become one of the most widely adopted (and expensive) school prevention programs in the nation.

As a research team, meanwhile, we did not feel satisfied with the extent or the sophistication of the analysis we had done under pressure of completing the required project report. Since we had longitudinal data, we wanted to try various forms of causal modeling and other statistical refinements so that we could publish our work in professional journals. In and around other professional obligations, therefore, we succeeded in obtaining the funds and the time to do such additional analyses, and our work was eventually published in the *Journal of Studies on Alcohol* (Hopkins, Mauss, Kearney, & Weisheit, 1988; Mauss, Hopkins, Weisheit, & Kearney, 1988). This time we pulled no punches: If the objective of school-based prevention programs is to prevent or modify alcohol- or drug-using *behavior,* then the program we evaluated did not work; nor is it likely that *any* such program could work, given the inherent limitations on school access to the most important causal variables. By now that same verdict has been pronounced by virtually all other evaluations of school programs in the professional literature (e.g., Moscowitz, 1989; Tobler, 1986), with the possible exception of antismoking programs—and these, of course, are carried on in the larger societal context of a nationwide antismoking campaign.

Once again, our research findings have not been well received by those whose interests are tied to the current social movement promoting drug- and alcohol-education programs in the public schools. Most states have now mandated such programs in their school curricula, and no one wants to hear that they don't work. When our findings were picked up by the media, the educators who had created the program in question (now in a private firm marketing the product) called a news conference in which they tried not only to refute our findings but also to question our motives, accusing us of having changed our earlier optimistic assessment of their program for the sake of gaining academic notoriety and promotions (despite the fact that the principal authors were full professors all along).

After that, the president of our university received letters from other public and private agencies involved in alcohol and drug education, demanding that our work be suppressed because of its manifestly destructive impact on their efforts to save the nation's children. Even the Johnson Institute's *Observer,* which printed a digest of our findings, later retracted and refuted them, published a complaining letter from the curriculum marketers, and refused us even a one-paragraph rebuttal. A little later our work was cited again in a *Wall Street Journal* article (Pereira, 1989), followed by another round of recriminations from journalists, curriculum developers, and educators convinced (despite the contrary evidence) that "education" is the answer to the nation's drug and alcohol problems. Once again we can see that a social movement and its struggle for material and political resources proves more important than science.

Conclusion

This chapter carries a rather cynical tone, especially in its last few pages. Yet it has not been my purpose to engender cynicism. Indeed, it is unlikely that any of my readers are naive enough to be surprised by any of my observations or experiences; many have had similar experiences in the world of alcohol- or drug-related research. Nor is it necessary to respond cynically to such experiences or observations. Knowing how the world works might actually be a deterrent to cynicism, for a politically knowledgeable researcher will not hold any unrealistic expectations about the likely impact of his work on the well being of humankind, and thus not be easily disillusioned.

Such knowledge might also make the researcher a bit more vigilant about maintaining scholarly independence in the face of the attractive career benefits that can come from cooptation by one or more of the recurring and politically fashionable social movements. Of course, sponsorship of our research by any entrepreneurial interest group, public or private, commercial or moral, need not mean that we serve only their interests as we conduct our work and publish it. However, our integrity will be much less in jeopardy to the extent that we make it our business to know who they are and what they are up to, no matter how sincere or well-intentioned they may be. Finally, if politics is the "art of the possible," then it is precisely the politically aware researcher, rather than the disillusioned scientific purist, who is the more likely to find ways to maximize the ameliorative impact of his or her work upon the surrounding society.

If the above two paragraphs on the functions of political realism seem gratuitous to my sophisticated readers, likewise it has not been necessary for me to point out how Bacon's 1943 research agenda has been so largely ignored during the intervening decades. That was not my purpose, either. Many others have already made the same observation, including Bacon himself (1984, 1987) and Paul Roman (1984). What I have tried to do instead is to propose an *explanation* for such a collosal failure on the part of such a sensible document as Bacon's agenda. I have argued that the agenda has been ignored mainly because social science research in the alcohol field, even more than in most other fields, has been driven and tossed by the shifting winds of recurring social movements and the concomitant shifts in the fads and fashions of funding. I do not argue that this is the *only* explanation, but I do believe that it is a *sufficient* one. Nor do I expect things ever to be otherwise in the alcohol or drug fields.

If Bacon called for us to appreciate the general *social and cultural* context in which alcohol use takes place, I am urging that we keep constantly in mind the general *political* context in which our research takes place. It is an unstable context of passing social movements and contending interest groups with their entrepreneurs. If we are ever able to give Bacon's early agenda the attention it deserves, we will have to learn how to do so in that political context. That is, we will have to learn better how to integrate the basic scientific research issues that he has raised with the movement-driven research agendas of our funding agencies. Perhaps the time has come for us to study Bacon's agenda once more with an eye toward accomplishing it more fully in the next 40 years than we have in the previous 40.

References

Armor, D.J., Polich, J.M., & Stambul, H.B. (1976). *Alcoholism and treatment* (NIAAA contract). Santa Monica, CA: Rand Corporation.

Bacon, S.D. (1943). Sociology and the problems of alcohol: Foundations for a sociologic study of drinking behavior. *Quarterly Journal of Studies on Alcohol, 4,* 402–445.

Bacon, S.D. (1984). Alcohol issues and social science. *Journal of Drug Issues, 14,* 7–29.

Bacon, S.D. (1987). Alcohol problem prevention: A "common sense" approach. *Journal of Drug Issues, 17,* 369–393.

Bibby, R.W., & Mauss, A.L. (1974). Skidders and their servants: Variable goals and functions of the skidroad "rescue mission." *Journal for the Scientific Study of Religion, 13,* 421–436.

Blocker, J.S. (1989). *American temperance movements: Cycles of reform.* Boston, MA: Twayne (Div. of G.K. Hall).

Blumberg, L.U. (1978). The institutional phase of the Washingtonian total abstinence movement. *Journal of Studies on Alcohol, 39,* 1591–1606.

Blumberg, L.U. (1980). The significance of the alcohol prohibitionists for the Washington Temperance Societies. *Journal of Studies on Alcohol, 41,* 37–77.

Chauncey, R.L. (1980). New careers for moral entrepreneurs: Teenage drinking. *Journal of Drug Issues, 10,* 45–70.

Cork Communicator (1981, Fall). *2*(4), entire issue, and various other issues from 1980 to present. San Diego, CA: Operation CORK.

Critchlow, B. (1986). The powers of John Barleycorn: Beliefs about the effects of alcohol on social behavior. *American Psychologist, 41,* 751–764.

DiCicco, L. (1978). Evaluating the impact of alcohol education. *Alcohol Health and Research World, 3*(2), 14–20.

Durkheim, E. (1950). *Rules of the sociological method.* New York: Free Press.

Fagan, R.W., & Mauss, A.L. (1978). Padding the revolving door: An initial assessment of the Uniform Alcoholism and Intoxication Treatment Act in practice. *Social Problems, 26,* 232–246.

Fagan, R.W., & Mauss, A.L. (1986). Social margin and social re-entry: Evaluation of a rehabilitation program for skid row alcoholics. *Journal of Studies on Alcohol, 47,* 413–425.

Fingarette, H. (1987). *Heavy drinking: The myth of alcoholism as a disease.* Berkeley: University of California Press.

Gusfield, J.R. (1963). *Symbolic crusade: Status politics and the American temperance movement.* Urbana: University of Illinois Press.

Gusfield, J.R. (1981). *The culture of public problems: Drinking-driving and the symbolic order.* Chicago: University of Chicago Press.

Heath, D.B. (1987). Anthropology and alcohol studies: Current issues. *Annual Review of Anthropology, 16,* 99–120.

Hopkins, R.H., Mauss, A.L., Kearney, K.A., & Weisheit, R.A. (1988). Comprehensive evaluation of a model alcohol education curriculum. *Journal of Studies on Alcohol, 49,* 38–50.

Hubbard, H. (1978). Focus on youth: The prevention model replication program. *Alcohol Health and Research World, 3*(2), 9–13.

Johnston, L.D., O'Malley, P.M., & Bachman, J.G. (1989). *Drug use, drinking, and smoking: National survey results from high school, college, and young adult populations, 1975–1988.* Rockville, MD: National Institute on Drug Abuse.

Kurtz, L.R. (1984). *Evaluating Chicago sociology: A guide to the literature, with an annotated bibliography.* Chicago: University of Chicago Press.

Levine, H.G. (1984). The alcohol problem in America: From temperance to alcoholism. *British Journal of Addiction, 79,* 109–119.

Lewis, J.D., & Smith, R.L. (1980). *American sociology and pragmatism: Mead, Chicago sociology, and symbolic action.* Chicago: University of Chicago Press.

Mauss, A.L. (1975). *Social problems as social movements.* Philadelphia: J.B. Lippincott.

Mauss, A.L. (1982). Salvation and survival on skid row: A critical comment. *Social Forces, 60,* 898–904.

Mauss, A.L. (1989). Beyond the illusion of social problems theory. In J.A. Holstein & G. Miller (Eds.), *Perspectives on social problems* (Vol. 1). Greenwich, CT: JAI Press.

Mauss, A.L., Hopkins, R.H., Weisheit, R.A., & Kearney, K.A. (1988). The problematic prospects for prevention in the classroom: Should alcohol education programs be expected to reduce drinking by youth? *Journal of Studies on Alcohol, 49,* 51–61.

McAndrew, C., & Edgerton, R.B. (1969). *Drunken comportment: A social explanation.* Chicago: Aldine.

Mooney, C., Roberts, C., Fitzmahan, D., & Gregory, L. (1979). Here's looking at you: A school-based alcohol education program. *Health Education, 10*(6), 38–41.

Moskowitz, J.M. (1989). The primary prevention of alcohol problems: A critical review of the research literature. *Journal of Studies on Alcohol, 50,* 54–88.

Musto, D.F. (1988). *The American disease: Origins of narcotic control* (expanded ed.). New York: Oxford University Press.

Noble, E.P. (Ed.). (1978). *Third special report to Congress on alcohol and health* (Technical support document, Chap. XII). Washington, DC: Government Printing Office.

Pereira, J. (1989). Alarming result: Even a school that is leading the drug war grades itself a failure. *Wall Street Journal, CCXIV*(93), A-1 & A-6.

Pittman, D.J., & Snyder, C.R. (Eds.). (1962). *Society, culture, and drinking patterns.* New York: John Wiley and Sons.

Reinarman, C. (1988). The social construction of an alcohol problem: The case of Mothers Against Drink Drivers and social control in the 1980s. *Theory and Society, 17,* 91–120.

Roman, P. (1984). The orientations of sociology toward alcohol and society. *Journal of Drug Issues, 14,* 1–6.

Room, R. (1975). Normative perspectives on alcohol use and problems. *Journal of Drug Issues, 5,* 358–368.

Room, R. (1976). Ambivalence as a sociological explanation: The case of cultural explanations of alcohol problems. *American Sociological Review, 41,* 1047–1065.

Rooney, J.F. (1980). Organizational success through program failure: Skid row rescue missions. *Social Forces, 58,* 904–924.

Rumbarger, J.J. (1989). *Profits, power, and Prohibition: Alcohol reform and the industrializing of America, 1830–1930.* Albany: State University of New York Press.

Schneider, J.W. (1985). Social problems theory: The constructionist view. *Annual Review of Sociology, 11,* 209–229.

Thompson, M.L., Daugherty, R., & Carver, V. (1984). Alcohol education in schools: Toward a life-style risk-reduction approach. *Journal of School Health, 54,* 79–83.

Tobler, N.S. (1986). Meta-analysis of 143 adolescent drug prevention programs: Quantitative outcome results of program participants compared to a control or comparison group. *Journal of Drug Issues, 16,* 537–567.

Wiener, C.L. (1981). *The Politics of alcoholism: Building an arena around a social problem.* New Brunswick, NJ: Transaction Books.

Williams, M., & Vejnoska, J. (1981). Alcohol and youth: State prevention approaches. *Alcohol Health and Research World, 6*(1), 2–13.

Zald, M.N., & McCarthy, J.D. (1987). *Social movements in an organizational society.* New Brunswick, NJ: Transaction Books.

CHAPTER 9

Working the Rules: Drinking Sanctions in College Residence Halls

EARL RUBINGTON

More than 45 years ago Selden Bacon (1943) spelled out a rationale for sociological studies of drinking. In "Sociology and the Problems of Alcohol," he listed ten components and seven aspects to be taken into account in the study of drinking. He called for a scientific approach to the study and solution of social problems of alcohol and pointed to the absolute necessity for making observations, developing and verifying hypotheses, and arriving at sociological generalizations. Since he wrote that piece, alcohol sociology has come of age. Many sociologists have piled bricks on the foundation he laid. And most of them apply what is now conventional wisdom behind alcohol studies, largely unaware of Bacon's contribution.

He said: replace opinions with facts; study drinking as institutionalized behavior; obsess not over pathology. During his Yale time he wore three hats: he professed sociology, directed the Yale Center of Alcohol Studies, and chaired the Connecticut Commission on Alcoholism. No one occupying all these positions could escape pressures from "wets " and "drys," politicians, Alcoholics Anonymous (AA) zealots, religious reformers, educators, and therapists of all kinds unless he developed a broad view of the alcohol field. Bacon saw that alcohol could and did make problems for sociology. The only way to solve these problems was to capitalize on them. Study how people make, define, distribute, and consume alcoholic beverages, and how they respond to the "problems" alcoholic beverages make for them, and you learn about society.

He noted how biased samples (if you could call them that) skewed perceptions on alcohol problems. He said most "theories" on drinking based their interpretations on the "exotic fraction," the obvious problem drinkers. This perceptual bias may still exist, albeit in a different form.[1] Many people reduced problems of alcohol to problems of alcoholism when they referred in footnotes to the *Quarterly Journal of Studies on Alcohol* or described Bacon's center as the Yale Center on Alcoholism Studies. Some people even thought the famous summer school was called the Summer

School on Alcoholism. Not a few also believed that the center and AA were synonymous, one and the same thing.

The last section of "Sociology and the Problems of Alcohol" lists methods; documents (past and current), field studies, participant observation, interviews, and questionnaires all come in for separate mention. The point was, Bacon concluded, to use these methods to observe and collect data, gather sociologically relevant information, and replace armchair philosophy with systematic sociological theory. Since he wrote those words, there have been hundreds of studies employing all these methods. An inventory of general sociological propositions on alcohol and society is yet to come. But, in one field which Bacon (with Robert Straus) helped to start, the study of college drinking (Straus & Bacon, 1953), perceptual bias continues to reign. Of the numerous studies on college drinking, most employ questionnaires and most reproduce a staggering array of descriptive generalizations. When it comes to the kind of sociological explanations that Bacon argued for, most of these studies come up short (Saltz & Elandt, 1987). The situation at present is the direct opposite of the one Bacon wrote about. Now we have an enormous amount of information on college drinking of questionable sociological relevance.

The economics of social research probably accounts for this state of affairs. Surveys yield more facts than all other methods combined. So now we know how many students by sex, social class, age, religious preference, etc., drink what, how much, how often, with what effects on grade-point averages, etc. But what the charter or rationale of college drinking is, its rules, sanctions, functionaries and personnel, and the way of drinking (to use some of Bacon's ten components) has not yet been well established. And how college students teach and learn the folkways (Bacon's term for the socialization process) draws heavily on one of two sources: past parental socialization or current peer pressure. The remainder of this chapter now turns to a consideration of an exploratory study of drinking in college, one that draws upon residential assistants (RAs) as informants on drinking in college, in an attempt to show that learning how to drink in college in part, at least, depends upon how functionaries and personnel (residents in college residence halls) interact with one another over the course of the academic year. Rule enforcement turns out to be a significant aspect of the socialization process, as Bacon pointed out.

The Research Problem and Theoretical Rationale

Age-specific prohibition produced the research problem this chapter explores. At the time of this study, the legal drinking age was 20 for students of City University (not its real name). The law bisected the population of students into those who could obtain and drink alcoholic beverages legally and those who could not. The nature of college life thrusts a diverse lot of people into numerous kinds of contacts in many different social contexts. Thus, people in both of these age categories meet and do a lot of different things together. Similarly, 90 percent of these college students had been drinkers in high school long before they entered college. If they came from a state that had 18 years as the legal drinking age, they drank legitimately. If they came from a state with 19, 20, or 21 years as the legal drinking ages, then they had drunk illegitimately. In their case they had obtained alcoholic beverages in one of three ways: People of legal age bought alcohol for them, people who sold alcoholic beverages did not ask for identification, or while students in high school they used false identification to obtain alcoholic beverages.

These sets of rules enjoined underage use and regulated legal age use. State law made it illegal to possess or consume alcoholic beverages if under the age of 20. Municipal law forbade open containers or drinking in public. University rules forbade underage drinking anywhere but permitted drinking in specified times, places, and circumstances for drinkers of legal age. The university forbade drinking in common areas (hallways, stairways, etc.), but permitted drinking in rooms if the student was of legal drinking age.

These social and cultural conditions generated the theoretical rationale for this preliminary study. In effect, these rules created for many students at City University a situation comparable to that of the years of Prohibition (1920–1933). Conflict over the norm exists with many people giving overt support but tacit opposition (people talk "dry" but act "wet"); high-ranking people flout the norm with impunity (Dollard, 1945); large numbers of people are forbidden to possess, use, sell or serve alcoholic beverages; and a very small number of persons are the designated agents of enforcement (Merz, 1981). All these social and cultural conditions combine to suggest a social distance proposition on the observance-enforcement of age-specific prohibition. Other things being equal, observance-enforcement of the prohibition varies with the severity of the sanction, the social distance between sanctioning agents[2] and objects of sanction, and consensus on the rule itself. (In "Sociology and the Problems of Alcohol" Bacon pointed out that low-ranking people could enforce the mores on people of higher rank, whereas high-ranking people could resist the enforcement of folkways by persons of lesser rank. Many of his students will recall him saying in class that the function of the police is to control minority groups. A few people will still remember that comment stems from his two-volume doctoral dissertation on the New Haven police.) Finally, with changes over time in the pattern of social interaction between enforcement agent and object of sanction, observance of and enforcement of the prohibition may change as the social distance between agent and object changes. The above considerations, taken together, became the basis for an exploratory study of how RAs at City University defined, interpreted, and enforced the underage drinking rules.

Setting and Method

The Main Street campus of City University (hereafter referred to as CU) is located in a commercial and residential section of an eastern city. University housing consists of 17 buildings which are either apartments, residence halls, or suites. Apartments have private bath and kitchen, suites have bath but no kitchen, while residence halls have common bathrooms and cafeteria. Residence halls, whether apartments, residence halls, or suites, may or may not be integrated by sex as well as academic class, as when freshmen and upperclassmen live in the same building. This study deals primarily with residential assistants who live and work in West Hall and East Hall. Both house freshmen and have men and women students living on alternate floors.

West Hall houses 400 student freshmen in three wings. On each floor an RA occupies a single room at the front end of a corridor shaped in the form of a square. Two single rooms, generally housing an upperclassman, adjoin the RA's room on either side. Each room holds two students. Between the north and south hallways lies the common lavatory and lounge. RAs in West hall generally supervise the 36 students who reside on their floor.

East Hall, a converted apartment house, also houses 400 students, predominantly freshmen, in a building with three wings. The RA lives in a single room and generally has charge of 35 students on his or her floor. Unlike West Hall, however, students live in doubles, triples, or quads. In addition, the floor layout differs in two important ways. The RA's single room lies at the base of a fork shaped by three hallways, each of which has its own fire door.

Main Street bisects the campus. The north side of the street, with two exceptions, has housing, both university and private, while the south side has administrative, classroom, departmental, gym, maintenance, and campus police buildings. Several large parking lots are also found on this side. Two drinking establishments cater to CU students. One, the Goal Line Stand, is on the south side of Main Street right next door to Barr Hall, a CU apartment building. The other, the Jug and Mug, is on the north side of the street, closer to the main body of student housing, both university and private. A block east of the Jug and Mug is Main Street Liquors, another establishment catering to CU students.

Violations of drinking rules made up a portion of news and feature articles in the *CU News*, the student weekly. During 1983–1984, the *CU News* ran a three-part series on student alcoholism. The weekly campus crime column reported over the course of the academic year approximately a dozen incidents in which students figured either as assailants or victims of assaults that took place after drinking. One woman student was allegedly raped while in an intoxicated condition, while another had to be rushed to the emergency room of a nearby hospital to be treated for alcohol poisoning. A male freshman in East Hall was said to have gone berserk after only two or three beers. It took five campus police officers to subdue the student, who was said to weigh only 130 pounds. In the process, the student broke the collarbone of one of the assisting officers. A feature article on the Jug and Mug noted that it had been suspended for three days for selling alcoholic beverages to underage students. The article went on to say that the Jug and Mug was now cooperating with the CU Community Relations Committee to enhance the image of Main Street.

CU houses the bulk of its incoming freshmen in four residence halls: North, South, East, and West. Since most of the students under the legal drinking age would be found in these residence halls, I centered the study on them. And since North, South, and West Halls have identical architecture, I interviewed RAs in West and East to see if variations in residence hall architecture had any bearing on violation and enforcement of the drinking rules. I interviewed East Hall's residence director (RD) and eight of its 13 RAs. I interviewed West Hall's RD, one of its two staff assistants, and nine of its 12 RAs. For comparative purposes, I also interviewed the RD of Arch Hall, a Carr Hall RA, an RA from an upperclassmen's residence hall, and three RAs from Barr, Elm, and Ford, all upperclassmen's apartments.

I explained to all interviewees that I was making an exploratory study of alcohol problems on campus and that I was primarily concerned with their experiences in enforcing CU rules and regulations on alcoholic beverages. I had a simple interview guide but departed from it frequently as I listened to the RAs talk about how they viewed their work, their floor, the rules, and their own experiences. I took notes during the course of 15 interviews, none in the other nine (the last interview was a joint one with two RAs); the grand total was 25 informants. Generally, I wrote up the interview notes (an average of five and one-half pages of double-spaced typewritten notes) immediately after the interview or the next day. I read the interviews over several times, developed a set of codes, and then analyzed the data. Some common themes appeared in the notes and indicated that changes over time took place in both the observance and enforcement of the drinking rules. A process of socialization between

RAs and their residents over the course of the academic year resulted in an etiquette of relations centering on the observance-enforcement issue. I turn now to a discussion of the evolution of this etiquette.

RAs are on duty on certain days and weekends during a given quarter. When on duty, they are responsible for the entire building. During their tour they make rounds three times a night. When making rounds, they patrol common areas on the lookout for vandalism, loud parties, drinking in hallways, open containers, and students coming into the building with cases of beer or bottles in bags. If they find students who are over 21 drinking in the hallways or walking around with an open container, they generally ask them to retire to their rooms. If students are underage, RAs ask them to "dump the beer." At loud parties, they check students' IDs if alcohol is being served in the room. If a party is illegal (too many guests and/or party not properly registered in advance with the Housing Office), they "break it up." Back in their rooms, they are still "on call." Residents of the building as well as of the neighborhood can and do call them up to advise them of disturbances. At the end of their tour of duty, they record what has happened in the "log." Students who repeat infractions or engage in more serious forms of misconduct may be "written up"; in such a case an RA files an incident report with the residence director. Repetition draws escalation of penalties and meetings with housing officials. The scale of sanctions begins with probation, moving troublesome students to another residence hall, and ends, if necessary, with banning the student from CU housing.

Two weeks before school starts in the fall, RAs participate in a week-long training session that includes seminars on the drinking rules and how to enforce them. One week before school starts, freshmen move in and the residence director orients them. Later, their RA acquaints them with CU rules for residence hall living at their first floor meeting. At this meeting, the RA pays particular attention to rules pertaining to the use of alcoholic beverages.

Evolution of the Etiquette

Changes take place over the academic year in contacts RAs have with residents of the whole building, as well as the residents on their own floor. The changes lead to status changes and ultimately, for some, changes in the definition and enforcement of drinking rules. Through a process of mutual socialization, RA and residents evolve an etiquette for handling rules on drinking.

At school's start, freshmen residents assign RAs to the master status of policeman (Hughes, 1945; Becker, 1963). This assignment makes many RAs personally uncomfortable. The passage of time, however, offers considerable opportunity to live down the stereotype and the negative feelings attached to it. In effect, as one informant explains below, the RA goes from being the incumbent of a status and an authority figure to becoming a person in his or her own right.

> *R-11:* Freshmen think all RAs are alike. At the beginning of the year, they think you're out to restrict their freedom. After they get to know you, they change their attitudes. There are certain trouble-makers who may have a thing about RAs. But when residents get to know you on a personal basis, they find out that RAs are people.... As they get to know you, they break down. And you're able to become more flexible.... I started strict, then I found it gradually breaking down. You get close to them, you get to know them.

In some instances, the humanization of the RA can take place in a matter of weeks.

> *R-12:* Residents on my floor came and told me after three or four weeks that they didn't like me at first; they thought I was prissy. But during those first days, I was wearing a dress, my glasses, and acting very formal, meeting parents, moving students in. Later they got to see me run around, play my music, cram, do the all-night studying, be human, cry, get upset. Other RAs won't show themselves, they act straight laced, even if they don't know all the answers. They won't ask.... I have even asked about courses to take. Other RAs act condescending. They say: "You're a freshman, but you'll get over it," as if they had never been a freshman themselves or gone through the same thing. If there are water fights or shaving cream on doors on the floor, I don't get upset. I sort of say to them after awhile, "C'mon guys." I'm human, I've even yelled at them at least once.

There are kinds of RAs that RAs do not want to be like, mainly because their behavior contributes to the negative social identity they seek to change. Those who are "sticklers for the rules," who are "hard assed," who just want to "write up" residents in the log, who just want to "catch" residents, who just want to "break up parties," are symbols of how not to be an RA. Several RAs, as R-12 in the excerpt above, commented on "other RAs," thereby contrasting themselves with these obvious examples of how not to be a successful RA. To be known as "fair, but firm" is preferable to being labeled "strict." In all this, the theme of reducing the authority gap between RA and resident comes through clearly.

At the onset, contact between RAs and residents are segmental, and given freshmen's initial introduction to college living, primarily focused on enforcement of the rules. RAs assert their authority most frequently when there are violations of drinking rules and regulations. These contacts themselves can become the basis for learning about residents as persons. In time, as RAs come to know more and more residents as persons, they begin to exercise that discretion required of all agents of social control. How this process evolves both through enforcement and drinking contacts is well illustrated by one RA.

Interviewer: Now tell me how the RD presented the rules at the first dorm meeting, how you presented them at the floor meeting, and then how you actually enforced those rules.

R-20: The RD cited incense, candles, but stressed the alcohol rules. At the first floor meeting, I said that state law prohibits underage possession or use and that CU supports the law. That means if minors are caught in common areas with open containers, the beer will be dumped and they will be written up in the log. Parties, in the room, again if all are underage, will be disbanded. I gave the impression that I would follow the rules to a T. I used discretion. When I first uttered the sentences, I took them as the word of God. I soon learned to ignore them.

Interviewer: Ignore them?

R-20: I was continually running to the log and writing incident reports. There were situations where people were drinking beer which were harmless and there were other situations where people were drinking beer which were harmful.

Interviewer: Distinction between the two situations?

R-20: In the second situation, people were drinking a lot of beer.

Interviewer: How did you know that a lot was harmful?

R-20: By the effects of alcohol, loud and boisterous, slurring speech, getting out of hand. The others were drinking beer, watching the basketball game, and were still sober.

From past history, I can usually determine if they are problem drinkers or not. Some tend to drink more than others. Some don't handle alcohol as well as others. Some get violent.

Interviewer: How did you learn all this?

R-20: I've seen some quite often, obviously intoxicated, one or two times. At the Jug and Mug, I drank with some of the residents. Some get in with legitimate driver's licenses owned by others, some change birthdates on their own licenses.

In a social environment that generates a high frequency of violations, RAs come to see that excessive application of sanctions only cheapens their value. Thus, RAs reduce sanctions as much as possible while at the same time developing a set of specific conditions for their application. These conditions arise, in time, out of the nature of the RAs duties as well as the range of social contacts they sustain in the course of residence hall life. When RAs are on duty, for example, they make their rounds over the entire building. At all other times, they are on call, but largely at the service of residents of their own floor. Thus, in the slow shift from status to personal contacts, they are much more likely to know more about residents of their own floors than residents of the entire building. Personal knowledge makes for differential enforcement of the drinking rules. As the following excerpts indicate, RAs sanction residents of their own floors differently than they sanction residents of other floors.

> *R-6:* You use your discretion. I treat guys on my floor different from the rest of the building.... I treat my guys good. I know them better than anyone else. This quarter I've noticed there's more smoking than drinking. When they drink, they become loud and violent. When they smoke pot, they're mellow, quiet; they just hang out. I tell them I smell smoke, suggest they stuff towels in the door, spray Glade. The residents were shocked to find that I let 'em smoke. All the other RAs bust residents for smoking pot. My floor has less vandalism this quarter, probably because the residents have been smoking pot instead of drinking.... If some of my guys are going to have a punch party, they tell me. I let them know who's on duty, I advise them to keep it quiet.

> *R-23:* I am less formal when I approach residents on my floor, more formal when I approach residents on other floors. With residents of other floors, I identify myself as an RA, I explain rules to them, then I make my request. With residents on my own floor, all I need to say is: "You know what you're doing."

> *R-22:* Experienced RAs use their judgment. If it's their resident, a dirty look may be enough. RAs are more protective of residents of their own floor, they're more lenient. They know more about the person from their floor, they know less about a person from another part of the building. When they know less about the person, they deal with the situation. Most residents have respect for their own RAs and are more likely to comply with their requests. Some students are more apt to obey RAs from other floors. A lot depends on the interactive chemistry of the floor.

> *R-20:* There are 350 students living in East Hall. There's been no violence on my floor. One resident, after drinking I think, tears postings off the wall. I'm closest to the 30 residents on my floor, as a person, than to the other 320 residents in the building. Those who come from other floors respect their RAs and/or their floors, but not necessarily me or my floor. In any case, I may know them slightly, if at all.

A reciprocal of differential enforcement of rules on the home floor is that residents can repay their own RA by breaking the rules on other floors. When the RA is not around, residents may be expected to stray from the housing rules. But when the RA is around, residents who want to stray can take themselves to other floors. Two women RAs independently of each other said in separate interviews that the residents of their floors were "not angels." By this they meant that their residents did get into trouble on other floors, not on theirs. These RAs learned of the trouble by reading about it in the log or by having other RAs tell them about it. Getting into trouble on other floors is a way of showing respect for one's RA as well as saving face. In part, the many comments that RAs made about theirs being a "quiet floor," not having a "problem floor," or one where there had been no vandalism may well be self-serving but also may reflect a truism. Many collective disturbances, such as loud parties coupled with underage drinking, may occur off the home floor. Thus, whenever residents export deviant behavior to another floor, they uphold the etiquette of their floor.

In their initial presentation of the alcohol rules to residents of their floors, RAs begin with either discretionary or scare tactics. RAs who enunciate a discretionary tactic at the outset indicate that they will be forced to punish those whom they catch. If some imply ways of avoiding capture, others spell them out. RAs who enunciate scare tactics at the outset imply that they will catch all who break the rules and that the penalties for such infractions will be severe. Those who begin with scare tactics end up with a discretionary stance that has become adapted to the realities of the floor. Some examples of discretionary tactics are as follows:

> *R-10:* If residents get out of line, stop them. If there's no noise, then there's nothing to do. It's not a problem if it's not in my sight, if they drink when I'm not around.
>
> *R-17:* I tell them all the rules. I tell them I don't want to see anything. I tell them not to let things get out of hand.
>
> *R-21:* At the first floor meeting, I inform the residents on what the housing handbook says. Then I tell them if you're drinking and don't call attention to yourself, then you will have no problems with me. Don't bother me or the people around you. In three years, I've had no problems.
>
> *R-24:* I told them at the floor meeting, you do what you do, but don't do it here.

Enunciations of the alcohol rules that spell out ways of evading capture are as follows:

> *R-8:* At my first floor meeting this fall, I state Housing Office Policy on alcoholic beverages. Then I explained how we were going to work the rules here. I told them to be smart, to do responsible drinking, that I was not going into rooms to see what they were doing.... You don't want to be a hard ass. I favor the soft or cool approach.

Scare tactics, by contrast, start out "hard," only in time to become "soft," as the quotes from interviews that follow demonstrate.

> *Interviewer:* How did your initial RA training deal with alcohol problems you might face?

R-12: During the year, we attend nine training sessions. And for a full week before school starts. They don't tell us anything about recognizing or dealing with students who may have drinking problems. They concentrate on telling us about the alcohol beverage rules and what to do when they get broken. I communicate the same set of rules when school starts in the floor meeting. I pass out 35 housing handbooks, then I go over the highlights. I tell them it's my job to enforce the rules. I tell them they've all had jobs and so they know something about enforcing rules they don't really like. CU is my employer. The Housing Office tells us to be strict in the beginning, act like a top sergeant, then later see whether to stay strict.

Interviewer: What are some of the things which still stick in your mind from your first week as a residential assistant?

R-17: On that first day, I kept thinking about what I'd say at the floor meeting. I wanted to start the year off right. About the housing rules, how I'm going to handle the floor. Some floors are hellish, some are angels. I was thinking about what I was going to tell them. I decided to tell them that I was going to be strict, sort of scare them. Common sense says it's easier to go from hard to soft than from soft to hard. It's harder to get tough. And some RAs had given me advice. I'm majoring in marketing. It's only common sense to start with a higher price, then mark it down.

R-11: In the beginning, it was a whole new situation.... When I had to break up a party, I told them that I was going to write them all up. We keep a log in the office. If anything happens, we make a note of it in the log. They didn't know what being written up meant. So it was fear of the unknown. I told them that people would be made aware of their actions. Anytime someone gets caught and doesn't know what will happen, it sets up a doubt, it intimidates them, they don't know how far they can go. If they don't know what's going to happen, they have fear. If they're written up, they could get tossed out. They don't know the procedure when they first come in, they don't know what they can get away with, what they can do.

The RAs in the following quotes depict the move from strictness to leniency and some of the social conditions under which they made the move.

R-6: None of the three RAs I had in my first years at CU were strict enforcers of the rules. If there was vandalism, of course, they reported it. But I've found that you get your best results if you play with the residents. It's a matter of give-and-take. In time, you give them a little more room, a little leeway.... You can't be strict all the time.

R-11: As they get to know you, they break down. You're able to become more flexible.... I started strict, then I found it gradually breaking down.... You have to give the freshmen a certain amount of leeway. You can't say you can't drink.

R-12: In the beginning, you're not popular. Residents see you as an authority figure. At the beginning of the year, you're really enforcing policy. In time, you learn how to roll with the punches.... In the beginning, I was strict. Later, I began to lay back as the residents learned how to work within the system.

With time, RAs and residents arrive at a mutual accommodation. Out of their collective experience with taking specific roles in situations of social control (that is, the enforcer and the one subject to rule enforcement), they constitute an agreement for how they will "work the rules."

Summary and Conclusions

In effect, RAs teach residents how to break drinking rules. Residents learn discretion in situations of social control. When an RA sees a student drinking in his or her room with the door open or walking in the hallway with a beer in hand, the RA suggests that the student drink in his or her room behind closed doors. Students with a "good attitude" accept the suggestion good-naturedly and retire to their rooms immediately. If in the future they never go out in the hallway with a beer in hand and always keep their door closed when drinking, they have learned "discretion."

The open corridors of West Hall amplify all sounds of student life. Consequently, more deviant behavior probably comes to the attention of West Hall's RAs. RAs say that when they try to quiet a loud party, they usually find illicit drinking. East Hall, with its numerous hallway fire doors, probably screens more deviant behavior from its RAs. Floor layouts occasion different patterns of interaction between RAs and residents. In a high visibility dorm like West Hall, RAs are apt to take more active roles in teaching discretion and "responsible drinking" (Weisheit, 1983). By contrast, the floor layout of East Hall encourages a more laissez-faire approach on the part of its RAs.

Finally, the version of responsible drinking that RAs preach and sometimes teach by example has little to do with moderating or limiting amounts. Rather, it stems from one of the implicit norms of urban life. This is the norm of "mind your own business" in a double sense. This norm requires discretion in deviance. Residents mind their business when drinking if they do so behind closed doors, keeping their voices and stereos down. In so doing they do not disturb their neighbors or come to the attention of the RAs. It seems that RAs who have somewhat more contact with residents, who themselves are drinkers, and who may on occasion drink with residents in or out of the building are more able to teach these norms by example and by selective enforcement. By contrast, RAs who are either light drinkers or abstainers, who have somewhat less social contact with their floor residents, and who observe the taboo against drinking with residents can only teach about responsible drinking by means of selective overenforcement.

Drinking, as Bacon noted, occurs as a part of many different human activities. Thus, drinking takes its meaning from the context in which it occurs. In college residence halls, that context involves three social units:

college administration (makers of alcohol policy), RAs (enforcers of that policy), and residents (observers of that policy). In the process of mutual socialization, relations on the floors shape the way RAs and residents alike define, interpret, and respond to those rules. "Moderate" drinking and its reciprocal "moderate" enforcement, this exploratory study suggests, evolves over time as both parties to the rules devise ways for "working" them. This collaboration yields positive sanctions and keeps the peace on the residence hall floor. Bacon (1943) explains the tentative findings when he points out that "negative sanctions of overt character have been the subject of a great deal of attention in relation to the drinking habits of specific individuals as well as of large groups.... In the great majority of instances they have failed"(p. 419).

Acknowledgment

This study was supported in part by a grant from the Northeastern University Research and Scholarship Development Fund.

Notes

1. Perceptual bias or systematic error (whether from alcohol use or study) attaches to Bacon's name. Over the years people have referred to him as Sheldon Baker and systematically misspelled his first name as Seldon.
2. After the 1951 publication of Edwin Lemert's *Social Pathology*, the term most sociologists would come to use for sanctioning agents was "agents of social control."

References

Becker, H.S. (1963) *Outsiders*. New York: Free Press.

Bacon, S.D. (1943). Sociology and the problems of alcohol: Foundations for a sociologic study of drinking behavior. *Quarterly Journal of Studies on Alcohol, 4*, 402–445.

Dollard, J. (1945). Drinking mores of the social classes. In *Alcohol, science and society* (pp. 95–104). New Haven: Quarterly Journal of Studies on Alcohol.

Hughes, E.C. (1945). Dilemmas and contradictions of status. *American Journal of Sociology 50*, 353–359.

Lemert, E.M. (1951). *Social pathology*. New York: McGraw Hill.

Merz, C. (1981). *The dry decade*. Seattle: University of Washington Press.

Saltz, R., & Elandt, D. (1987). College drinking studies, 1976–1985. *Contemporary Drug Problems, 13*, 117–159.

Straus, R., & Bacon, S.D. (1953). *Drinking in college*. New Haven: Yale University Press.

Weisheit, R.A. (1983). Contemporary issues in the prevention of adolescent alcohol abuses. In D.A. Ward (Ed.), *Alcoholism: Introduction to theory and treatment* (pp. 253–266). Dubuque, IA: Kendall-Hunt.

CHAPTER 10

Social Policy and Habitual Drunkenness Offenders

DAVID J. PITTMAN

This chapter focuses on one of the most ignored and disesteemed groups in the United States: habitual drunkenness offenders. My frame of reference is to examine the social policy changes toward this category of individuals over approximately the last half-century. The following areas are discussed: (1) the seminal works on habitual drunkenness offenders by Selden Bacon in the 1940s and by David Pittman and Wayne Gordon in the 1950s, (2) the social movement to decriminalize the public drunkenness offense in the 1960s and 1970s, and (3) barriers to decriminalization and treatment of these individuals in the 1970s and 1980s.

The Drunkenness Offender After Prohibition's Repeal

More than a half-century ago the noble experiment of Prohibition ended with the ratification of the 21st Amendment to the U.S. Constitution in 1933. Repeal reopened the door to the scientific study of various alcohol problems and experiments with new treatment modalities for individuals who were alcohol dependent. In 1940, the Section on Alcohol Studies of the Laboratory of Applied Physiology at Yale University was founded, as was the *Quarterly Journal of Studies on Alcohol*. The major influences of the Yale Center of Alcohol Studies (now at Rutgers University) in the 1940s were: (1) to disentangle alcohol issues involving research, education, and treatment from the straightjacket of the wet/dry controversy that had characterized American thought since the early 1800s; (2) to create a climate in which social and cultural factors in alcohol use and misuse could be examined within a scientific framework; and (3) to spawn social movements (e.g., the National Committee for Education on Alcoholism, which later became the National Council on Alcoholism, summer schools on alcoholism, Yale Plan Clinics for the rehabilitation of alcoholics, etc.) to increase the understanding and expand the treatment of alcohol-dependent people. A young sociologist,

Selden Bacon, who later became director of the Yale and then the Rutgers centers, was active in all these areas.

Bacon's study, "Inebriety, Social Integration and Marriage" (Bacon, 1944–1945), was based on all persons arrested for drunkenness in a five-week period in five large and three small Connecticut towns in 1942, as well as those arrested on any charges at the same time who were used as a comparison group. As Bacon noted, these inebriates represented what at the time were referred to as "public nuisance" drunkards by the police. Some 1,200 men were interviewed; they were predominantly a middle-aged group who participated to a limited extent in community organizations. Bacon's research focus was on the "undersocialization" of these men, which was reflected in the fact that a majority of the sample was single. A major contribution was his emphasis upon the importance of associational factors, only hinted at by researchers who preceded him, in the genesis of inebriety; however, his concern with marital status as being the most significant interpersonal experience overlooked the possibility of other social networks that could provide social integration for chronic drunkenness offenders.

Bacon's research on inebriety provided part of the theoretical framework for the research design of Pittman's joint study with C. Wayne Gordon in the early 1950s on the chronic police case inebriate (Pittman & Gordon, 1958). In fact, we used the same undersocialization hypothesis in the analysis of our data, which were obtained by interviewing 187 men in the county jail who had been sentenced at least twice to a short-term correctional institution on a charge of public intoxication. The demographic characteristics of this New York State sample in many respects were similar to those Bacon had studied a decade earlier. These men, whose average age was 48 years, were older on the whole than other criminal law violators and those found in alcoholism treatment facilities. Forty-one percent of the men had never married. Of those married, 96 percent had been separated, divorced, or widowed. Practically all of them had either failed to establish families or those they had established were broken. Only 2 percent were living with their wives at the time of arrest; in the county studied, only 11 percent of the men in the same age brackets had broken marriages. Seventy percent had not gone beyond grade school, compared to 40 percent in the county at large. Sixty-eight percent were unskilled workers, 22 percent skilled, and only 3 percent professional, compared to 13, 46, and 22 percent in the general population.

Of particular relevance for understanding the habitual drunkenness offender was the finding in our study that these men's careers could be classified as "early skid" and "late skid." The early skid career pattern included about half of the offenders. Two-fifths had been jailed twice in their twenties, and the others by their early thirties. The early skid pattern represented serious social maladjustment in youth carrying over into middle adulthood; poor work records outside of institutions; and poor marital adjustment, if any. The late skid men did not experience their second public intoxication conviction until they were in their forties or even fifties. During the time they were developing dependence upon alcohol, they often had stable jobs and families for long periods. Apparent in the late skid career pattern was physical decline and great difficulty in maintaining economic stability in physically taxing marginal jobs. Younger men replaced the late skid person on the casual day-laborer jobs. His drinking increased, and finally his tolerance for alcohol declined. The early skid person probably drank because of early social, economic, or psychiatric crippling; the late skid individual probably drank because of physical decline and failure to maintain employment as he approached later middle age.

Bacon's study of public inebriates was undertaken in the social context of wartime America (1941–1945), which was receptive to innovations in many areas, including alcohol problems. Every individual was needed in the war effort to defeat the enemies of democracy; thus, the first formal program to rehabilitate alcoholic workers was founded by the DuPont Corporation in 1943. Moreover, society could not afford the luxury of losing individuals to jail for alcohol problems; in fact, the Connecticut War Council cooperated in the execution of Bacon's research. Thus, the need to rescue individuals from deviancies such as drunkenness and other alcohol-related problems accelerated the support for the establishment of Yale Plan Clinics in 1944 and other similar institutions to prevent inebriety and rehabilitate individuals with alcohol problems, including habitual drunkenness offenders.

Although Bacon's study made no specific social policy recommendations concerning what changes should occur in the disposition of public inebriates, it should be noted that he was a member of the original board of directors of the National Committee for Education on Alcoholism founded in 1943; the major purpose of the organization was to educate the public that alcoholism was a disease and that the alcoholic was a sick person who not only was worth helping but could be helped with his or her problem. The wartime experience set the stage for a liberalization of societal attitudes toward alcohol use, as reflected by many jurisdictions removing restrictions on "liquor-by-the-drink," Sunday sales, etc. Societal concern with alcohol-related casualties was reflected in the passage of the first congressional alcoholism legislation since Prohibition with the enactment of the District of Columbia Alcoholism Act in 1947, which in effect decriminalized public drunkenness in that jurisdiction; no funds were provided for facilities to implement the legislation. With the return of American military personnel to civilian life in the latter part of the 1940s, innovations in reference to alcohol problems almost ceased, especially programs for the habitual drunkenness offender.

The social context in which the *Revolving Door* study by Pittman and Gordon (1958) took place was quite different than the one in which Bacon worked. Rochester, New York, the locale of study, in the early 1950s was a place where innovative ideas in alcoholism were widely accepted, although one might say that the attitude toward jailed inebriates was one of noblesse oblige when viewed from today's perspective. However, there was a real concern for the physical and social welfare of these individuals. There was a very active organization, the Health Association of Rochester and Monroe County, which had an alcoholism committee; they were concerned with the jailed population at the Monroe County Penitentiary, really a regional jail. The individuals most active in this group were the late Jack Norris, who was the medical director of Eastman Kodak and also later a nonalcoholic trustee of Alcoholics Anonymous, and the Reverend Thomas Richards, who ran a mission for skid row denizens. Ray McCarthy, who was the director of alcoholism research for the New York State Commission on Mental Health, believed that a study should be done on this population and helped to secure the funds. Much to the credit of the health association, and especially to Jack Norris, they decided to ask for sociological direction of this study, and I was assigned by the chairman of sociology at the University of Rochester, Earl Koos, along with a colleague of mine, Wayne Gordon, to participate.

In our study we found that the skid row individual incarcerated for public drunkenness was and perhaps always has been at the bottom of the socioeconomic ladder; he was isolated, uprooted, unattached, demoralized, homeless, and stigmatized. In this situation he frequently drank to excess. Admittedly, through his own behavior, he was one of the least respected members of the community, and his treatment by the community had been at best negative and expedient. Furthermore, we concluded that many individuals arrested for public drunkenness on skid row were alcoholics, but treatment for their alcoholism was clearly not part of the correctional regimen. The process of arresting inebriates, detaining them for a few hours or a few days, and then rearresting them was termed by us a "revolving door." Some individuals had been arrested 100 to 200 times and had served 10 to 20 years in jail on short-term sentences, which in reality was life imprisonment on the installment plan. The recidivism rates for public drunkenness clearly indicated the futility of the criminal justice system in dealing with the underlying social, economic, and medical problems involved. Our major social policy recommendation in 1958 was:

> A Treatment Center should be created for the reception of the chronic public inebriate. This means that they should be removed from the jails and penal institutions as the mentally ill in this country were removed from the jails during the last century. Given the present state of knowledge concerning alcoholism, the time is ripe now for such a change. The present system is not only inefficient in terms of the excessive cost of jailing an offender 30, 40, or 50 times, but is a direct negation of this society's humanitarian philosophy toward people who are beset by social, mental, and physical problems. (Pittman & Gordon 1958, pp. 141–142)

Thus, the *Revolving Door* study documented that imprisonment of alcoholics not only reinforced their dependence on alcohol but that it was a waste of money in that it is the rare alcoholic who would be deterred from drinking by a jail term. As a consequence of the study, a halfway house for a limited number of jailed alcoholics was established in Rochester, New York. In summary, although our work on inebriates was influenced by Bacon, we went further than he by making specific social policy recommendations that we believed would aid these individuals' return to a productive life.

The Decriminalization Movement

The 1960s of President John Kennedy's "New Frontier" and especially President Lyndon Johnson's "Great Society" and "War on Poverty" programs were in an era in which the social policy orientation was that the federal government could solve all America's social and economic problems, including alcoholism. By 1964, a high level committee on alcoholism had been established in the Department of Health, Education and Welfare by Secretary Anthony Celebrezze to study the effects of alcoholism on the nation's well-being. This committee sponsored several national conferences on the chronic drunkenness offender and accidental injury in Washington in 1964 and 1965. In this milieu of domestic reform the social movement to decriminalize the public drunkenness offense occurred.

Historically in North America and Great Britain, public drunkenness in almost every legal jurisdiction had been defined as a criminal violation. Laws existed on state and municipal levels prohibiting public drunkenness. Although disorderliness was a prerequisite for arrest under some laws, the homeless skid row inebriates faced repeated arrest for disorderly and nondisorderly drunkenness. There were, however, variations from city to city on how severely the drunkenness statutes were enforced by the police. At one extreme were the practices in Atlanta and Washington, D.C., where enforcement procedures were very strict: in Atlanta, in 1965, of a total of 92,965 arrests, 52.5 percent were for drunkenness; while in Washington, D.C., the corresponding figure for 86,464 arrests was 51.8 percent for drunkenness. At the permissive end was St. Louis, where the police were more tolerant and had a sociomedical orientation to alcoholism; there in 1965, of all 44,701 arrests, 5.5 percent were for drunkenness (President's Commission, 1967, p. 254).

A large number of these actions involve the repeated arrest of the same men. To illustrate, let us take the case of Portland, Oregon, for 1963; in this year there were 11,000 law violations involving drunkenness or the effects of drinking, but only around 2,000 different persons accounted for these arrests (R.R. Wippel, personal communication, 1964).

Persons arrested and held for prosecution for public drunkenness were almost never represented by counsel and almost always found guilty. In 1969, reports to the Federal Bureau of Investigation (FBI) from 2,640 cities representing a population of 66,155,000 showed that 86.2 percent of all persons charged with public drunkenness were found guilty (FBI, 1970). Moreover, these offenders frequently found themselves incarcerated. Indeed, there was strong evidence that chronic inebriates constituted one of the largest groupings of individuals incarcerated in short-term correctional institutions. Alcohol-related offenses accounted for 35 percent of the incarcerations in the St. Louis City Workhouse for the period 1957–1959. Benz (1964) completed a study of the penal population in the Monroe County (Rochester, New York) Jail, which showed that alcoholic offenders accounted for 62.5 percent of the prisoners and 73.1 percent of the total commitments in the year 1962.

As the Bacon and the Pittman and Gordon studies noted, the very nature of public drunkenness cases excluded most middle-class and upper-income alcoholics and excessive drinkers who typically drank in private or semiprivate settings. Thus, advocates of decriminalization of the public drunkenness offense pointed out that these laws affected mainly the lower class who drank in public and, in effect, were class laws.

Decriminalization of the public drunkenness offense in the late 1960s and early 1970s in various political jurisdictions was the result of: (1) court decisions related to the legal status of the intoxication offender; (2) the Presidential Crime Commissions; and (3) major innovations in processing drunkenness offenders.

Court Decisions

In 1966 two major legal decisions affecting the public intoxication offender in the United States were rendered. First, in January the United States Court of Appeals for the Fourth Circuit in Richmond found in favor of the appellant, Joe B. Driver of North Carolina, who had been arrested more than 200 times for public intoxication. In a unanimous decision, Judge Bryan stated for the court:

> The upshot of our decision is that the State cannot stamp an unpretending alcoholic as a criminal if his drunken public display is involuntary as a result of disease. However, nothing we have said precludes appropriate detention of him for treatment and rehabilitation so long as he is not marked a criminal. (*Driver v Hinnant*, 1966)

In the same vein, the United States Court of Appeals for the District of Columbia in March 1966 ruled unanimously in favor of the appellant, De-Witt Easter, who was contesting his conviction for public intoxication on the grounds that he was a chronic alcoholic. The decision stated: "Chronic alcoholism is a defense to the charge of public intoxication and therefore is not a crime" (*Easter v District of Columbia*, 1966).

These court cases dealt only with the chronic alcoholic and one manifestation of the disease—public intoxication. This is illustrated in the following excerpt from the legal brief filed on behalf of DeWitt Easter by his attorneys. It states:

> There is no quarrel here with the principle of the Harris case that Section 25–128 of the D.C. Code [public intoxication statute] may be used to punish the normal individual who goes on a "binge" from time to time, or the common drunkard whose intoxication results from indolence but not from addiction. Such persons could, if they wished, control their drinking, and therefore are criminally responsible for their actions. Nor does Appellant [Easter] argue that a chronic alcoholic who drives a car, or commits murder, is *ipso facto* not guilty. A finding that a chronic alcoholic has no capacity to avoid appearing in public in an intoxicated condition is *not* a finding that a chronic alcoholic may with impunity commit murder... that is an entirely different question of criminal responsibility.... (*Easter v District of Columbia*, 1965)

The *mens rea* approach is aimed at helping only the chronic alcoholic, and does not encompass all chronic drunkenness offenders nor cope with the great range of alcohol-related offenses, such as vagrancy, common assault, and disorderly conduct.

In 1967, the Supreme Court of the United States was asked to rule, in *Powell v Texas* (1968), on the constitutionality of the use of a public intoxication statute in cases involving chronic alcoholics. It was the contention of medical, legal, and other professional groups supporting Powell that chronic alcoholism was a positive defense to the charge of public intoxication, and that these individuals should not be incarcerated but should receive medical and social treatment. In June 1968, the Supreme Court, in a narrow five-to-four decision, held that a chronic alcoholic could be convicted under a state law against public drunkenness. The majority decision was by four members of the Court who joined in one opinion, which noted that there was disagreement among medical authorities that alcoholism was a disease, and by a fifth justice whose opinion agreed with the result reached by them, but who took a narrower position in doing so. However, this decision did not negate the lower-court decisions, given the absence of a clearcut majority of five justices agreeing in one decision. Thus, local and state laws (which have been enacted subsequent to the *Powell* decision) that allow chronic alcoholism as a valid defense to public intoxication criminal charges remain constitutional.

The advocates of decriminalization by virtue of the *Powell* decision, which failed to render public intoxication statutes unconstitutional, were left the recourse of legal reform on a state-by-state and/or community-by-community basis. In fact, this is what happened. In 1971, the National Conference of Commissioners on Uniform Law endorsed the Uniform Alcoholism and Intoxication Treatment Act for enactment by the various states. Basically, the Uniform Act had as its major thrust to remove the alcoholic and intoxicated person from the criminal justice system and to provide him or her, as a sick person, with a continuum of care. When the Congress renewed the 1970 Comprehensive Alcohol Abuse and Alcoholism Prevention, Treatment, and Rehabilitation Act (PL91-616) in May 1974 with PL93-282, the legislation provided special grants to those states that had repealed the criminal statutes and ordinances under which drunkenness was the "gravaman of a petty criminal offense." Despite these incentives, by the early 1980s only 32 states enacted the basic provisions of the Uniform Act.

The Presidential Commissions

Important in the decriminalization movement were President Johnson's appointment of the President's Commission on Law Enforcement and Administration of Justice and the President's Commission on Crime in the District of Columbia, both in 1965. These commissions had as one of their specific foci the public drunkenness offender. Their interest was based on the fact that in the mid-1960s one-third of all arrests in the United States—about 2 million of the 6 million—were for public intoxication. Of note was the President's Commission on Law Enforcement and Administration of Justice's (1967) recommendations concerning alcoholism and the public drunkenness offense:

1. Drunkenness should not in itself be a criminal offense. Disorderly and other criminal conduct accompanied by drunkenness should remain punishable as separate crimes. The implementation of this recommendation requires the development of adequate civil detoxification procedures.
2. Communities should establish detoxification units as part of comprehensive treatment programs.

3. Communities should coordinate and extend aftercare resources, including supportive residential housing.
4. Research by private and governmental agencies into alcoholism, the problems of alcoholics, and methods of treatment should be expanded.... Consideration should be given to providing further legislation on the Federal level for the promotion of the necessary coordinated treatment programs. (pp. 256–257)

These were forward-looking recommendations for the late 1960s, some of which were to be achieved in the 1970s and 1980s; however, these recommendations have never been fully implemented in the United States.

Innovations in Processing Drunkenness Cases

Bold approaches to handling the problem of public drunkenness within a sociomedical context first occurred, interestingly enough, in Eastern Europe—namely, Czechoslovakia and Poland—with the establishment of "sobering-up stations" or detoxification centers. These stations, instead of jails, were used to process drunkenness cases.

Sobering-up stations have become an integral part of the network of alcoholism services in Poland and Czechoslovakia. For example, in Warsaw any person found drunk on the street or lying in a doorway is taken by police to the sobering-up station. In Czechoslovakia, patients from the sobering-up station are referred to lectures on alcoholism and its effects (called "Sunday Schools" since the lectures are held on Sunday). Generally, when the individual appears a second or third time at the sobering-up station, a full-scale medical and social evaluation begins and a plan for therapeutic intervention is worked out, involving voluntary approaches at first. If the patient does not proceed with voluntary treatment, then compulsory treatment is begun.

The first detoxification center to open in North America for persons detained by the police for public intoxication was in St. Louis in November 1966, sponsored by that city's Metropolitan Police Department in cooperation with St. Mary's Infirmary and the Social Science Institute, Washington University (which this author directed at the time), under a grant from the Office of Law Enforcement Assistance of the U.S. Department of Justice, supplemented by local and state funds.[1]

The elements for social change in St. Louis were found in the emphasis the city's power structure—government, police, civic leaders, and social agencies—placed on the rehabilitation of the alcoholics. Much educational work with these groups and the area's mass media had prepared the community to accept the idea that public drunkenness offenders were sick individuals who needed sociomedical care instead of a fine or jail terms. For example, in 1962 and 1963 many key St. Louis personnel visited the Alcoholism Treatment and Research Center and held numerous informational conferences with staff members (Pittman, 1967).

As a result of these conferences and further studies, the St. Louis Board of Police Commissioners in 1963 instituted a major policy change in reference to intoxicated persons on the street. The St. Louis Metropolitan Police Department made it mandatory for all individuals "picked up" from St. Louis streets to be taken to the emergency rooms of one of the two city hospitals for physical examination. This meant that routine physical evaluation was provided all alcoholics processed by the police; if these individuals were in need of medical care, they were to be hospitalized instead of being jailed. If medical care were deemed unnecessary, the intoxicated person was "held until sober"—not more than twenty hours—and released to the community.

St. Louis was one of the few American cities in which this early innovation in the handling of the public intoxication case occurred. It squarely placed the locus of responsibility for the alcoholic in the treatment sphere and was in keeping with modern practices toward the publicly intoxicated person in a number of European countries. However, the Board of Police Commissioners was dissatisfied that large numbers of alcoholics were not admitted to the hospitals for medical care. At times more than 90 percent of the "examined" public intoxicants were returned by the physicians to the police for processing.

Thus, the St. Louis Metropolitan Police Department was one of the few agencies to apply for a grant for the operation of a detoxification center when such centers became eligible for support under the Law Enforcement Assistance Act of 1965. The St. Louis Detoxification and Diagnostic Evaluation Center (originally at St. Mary's Infirmary, and in December 1968 transferred to the St. Louis State Hospital) was a 30-bed unit with 24-hour medical and nursing coverage. The goals of the center were:

1. To remove chronic inebriates to a sociomedical locus of responsibility that would markedly reduce police processing time.
2. To remove chronic inebriates from the city courts and jails.
3. To provide sociomedical treatment for them.
4. To begin their rehabilitation.
5. To refer them to an agency for further rehabilitation, with the goal that they will return to society as productive persons.

This detoxification center was the first systematic attempt in North America to provide treatment for the alcoholic at the moment the police intervened in the process. The center operated as follows:

1. A police officer brought the "intoxicated" person to the reception room.

2. Center personnel completed a medical examination of the patient.
3. The patient was showered, given clean clothing, and assigned a bed.
4. Special nursing care and diets were provided.
5. Therapeutic activities—films, group meetings, discussions, and lectures—were provided.
6. Each patient was counseled individually.
7. The patient, when necessary, was referred to other social, health, and governmental services for further help.
8. The average length of stay was seven to 10 days.

An evaluation study of the effectiveness of the St. Louis detoxification center was conducted by Jim Weber, a police officer and sociology graduate student, under the direction of this author. The goals of the study (St. Louis Detoxification and Diagnostic Evaluation Center, 1970) were to measure the effect of the new process for handling public inebriates in time-savings for the police and indirectly for the courts and penal institutions, as well as the impact of the short-term treatment in the detox center on the subsequent behavior, including drinking, of the inebriates. In short, Weber's goal was to determine whether the revolving-door cycle for public drunkenness offenders could be changed. Weber interviewed a random sample of the male patients approximately four months after their discharge from the center. His results were encouraging in that they demonstrated that the police used less time to handle drunkenness cases, and that at follow-up approximately one-half of the patients showed improvement in their life situation. The most positive aspect about the study was that many of these individuals could be helped and rehabilitated. However, the success rate, as it has commonly been termed, was not as high as it would have been with upper- and middle-income patients who have more social and financial resources at their disposal. Thus, the St. Louis center became a model for many other communities that developed detoxification centers to handle their chronic drunkenness offenders.

Throughout the period from the late 1960s to the early 1980s, numerous states and municipalities decriminalized the public drunkenness offense. For example, a Missouri law enacted in 1978 stated:

> No county or municipality, except as provided in section 67.310 [dealing primarily with driving while intoxicated] may adopt or enforce a law, rule or ordinance which authorizes or requires arrest or punishment for public intoxication or being a common or habitual drunkard or alcoholic. (Mo. Ann. Stat., 1981)

Conversely, a number of states, such as California, have never decriminalized the public drunkenness offense. We will now explore the major reasons why barriers were created to the implementation of the Uniform

Alcoholism and Intoxication Treatment Act, which not only decriminalized this offense but provided rehabilitation services for habitual drunkenness offenders.

Barriers to Decriminalization and Treatment

The optimistic beliefs of the late 1960s and early 1970s that all public drunkenness offenders would be removed from jail cells and receive adequate treatment never fully materialized. A number of reasons for this development can be enumerated. By the mid-1970s a more conservative trend emerged in the United States that accelerated with the end of the Vietnam War in 1975. After the initial enthusiasm generated by the passage of the Hughes-Javits Alcoholism Act in 1970, the interest in the public drunkenness offender and the skid-row alcoholic began to fade. The rush to enact state decriminalization laws in the 1970s and the tendency to reduce the legal purchase age for alcoholic beverages began to wither away by the late 1970s. To paraphrase the sociologist Joseph Gusfield, American society, throughout the 1980s, began to view people with problems, such as habitual drunkenness offenders, homeless individuals, skid-row derelicts, etc., as problem people.

But there were more specific reasons for the problems the various states experienced after their enactment of the Uniform Alcoholism and Intoxication Treatment Act.[2]

Planning

A number of states passed the Uniform Act without any preliminary planning or a realization that the new system would be difficult to operate and without full awareness of what the movement from a criminal justice approach to a sociomedical one for public inebriates would entail. This lack of planning meant that law enforcement officials expected to be relieved of processing public drunkenness offenders. However, adequate treatment centers to provide care for these individuals rarely were established. In the absence of facilities for emergency care or even treatment on an outpatient basis, many inebriates were still placed in protective custody in the jail. In some jurisdictions public inebriates continue to be arrested but on a different charge, for example, disorderly conduct.

Funding

Some states enacted the Uniform Act without providing additional appropriations to fund essential programs. These legislatures operated on the assumption that savings from the diversion of inebriates from the criminal justice system could be reallocated for the Uniform Act's purposes. Rarely were there any revenues to reassign; in those cases in which there were savings in police, court, or correctional areas these funds were expended for other pressing needs involving crime control. Since indigent alcoholics received minimal or no care before decriminalization, the effect was to bring a population group into the health delivery system who needed major medical and social services. Thus, costs increased without adequate monies being available. Resistance to providing resources for this group of individuals derived to a large extent from the low priority that is attached to the needs of public inebriates.

Health Care

The decriminalization movement was constructed on two fundamental assumptions: (1) alcoholism was a disease; and (2) therefore, the offender's public intoxication was involuntary and symptomatic of his or her disease. Thus, detoxification centers such as the one in St. Louis were based on a medical model. Therefore, the staff included physicians available 24 hours a day, registered nurses, paramedical personnel, alcoholism counselors, social workers, etc.

But there was major resistance in most American states to the provision of this type of service for habitual drunkenness cases, many of whom were not only homeless poor people but also were in the chronic stage of alcoholism. Therefore, some American communities developed what they referred to as nonmedical detoxification centers; these facilities provided a context that emphasized "talking a patient down" from an intoxication episode, the nonuse of drugs to detoxify the patient, and the provision of care for alcoholics through counseling, illustrated lectures, and other techniques to motivate the patient to achieve sobriety. By the early 1980s the term *nonmedical* had been dropped in favor of the social model of detoxification; this term acknowledged the fact that medical care was available if necessary in complicated cases of withdrawal from alcohol. Currently, it is the exception to provide medical detoxification centers, even in St. Louis, for habitual drunkenness offenders. Nonmedical and later social detoxification centers conveyed to the public that there was an inexpensive way to provide services for poor alcoholics without health insurance despite the fact that many of these individuals had other medical problems, such as liver cirrhosis, peripheral neuropathy, Wernicke-Korsakoff syndrome, etc.

The early medical detoxification centers for public inebriates differed little from today's alcoholism treatment centers, which are a ubiquitous part of the American medical scene. In reality, there are two systems of health care for American alcoholics today. Despite the fact that the United

States spent 12 percent of its gross national product on health care in 1988, it is the only nation in Western society, with the exception of South Africa, that does not have a system of universal health insurance for its citizens. For those with private and public health insurance coverage, treatment for alcohol dependency typically takes place in medically oriented inpatient centers. For those without this insurance (unless covered by Medicare), such as habitual drunkenness offenders, treatment for alcoholism dependency is haphazard—at best they may receive care in a social detoxification center or in the emergency rooms of public hospitals. At worst, they receive no medical care. In short, there appears an unwillingness on the part of American society to provide adequate health care for all its citizens, regardless of resources.

Conclusion

The social movement to decriminalize the public drunkenness offense, which was constructed on the empirical studies of Bacon and Pittman and Gordon among others, was only partially successful. There were a number of reasons for this not occurring—one of which is that low-income public drunkenness offenders do not have a high priority in the public legislative agenda. No detoxification center had full access to the kinds of transitional resources that were necessary for the adequate treatment of these individuals, such as the provision of halfway houses, domiciliary care, or even custodial care. American society has never been willing to invest the kinds of resources needed to cope with this problem. Further, advocates of decriminalization were accused of moving the "revolving door" from the jail to the hospital, because these centers constantly accepted persons who relapsed. Moreover, the emphasis has been on upper- and middle-class alcoholics with health insurance to the exclusion of lower-income ones, such as habitual drunkenness offenders. However, there is always the possibility that any citizen may be arrested for public drunkenness or, in those states that have decriminalized, on a related charge; the consequences may be lethal, as this August 17, 1988 story in the *New York Times* indicates:

> ________was arrested Sunday night on a charge of public drunkenness and placed alone in a receiving cell. . . . [He] hanged himself with his shirt, which he had tied to a window grate. (p. 18)

Notes

1. For the author's role in this endeavor, see Pittman (1987), 11–21, 121–126).
2. For detailed accounts of the American experience with decriminalization of public drunkenness, see Aaronson, Dienes, & Musheno (1982, 1984).

References

Aaronson, D.E., Dienes, C.T., & Musheno, M.C. (1982). *Decriminalization of public drunkenness: Tracing the implementation of a public policy*. Washington, DC: National Institute of Justice, U.S. Department of Justice.

Aaronson, D.E., Dienes, C.T., & Musheno, M.C. (1984). *Public policy and police discretion: Processes of decriminalization*. New York: Clark Boardman.

Bacon, S. (1944–1945). Inebriety, social interaction and marriage. *Quarterly Journal of Studies on Alcohol, 5*, 86–125, 303–339.

Benz, Elizabeth. (1964). *Man on the periphery*. Rochester, NY: Rochester Bureau of Municipal Research.

Driver v. Hinnant, 356 F.2d 761 (4th Cir. 1966).

Easter v. District of Columbia, 209 A.2d (D.C. Ct. App. 1965).

Easter v. District of Columbia, 361 F.2d 50 (D.C. Cir. 1966).

Federal Bureau of Investigation (FBI). (1970). *Crime in the United States: Uniform crime reports—1969*. Washington, DC: U.S. Government Printing Office.

Mo. Ann. Stat. 67.305 Supp. 1981.

Pittman, David J. (1967). Public intoxication and the alcoholic offender in American society. In President's Commission on Law Enforcement and Administration of Justice, *Task force report: Drunkenness* (pp. 7–28). Washington, DC: U.S. Government Printing Office.

Pittman, David J. (1987). Conversation with David J. Pittman. *British Journal of Addiction, 82*, 11–21, 121–126.

Pittman, David J., & Gordon, C.W. (1958). *Revolving door: A study of the chronic police case inebriate*. Glencoe, IL: Free Press.

Powell v. Texas, 392 U.S.514 (1968).

President's Commission on Law Enforcement and Administration of Justice. (1967). *The challenge of crime in a free society*. Washington, DC: U.S. Government Printing Office.

St. Louis Detoxification and Diagnostic Evaluation Center. (1970). *Final report*. Law Enforcement Assistance Administration (LEAA), United States Department of Justice #0-373-790. Washington, DC: U.S. Government Printing Office.

CHAPTER 11

Problem Definitions and Social Movement Strategies: The Disease Concept and the Hidden Alcoholic Revisited

Paul M. Roman

Much of the enterprise surrounding the sociological study of alcohol issues is related to the disease or medicalized conception of alcoholism. The specification and diffusion of this concept have been a fundamental characteristic of the "modern" alcoholism movement (Roman, 1988a; Schneider, 1978). Sociologists have had mixed relationships with the medicalized conception. In some instances they have been critics (cf. Mulford, 1984, 1990; Roman & Trice, 1968; Seeley, 1962), although the most recent detailed and strident critiques have been authored by a philosopher (Fingarette, 1988) and a psychologist (Peele, 1989). Much sociological work has implied support (and in some instances provided both direct and indirect support) for the medicalized conception by basing research programs within its paradigm; that is, studies of etiology, prevention, and treatment, together with their accompanying models and language, all more or less affirm medicalized conceptions of alcohol problems. Without doubt, many sociologists have benefited from research and other support from agencies that are strongly aligned with the medicalized concept, such as the National Institute on Alcohol Abuse and Alcoholism (NIAAA), the Christopher D. Smithers Foundation, and the National Council on Alcoholism and Drug Dependence.

Aspects of ambivalence toward the medicalized conception can be found in the work of Selden Bacon. There can be little doubt that in his significant leadership roles at the Yale and Rutgers Centers of Alcohol Studies, Bacon's professional identification as a sociologist was a singularly important influence in legitimizing sociological study of alcohol issues. An overview of Bacon's work indicates a continuing pattern of support for the medicalized conception of alcohol problems, but most of this support is cautious and inferential rather than direct. Across Bacon's wide-ranging publications, one cannot find statements that could be described as all-out embracing of the medicalized conception, yet also there is scant evidence of intense criticism. This ambivalence had organizational supports. Like many scholars who are both researchers and administrators, Bacon knew the sources of support and additional resources. His roles required continuing rapport with influential individuals and constituencies who were unequivocally and even blindly committed to the validity of the notion that alcoholism is a "disease like any other" and that other alcohol problems are most effectively dealt with through a medicalized approach.

As is indicated in the introductory chapter, alcohol issues and alcohol studies have become a large-scale enterprise over the past two decades, contrasting sharply with the previous three decades following the repeal of Prohibition. Within this enterprise, the dominance of the medicalized conception, especially in the rather primitive forms in which it is portrayed in the mass media, is changing rapidly. These changes are important in that they may foreshadow changes in patterns of support for public policies, and affect the organization of constituencies and components of what might be regarded as alcohol-linked social movements.

The purpose of this chapter is twofold. First, a discussion of two patterns of cultural and organizational change that are alleged to be undermining the "classic" medicalized conception of alcohol problems is presented. These macrolevel changes concern (1) the continuing and increased imprecision in the definition of alcohol problems, and (2) the challenge to the disease concept's dominance of social scientific discourse about alcohol problems by those who study the effects of alcohol availability on alcohol problems. This discussion supplements an earlier analysis (Roman & Blum, 1991) of the growing "murkiness" of the medicalized conception of alcohol problems. In that essay we described the growing influence of psychotherapy in alcoholism treatment and the introduction of alcohol warning labels as contributors to the murkiness of the medicalized conception.

The second goal of this chapter is to consider the implications of a repeated theme in Bacon's sociological analysis of alcohol problems, namely attention to the "hidden alcoholic." This concept is analyzed in terms of its utility in the past and at present as a "claims-making" device by the social movements associated with medicalized conceptions of alcohol problems.

Continuing Definitional Confusion Surrounding Alcohol Problems

The medicalized concept of alcoholism may be viewed as an American cultural innovation introduced (or reintroduced [Levine, 1978]) with intentional and deliberate social vigor in the 1930s and 1940s (Schneider, 1978). Its introduction was intertwined with the discovery of the strategy of Alcoholics Anonymous (AA) in providing a remedy for chronic alcohol abuse that was compatible with the basic value orientations of American culture (Roman, 1988a; Trice & Roman, 1970). The vigor accompanying the introduction of the medicalized concept, coupled with societal interest and organizational resource investments, increased over the following three decades.

The innovation of the medicalized concept of alcohol problems, while initially conceptualized many decades earlier (Levine, 1978), embodied solutions to two cultural anomalies that were produced by the repeal of Prohibition in 1933:

1. Popular concern in the 1930s and 1940s over recently relegalized beverage alcohol was reduced by supporting an image of alcohol as a potent but manageable cultural item. The repeal of Prohibition effectively undermined official societal support for the perspective that alcohol problems could be explained as a singular result of alcohol consumption. Individual abnormalities accompanying the ingestion of alcohol were substituted for alcohol as the principal cause of chronic deviant drinking, which was labeled alcoholism. As an official perspective, this sharply reduced attention to the causal role of drinking per se in the development of alcoholism. This approach coincidentally spawned an organizationally based movement that centered on alcoholism (rather than the myriad consequences of drinking) as the most serious and important societal problem associated with drinking.
2. This shift in focus from alcohol to alcoholism was in no small way due to the discovery of a viable "solution" for alcoholism via Alcoholics Anonymous, while evidence of the insolubility of the "drinking-as-the-alcohol problem" remained very much evident in the cultural memories of Prohibition. The repeal of Prohibition led to disassembling the massive law enforcement mechanism that Prohibition had demanded. This, together with the "new" official definition of alcohol's limited menace, made anomalous the primacy of law enforcement approaches to alcohol issues. AA, in turn, provided the basis for a new and different means for dealing with America's alcohol problems through compassionate help and constructive intervention rather than through legal definitions, law enforcement, punishment, and social exclusion.

Solution to these two anomalies produced by the repeal may be seen as the cultural foundation of the medicalized conception of alcohol problems. At present, there are at least four specific and major cultural and organizational changes that are undermining the definitional clarity of the disease concept of alcoholism, encouraging confusion in the public mind about the meaningfulness of this conception, and acting to discourage referrals to treatment.

Scope of Definition

First, while it was initially defined to apply strictly to the several varieties of alcoholism (Jellinek, 1960), the disease model has come to be used very loosely to refer to a vast range of alcohol problems. These applications include many instances where there is no evidence of loss of individual control over drinking, evidence that once was believed to be the "heart" of the disease designation of alcoholism. It is suggested that popular support for the medicalized conception of alcohol problems will be reduced or undermined unless those who are given medicalized labels can be shown to genuinely lack control over their drinking behavior.

In the typical definition of drinking problems that has come to dominate clinical practice, "control" over drinking is defined interactionally rather than individually, that is, the control "problem" is in the eyes of interactional partners who define drinking problems in terms of impaired role performance (Cahalan, 1970). This definition is, ironically, a sociological rather than medical definition; in its crudest form, this definition accedes diagnostic authority to the "person on the street" rather than requiring a medical judgment (assuming that the person on the street has a role relationship with someone that she or he believes to drink routinely in ways that interfere with what that someone ought to be doing, i.e., is a "significant other" in relation to the drinker).

In its initial formulation, there were elegant advances offered by this sociological approach to defining problem drinking. First and foremost, it sidestepped the ambiguity of using the quantity and frequency of alcohol consumption as the sole basis for the definition of an alcohol problem, thus disowning a potential heritage from the temperance-dominated era. Further, it did not require evidence of physical consequences or organ damage for a label of problem drinker to be applied. But its fundamental quality was to describe a process rather than to provide a new list of symptoms. This sociological process was one in which an individual's drinking and its association with role impairments and other troubles led to social reactions whereby the drinker's role partners defined the individual as substandard, inadequate, and/or deviant in her or his role performances.

The most prominent early use of this concept was as an epidemiological device, to provide estimates of different categories of drinking behavior in general populations without being limited to a trichotomy of abstainers, drinkers, and alcoholics (Cahalan, 1970; Cahalan & Room, 1974). In these survey studies, inventories of role performance problems and others' judgments were indeed used in a manner similar to symptom lists. There is little evidence, however, that the field researchers intended for this epidemiological device to be translated into a clinical device.

In fact, it can be observed that in this clinical translation, the designation of problem drinker does not even require a significant other, either on or off the street, but can be applied by a clinician who projects that a hypothetical significant other would likely find the individual's drinking-related behaviors to be problematic.

Herein lies a general explanation of the "loss of control" over the medicalized definition of alcohol problems. The role impairment definition of alcoholism basic to clinical practice that alcoholism specialists have worked so hard to diffuse is utilized so loosely that it can be applied to practically any drinker who exhibits problems in some arena of life functioning, whether or not these problems are causally linked to drinking behavior.

As mentioned, this definition disowns heritage from the temperance era by the "distance" that it moves drinking problems from drinking per se. In a complex way, the cultural adoption of the "problem" definition can be traced back to the contrasts that the new alcoholism movement was attempting to make with the previously dominant temperance ideologies. The temperance definition of alcohol problems centered on consumption, with "hard-liners" viewing any consumption as the equivalent of a problem. It is important to recognize that the role impairment definition of alcohol problems requires routine alcohol consumption and that quantity/frequency of consumption by itself is not a good predictor of the drinker's behavior over social space and time. Nevertheless, it may be that reliance on any definition of drinking problems that is primarily drinking-based may be "too close for comfort" to temperance imagery of alcohol problems. Regardless of this derivation, it is clear that the role-impairment definition of alcohol problems and, to some extent, alcoholism, encompasses a vast range of drinking behaviors and associated behavior patterns that enhances confusion over the "right" definition of both individual and collective "problems."

New Intervention Strategies

A second major contribution to inconsistency in the definition of alcohol problems is found in newly institutionalized structures of intervention that essentially define all alcohol problems within a medicalized frame of reference. Especially prominent are employee assistance programs (EAPs) (Blum & Roman, 1989; Roman, 1988b) wherein job performance impairments associated with drinking are defined and managed within medical frames of reference. Closely parallel are the practices and assumptions within student assistance programs, which are modeled after EAPs. To some extent, similar phenomena are also found in drinking driver programming, but here the attention is limited to EAPs as the main example.

EAPs are an outgrowth of industrial alcoholism programs. Part of their basic design comes from these earlier programs, particularly the use of job-based leverage to motivate employees with drinking problems to desire to change their behavior and to seek assistance. This is accomplished through the documentation of job performance decrements that are used as the basis for a confrontive discussion with the employee. This discussion is geared to provide the employee with evidence of a performance problem, coupled with the demand that performance be brought back to an acceptable level, but emphasizing that assistance is available through company auspices if the individual believes that a personal problem, such as substance abuse, is underlying the performance problem. A diagnostic decision about the nature of the problem is made by an EAP counselor, either employed by the company or working for an external EAP service contractor. Thus, rather than attempting identification of the signs and symptoms of alcohol problems "on the shop floor," the initial evidence of an alcohol problem is a job performance problem.

While EAP counselors do not attach the diagnosis of "alcoholism" to every observed instance of an employee alcohol problem, they work under the direction of policies and procedures that prescribe that alcohol abuse (as well as any other presenting problem that comes to the EAP's attention) is to be treated in a medicalized fashion. For example, many policy statements indicating that alcohol problems (not only "alcoholism") are to be dealt with like "any other health problem."

Thus, these program designs require that any level of alcohol problem that is perceived to affect work performance (by either the employee, supervisor, or work peers) must be dealt with "as if" it were a medical problem. As might be expected, language that differentiates alcoholism and alcohol abuse is not always used with precision, especially in settings dominated by concerns with practicality. From the present authors' observations, such imprecision leads to the perception in many EAPs that all alcohol problems represent some level of alcoholism or addiction, ranging from early to late stage.

EAPs have diffused very widely and rapidly over the past 20 years. As a workplace entity with high visibility to the workforce in most instances, they are important as transmitters of information about the nature of behavioral problems and their proper management. Since EAPs are bound by their guiding policies to treat all alcohol-related problems as health problems, they are major (and growing) contributors to the confusion produced by a lack of differentiation within the vast array of alcohol-related behaviors.

Intermingling Alcohol and Drugs

A third contribution to definitional confusion about alcoholism and alcohol problems lies in the definitional overlap between alcohol and drug problems at the macro level. Due to apparent desires to both capture public attention and capture a share of public dollars, there have been deliberate efforts to intermingle the definition of alcohol problems and problems with illegal drugs. Most recently such a motivation has been evident in the change of the name of the National Council on Alcoholism (the organization that was the very first to have as its top priority the systematic diffusion of the concept of alcoholism as an illness, treatable illness, and a disease "like any other") to the National Council on Alcoholism and Drug Dependence. This trend may be traced to the early 1970s when the National Institute on Alcohol Abuse and Alcoholism's informational campaigns began the emphasis on "alcohol as a drug." This soon led to language such as "alcohol and other drugs."

This raises multiple concerns about the effects on public attitudes and behaviors. One concern is that intense negative feelings toward illegal drugs fostered among the general public by all levels of government can spread to attitudes toward alcohol problems. For the present analysis, it is evident that a major problem lies in the fact that the illegality of many nonalcoholic drugs leads to the definition of any use of an illegal drug as a drug problem. This extremely loose definition even contrasts with the very loose combination of drinking and role impairment that can be the basis for the definition of alcohol problems, as described previously.

The definitional issue alone clearly describes the conceptual confusion that is created by this intermingling, despite the short-term payoffs that might result through the alcohol field's sharing in public appropriations for research and treatment. One set of substances is viewed within a prohibitionist framework, with use of the substances seen as the fundamental problem. In the case of alcohol and medicalized conception, the problem has been transferred from the alcohol user to the alcohol abuser and the alcoholic, but we may be observing trends toward defining any problematic use of alcohol as an alcohol problem worthy of medicalized intervention.

In addition, while there has been substantial resistance to merging alcohol and drug interests at the federal level, most state agencies have now combined these concerns. This has enhanced the "integration" of alcohol and drug dependencies. While the treatment of opiate addictions does not appear to have become integrated with alcoholism treatment (and is numerically a minor problem from a relative point of view), private alcoholism treatment facilities have shown a high degree of readiness to apply their technologies to drug dependencies other than alcohol, especially cocaine dependence.

Polydrug Abuse

A fourth compounding factor in definitional confusion is perhaps the most serious and fundamental, namely the apparent disappearance of the "typical" or "pure" alcoholic who is the foundation for the disease concept of alcoholism. This disappearance is manifest in treatment settings. Evidence from our own recent field survey of 125 private inpatient alcoholism treatment centers indicates a sharply increasing proportion of "polyabusers" in the client population, that is, those whose personal troubles are associated with the use of alcohol and a range of psychoactive substances.

The prevalence of polyabuse is similar to but distinctive from the intermingling of alcohol and illegal drug issues at the policy and organizational levels. While many of these polyabusers use alcohol to excess, they tend to identify themselves primarily as drug abusers rather than as alcoholics. This is sometimes encouraged by treatment personnel who readily define any use of illegal drugs as indicative of a drug abuse problem, with the cultural intensity of this label evidently prioritizing it above an alcohol problem.

Polyabusers' personal labels of their problems have in some instances created difficulties with attempted affiliation with AA, where the stated ideology is centered exclusively on alcohol problems. AA has not "changed" in response to the increased number of polyabusers among its affiliates, and it may be that resistance to those who have problems with drugs other than alcohol may be restricted to "old timers." This highlights the strong association between age and polyabuse, with such patterns associated with younger age groups, suggesting pressures for age stratification within AA groups.

Yet from all indications AA remains vigorous and continues to grow. Thus, indeed, the fellowship may have the flexibility to adapt to the change in the nature of population in need of its services. However, it would appear that an unavoidable contradiction between stated ideologies around the exclusive focus on alcohol problem issues and actual practices in supporting recovery from polyabuse will eventually have to be addressed.

From a theoretical point of view, all these changes in the "using" patterns of the clientele who were formerly "exclusively alcoholic" would appear to demand that there be a common conception of alcohol and drug dependencies, but the medicalized conception of alcohol problems does not seem to have moved in this direction. If anything, there appears to be resistance among alcohol problem researchers to a "unified" theory of addictions, affected at least in part by the stigma and legal definitions that are associated with many nonalcoholic drugs. In our own research on public attitudes toward alcohol and drug problems among adults in Georgia (Blum, Roman, & Bennett, 1989), we have found surprisingly high levels of acceptance of a medicalized conception of cocaine dependence, yet at the same time we are impressed by the ambivalence indicated by the large proportion of respondents found in the "don't know" categories of these survey items.

This set of four major definitional problems is related to the two social anomalies that underlay the initial waves of diffusion of the medicalized conception of alcohol problems. We can uncover two major vulnerabilities that are especially evident in the contemporary definitional ambiguity and confusion. These vulnerabilities have only recently become evident as alcohol problems have moved up the social problems "attention cycle."

First, the medicalized conception is alcoholism-oriented, and has great difficulty logically accommodating nonalcoholic alcohol problems. Such problems are included within the scope of the alcoholism movement's "claimed territory" or turf, but there are logical and strategic problems that are frequently glossed over. To refer to the discussion above, a principal problem is use or advocacy of a medicalized approach to deal with nonalcoholic alcohol problems where, by definition, the key ingredient of personal "loss of control" is not present. In other words, the alcoholism movement's definition of nonalcoholic alcohol problems calls on the one hand for sick role occupancy by the alcohol-troubled individual, yet on the other hand the individual does not meet the basic loss-of-control criterion that is the cornerstone of the disease concept.

Up until now, this lack of fit has not appeared to have created major problems or attention. However, it is clear that as expenditures for alcohol problem treatment continue to grow, along with health care cost containment concerns, scrutiny of alcoholism treatment will continue and will probably escalate. While most such attention to date has been centered on treatment efficacy, there is no doubt that it is increasingly leading to scrutiny of diagnostic criteria through various mechanisms of "managed health care."

A second vulnerability lies in the fact that the alcoholism movement does not offer alternatives to its humanitarian and humane remedies for dealing with alcoholism. Since all alcohol problems are tacitly regarded as some form of alcoholism, nascent or otherwise, medicalized approaches for dealing with these problems are all that are offered. Consequently, there is no latitude for addressing those instances of "badness" where the offended and observing public demands that some form of punishment or retribution occur. Such a rigidity of intervention choice on the part of alcoholism specialists undermines their potential constituency support, and makes them appear more naive in terms of problem differentiation than the supposedly ignorant public.

An arena for conflict around this issue is the management of the drinking driver (Jacobs, 1989). Through both voluntary organizations and intense media support, many in the public have come to demand strong penalties imposed on those who cause injury while drinking and driving, with these demands extended to some extent to cover those drivers who are simply found to have been drinking. The judicial system, by contrast, appears increasingly disposed to refer these individuals to treatment rather than incarcerate them. Such decisions reflect the strong influence of the medicalized conception of alcohol problems, the belief that treatment is the most effective means of changing behavior, and the attitude of the criminal justice system that it does not want to resume the custody of alcohol-troubled individuals that was happily surrendered with the decriminalization of public drunkenness. While these practices of offering treatment rather than retribution generally have a low social visibility, the potential for conflict is there.

Loss of Exclusive Identity with Science

Returning to the intentions of the definitional transformation of alcohol problems that began in the 1930s, it is clear that a major goal was support for what was then a new approach to analyzing, interpreting, and understanding alcohol and alcohol-related phenomenon. This new approach was tacitly objective and "scientific." It is our contention that a most important feature in determining the diffusion and acceptance of the medicalized conception of alcohol problems has been this foundation in "science."

The attractiveness of this approach in the 1930s and 1940s was greatly enhanced by its contrast with the moral bases and patterns of reasoning that appeared as the foundations of the temperance, prohibition, and antisaloon movements. By tapping the fundamental American cultural value orientation toward science and rationality (Roman, 1988a, 1988b; Williams, 1970), the "modern" alcoholism movement eventually made its way into acceptance by the scientific community. To a lesser but nonetheless significant extent, scientific trappings have also facilitated acceptance by the community of medical practitioners.

Until the 1970s, the medicalized conception had a "monopoly" on its identity with and attachment to science. The loss of this monopoly is a major contributor to the confusion and competition over governing images of alcohol problems contemporarily evident in the United States. This monopoly ended when advocates of images of alcohol problems and solutions to alcohol problems that are alternative to and/or competitive with the medicalized conception began utilizing the same "royal road" of science in their diffusion efforts within American culture and society.

Arguments emphasizing the economic and societal impacts of alcohol problems are based on complex statistics and detailed formulae, drawn from what might be regarded as the central armory of social science technology (for the initial major example of this strategy, see Berry & Boland, 1977). The advocates of dealing with alcohol problems through reducing consumption present detailed scientific information to bolster their claims about the positive social and health consequences of reducing drinking (Mosher & Jernigan, 1989). The impact of alcohol on the nation's highways is presented in both simple and complex statistical language (Jacobs, 1989), with the questionable evidence that alcohol consumption is a major causal factor in auto accidents and accident fatalities apparently convincing to much of the public (cf. Gusfield, 1981). A scientific basis for the warning labels that are mandated on alcohol beverage containers was distinctively presented in support of the enabling legislation in the forms of scientific evidence of the impact of drinking on developing fetuses, evidence of escalated risk of motor vehicle and workplace accidents caused by drinking, and a vast amount of data on the adverse consequences of alcohol consumption on virtually every organ system (Department of Health and Human Services, 1987).

The key point is that the advocates of the medicalized conception of alcohol problems are no longer alone in their use of the scientific approach. It is likely no exaggeration to observe that these alternatives have gained popularity and interest in response to two changes: the increase in societal resources available to support interventions and research related to alcohol problems, and the consequent upward mobility of alcohol studies on the ladder of scholarly respectability. The organizations representing the "competition" of the medicalized conception of alcohol problems have adopted what is essentially the same strategy of generating public and legislative support on the basis of scientific evidence. Further, it is probably a safe generalization that across all these models, parallel issues regarding definitions, data adequacy, reliability, and validity can be raised such that none is scientifically "best."

This contrasts with the post-Prohibition era when there was a sharp contrast between the "modern" alcoholism-oriented approach based in science and the alternative alcohol-oriented perspective that was characterized as being based primarily on moral and religious beliefs. This leads to the tantalizing question of whether the long-term acceptance of the medicalized approach to alcohol problems was based more on its "form" as a scientifically oriented explanation than on its "content" as an intervention strategy.

In 1990, the competing models of the basis of and solutions for the alcohol problem are all based on scientific approaches. There is little doubt that the backgrounds and motives of the researchers and practitioners who focus on the availability of alcohol as a principal cause of alcool problems (and who are pejoratively dubbed "neoprohibitionists" by some advocates of medicalized interests) differ sharply from those of the temperance workers of the 19th and early 20th centuries. In a very real sense, the "battleground" once occupied by the "wets" and "drys" has moved from the evaluation of ideology to the evaluation of evidence. But in another sense, the battlefield is very foggy, for it is not clear what "problems" are really being addressed or what solutions are really being sought.

The Fading Utility of the Hidden Alcoholic Concept

A theme that is found in several of the more widely diffused journal articles with which Selden Bacon is identified (Henderson & Bacon, 1953; Straus & Bacon, 1951) is "hidden alcoholism." There is no certainty as to the origin of this concept, but it may be seen as considerably important in organizing theory and practice about alcohol problems since the repeal of Prohibition.

The idea that major and significant alcohol problems are socially "hidden" can be contrasted to the image of alcohol problems during the temperance era, roughly from the 1830s through the enactment of Prohibition. Here it seems that the obnoxious presence and blatant visibility of alcohol-related problems were a principal component in their cultural characterization. In the post-Prohibition period a massive amount of effort has been centered around a reconstruction of the image of alcoholism, an effort that has elsewhere been described as the construction of a "new epidemiology" (Roman & Blum, 1987). Part of this effort has required dismissal of the "skid row image" of the alcoholic, an image that may be a legacy of the temperance era, with the skid row bum portraying the long-term consequences of capture by King Alcohol. Many researchers would trace the first major scientific step in this image transformation to the study of alcoholism and social stability reported by Straus and Bacon (1951).

The contention here, however, is that while the concept of hidden alcoholism may have originally served political purposes in capturing both legislative and public imaginations, I suggest that it no longer may be an asset for social scientific study of alcohol issues. The arguments developed here are abbreviated, but hopefully may serve to stimulate others' interest in the sociology of alcohol-related science and knowledge.

Beyond the arena of alcohol studies, the concept of hidden deviance is persistent in social scientific work, despite its grounding in notions of functionalist sociology, which have otherwise lost much popularity as approaches to sociological analysis. In other words, the conception of hidden deviance implies an imbalance in the functioning systems of identification and detection, that presumably can be corrected with an enhancement or sharpening of these organizational strategies, with the hidden problem ultimately uncovered and eliminated in a "balanced" system. Within a conflict-oriented sociological analysis, "hidden deviance" may be seen as a description of definitions of conformity, propriety, and life styles differing across social segments that possess different degrees of social power. It would follow that "hidden deviance" in the vision of one group would be conforming and appropriate behavior to another.

Such an approach may fit well with a conflict theorist's view of some types of property crime as a means for creating a more just distribution of wealth and resources. It is difficult if not impossible, however, to find published advocacy of a perspective that hidden alcoholics should be ignored and allowed to go about what they may regard as an appropriate and normal life style. Indeed, the opponents of the alcoholism-as-a-disease perspective seem to ultimately offer their own suggestions for intervention and control. Certainly these suggested strategies are typically of a different magnitude and orientation than the dominant perspective, yet they implicitly recognize that "something" must be "done" to deal with this set of obviously destructive social and personal behaviors.

The persistent hidden alcoholic concept has two principal dimensions: a dominant notion associated with epidemiology and a less emphasized but equally important dimension associated with the principles and practices of Alcoholics Anonymous.

The dominant notion of hidden alcoholism centers on a very simple social fact that is probably well diffused among the general public: the numbers of alcoholics known to experience formal treatment and/or affiliation with Alcoholics Anonymous are tiny compared to the estimated prevalence of alcoholism and alcohol problems (Rubington, 1972). The gap between these two sets of figures (with some estimates listing as much as 75 percent of the alcohol-troubled population as untreated) is filled by alcoholics who are functioning in social roles and are essentially "on the hoof" in society.

Despite their apparent social integration, the hidden alcoholic imagery indicates that these people are in the throes of a life-threatening disease but are characterized by "denial." At the same time the consequences of the hidden alcoholics' drinking behavior create substantial costs for those in the families, social networks, workplaces, and communities where they are still actively performing social roles. Thus, at best, hidden alcoholics are menaces to themselves and to others.

Some form of epidemiological survey data underlies practically every contention of hidden alcoholism and, more recently, hidden abuse of illegal drugs. In other words, the "data" show a much higher prevalence of the problem than is evident in the records of treatment centers or estimates provided by AA. At the same time it is crucial to keep in mind the characterization of the bulk of "hidden" alcoholics and drug abusers as persons who do not want to be detected, and who use every means at their disposal to deny, manipulate, cover up, and otherwise protect their engagement in patterns of substance use that they have found rewarding.

Following this logic, it is important to ask how hidden alcohol and drug problems are measured in epidemiological surveys. While it might be ideal to search for "markers" of such behavior in surveys of untreated populations, such markers are equivocal and are in any event not conveniently utilized in surveys. Thus, the means for identifying alcohol and drug-related behaviors in field surveys is to ask directly about them, that is, how often and how much one consumes of certain substances and what consequences ensue from this consumption.

Such an approach does not square with the characterization of the substance-dependent individual embedded in a pattern of denial and cover-up, for why can one assume that such individuals would be candid with field interviewers? Thus, within this pattern of logic, it may be suggested that the surveys that are used by those in policy-oriented circles to demonstrate the existence of hidden problems are finding "something" out there in the untreated population, but this "something" does not closely fit these same proponents' characterization of hidden problems that could not possibly be revealed by this methodology.

A mirror image of the hidden alcoholic is offered by Rubington (1972), who makes the obvious but profound point that hiding is a reciprocal process, requiring the presence of seekers, not unlike the children's game. Such a conception was later elaborated as the social constructionist approach to social problems (Spector & Kitsuse, 1977). Thus, one might expect a rise in the numbers and proportions of the projected hiders as the numbers of seekers rises. If this is true, the hidden group should have become especially prominent as the population of seekers has grown over the past two decades in the United States.

Presumably, the huge numerical gap between identified and unidentified alcoholics could be used as an index of the effectiveness of the seekers' strategies. There can be little doubt that the notion of "hidden" deviance of any sort is related to political organizational goals. Those who propose that substantial segments of a problem category are hidden are also proposing means to ferret out these hidden deviants. In fact, it is likely that most descriptions of hidden deviance appear as part of expositions that describe means by which portions of this hidden sector can be "discovered" and processed in some fashion believed to be appropriate. A dilemma centers on a necessary balancing act: when making the case for organizational or programmatic effectiveness by describing gains in the uncovering of portions of the target population, group leaders are usually at the same time requesting additional resources to uncover the substantial proportions of the target group that are still hidden.

The second dimension of the hidden alcoholic concept is notably different. It is centered in Alcoholics Anonymous, with this usage of the concept reflected in the influential work of Beauchamp (1980). By definition, the anonymity of recovering alcoholics is guaranteed by AA, the traditions of which essentially forbid individuals from revealing their identity as AA affiliates per se. This anonymity has several implications.

It is clear that unless they choose self-disclosure, members of AA may be "hidden (recovering) alcoholics," possibly present anywhere and everywhere but known only to some segment of this "underground" community. Further, there is no doubt that a continuing emphasis on anonymity supports and encourages stigma. If the label of alcoholic were not stigmatic, then why cover it up? Clearly, there are many in AA who question the current relevance (and possible stigmatizing side effects) of the continuing emphasis on anonymity within this fellowship, yet there is no evidence of movement to alter the traditions or to introduce procedural changes such that "closed" meetings would be eliminated.

The broader significance of the AA-based version of hidden alcoholism is found in the array of "confidentiality" regulations that surround alcoholism interventions and treatment. On the one hand, such regulations might be seen as reflecting medicalization processes wherein confidentiality guarantees are part of medical treatment contracts. On the other hand, however, the emphasis on confidentiality within the organizational processing of alcoholic patients and its institutionalization into legal regulation could be seen as a direct transfer from the principles of AA. But there are multiple ironies in the fact that the social movement promoting "alcoholism as a disease like any other" (presumably with the goal of increasing humanitarian reactions and decreasing stigmatization) has also been behind the legislative enactment of confidentiality regulations.

The persuasion that the skid row bum was not representative of the American alcoholic, and that much or most of the American alcoholic population was hidden, seems to have succeeded by the 1990s. Hidden alcoholism and the persistent gap between the identified and hidden populations continue as bases for mandates that more needs to be invested in the efforts of the seekers. At the same time, those pressing for more resources for the seekers have to persuade decision makers that progress is being made, or that the gap between the two populations would be even greater if it were not for the seekers' continuing but underfunded efforts.

There are at least two ways in which the hidden alcoholism concept creates confusion and reduces credibility of the whole alcohol-problem endeavor. First, there is definitional ambiguity associated with alcohol problems, an issue partly addressed earlier in this chapter. Since there is no biological marker for the presence of alcoholism, its diagnosis must be generated by self-reports or combinations of self-reports and observations of behavior. It would seem that quantity-frequency measures of alcohol consumption would be adequate to define alcohol problems, but this is not generally the case. The inadequacy of drinking per se as a criterion rests more on people who do not drink large amounts but get into a lot of trouble than on people who drink great amounts and do not get into trouble.

Thus, defining alcoholism has come to be centered on impairments in role performance that are found among individuals who drink at least a modest amount. Role impairment measures are probably "harder" data than quantity-frequency of consumption if we take into account variations in body size, metabolism, tolerance, etc. The use of role impairment definitions allows interventionists to "break through" denial and avoid negotiation over the "how-much-is-too-much" issues surrounding consumption.

The means by which hidden alcoholism undermines credibility is through this sociological definition of alcoholism with its central focus on role impairment. This brings us back to a mini-version of sociological functionalism. The characterization of hidden alcoholics rests on the notion that they are still "functioning" in the community. Presumably, if they are not functioning, action of some sort would be necessitated. While it is possible to conceive of a small proportion of individuals who are not able to perform assigned social roles due to drinking and toward whom no action is taken, credibility is genuinely stretched if one is to believe that most of those in the "gap" group are alcoholics who are not functioning at levels acceptable to their significant others.

It can be argued from a public health perspective that excessive drinking is a significant health hazard that should be curbed if at all possible. "Heavy" drinkers without evidence of role impairment would likely be the target of some interventionists' claims that such individuals are "deeply into denial." Given our current level of knowledge of social processes, the role impairment definition of alcohol problems that we have just reviewed, and the typical lack of interest in prevention that is reflected in the health care system, the likelihood that such individuals would come to the attention of interventionists is not high.

The second mechanism by which the hidden alcoholism concept undermines the credibility of the alcohol-problem endeavor is through its implied characterization of the disease that afflicts the hidden alcoholic. As mentioned, among clinicians "denial" is a favorite descriptor of the active alcoholic in American society. Like the concept of the hidden alcoholic, which implies seekers, denial implies the existence of truth. Denial is conceptually close to lying. It implies that some undesirable attitude, trait, behavior, or condition is being denied—that is, the "truth." Thus, it follows from the attribution of denial that deniers "really" know that they have a drinking problem and may even "really" know that they are alcoholics.

Hidden alcoholism is not without its bright side for the interventionist community, and perhaps even for the research community. To account for the massive population of hidden alcoholics, the concepts of enabling and codependency have been in steep ascendancy among interventionists throughout the 1970s and 1980s. Those around the active alcoholic are described as reinforcing and perpetuating the disorder, thus explaining much of the hidden alcoholic phenomenon. In many instances these behaviors by the nonalcoholic significant others allegedly create needs for intervention and treatment.

Enabling and codependency are yet to be the subject of solid scientific investigation. Yet in the urgency of claims making and the marketing of treatment services, clinicians repeatedly affirm the reality of these phenomena as if they had been scientific discoveries. In order to be operative, enabling and codependency hinge upon the existence of hidden alcoholism. The massive diffusion of these logically problematic folk ideas, incorrectly cast as the results of scientific discovery, comprises the broadest example of the undermining of the "modern alcoholism movement" by the notion of hidden alcoholism and its accompanying baggage of ideas.

To conclude, the time may have arrived to retire the concept of hidden alcoholism, recognizing it to be a political tool that is not generally useful in advancing scientific understanding of patterns of chronic excessive drinking. Empirical reality seems better described as comprised of (1) persons without drinking problems, (2) persons whose drinking may exceed some objective standard but who are socially integrated, (3) persons whose drinking behavior is leading in the direction of social reactions that will precipitate possible referral and labeling, and (4) persons whose drinking behaviors have led others to take actions toward them that have resulted in the formal label of alcoholic. While Bacon's work can be associated with the notion of hidden alcoholism, he offers a significantly more important legacy in his urgings that alcohol studies need clean and clear definitions in order to be valuable to both scientific and interventionist enterprises.

Acknowledgment

Partial support from Research Grant R01-AA-07218 and Training Grant T32-AA-07473 from the National Institute on Alcohol Abuse and Alcoholism during preparation of this manuscript is gratefully acknowledged.

References

Beauchamp, D. (1980). *Beyond alcoholism.* Philadelphia: Temple University Press.

Berry, R., & Boland, J. (1977). *The economic costs of alcohol abuse.* New York: Free Press.

Blum, T.C., & Roman, P.M. (1989). Employee assistance programs and human resources management. In K.M. Rowland & G.R. Ferris (Eds.), *Research in personnel and human resources management* (vol. 7, pp. 259–312). Greenwich, CT: JAI Press.

Blum, T.C., Roman, P.M., & Bennett, N. (1989). Public images of alcoholism: Data from a Georgia survey. *Journal of Studies on Alcohol, 50,* 5–14.

Cahalan, D. (1970). *Problem drinkers.* San Francisco: Jossey-Bass.

Cahalan, D., & Room, R. (1974). *Problem drinking among American men.* New Brunswick, NJ: Rutgers Center of Alcohol Studies.

Department of Health and Human Services. (1987). Review of the research literature on the effects of health warning labels: A report to the U.S. Congress. (Contract No. ADM 281-86-0003). Washington, DC: National Institute on Alcohol Abuse and Alcoholism.

Fingarette, H. (1988). *Heavy drinking: The myth of alcoholism as a disease.* Berkeley: University of California Press.

Gusfield, J.R. (1981). *The culture of public problems.* Chicago: University of Chicago Press.

Henderson, R.M., & Bacon, S.D. (1953). Problem drinking: The Yale plan for business and industry. *Quarterly Journal of Studies on Alcohol, 14,* 247–262.

Jacobs, J. (1989) *Drunk driving.* Chicago: University of Chicago Press.

Jellinek, E.M. (1960). *The disease concept of alcoholism.* New Brunswick, NJ: Rutgers Center of Alcohol Studies.

Levine, H. (1978). The discovery of addiction: Changing conceptions of habitual drunkenness in American history. *Journal of Studies on Alcohol, 39,* 143–174.

Mosher, J.F., & Jernigan, D.H. (1989). New directions in alcohol policy. *Annual Review of Public Health, 10,* 245–279.

Mulford, H.A. (1984). Rethinking the alcohol problem: A natural processes model. *Journal of Drug Issues, 14,* 31–43.

Mulford, H.A. (1990). The extent and patterning of job-related drinking problems. In P. Roman (Ed.), *Alcohol problem intervention in the workplace: Employee assistance programs and strategic alternatives* (pp. 125–140). Westport, CT: Quorum Press.

Peele, S. (1989). *The diseasing of America.* Lexington, MA: D.C. Heath.

Roman, P. (1988a). The disease concept of alcoholism: Sociocultural and organizational bases of support. *Drugs and Society, 2,* 5–32.
Roman, P. (1988b). Growth and transformation in workplace alcoholism programming. In M. Galanter (Ed.), *Recent developments in alcoholism* (vol. 6, pp. 131–158). New York: Plenum.
Roman, P.M., & Blum, T.C. (1987). Notes on the new epidemiology of alcoholism in the USA. *Journal of Drug Issues, 17,* 321--332.
Roman, P.M., & Blum, T.C. (1991). The medicalized conception of alcohol-related problems: Some social sources and some social consequences of murkiness and confusion. In D. Pittman & H. White (Eds.), *Society culture, and drinking patterns reexamined.* New Brunswick, NJ: Rutgers Center of Alcohol Studies.
Roman, P., & Trice, H. (1968). The sick role, labeling theory and the deviant drinker. *International Journal of Social Psychiatry, 14,* 114–136.
Rubington, E. (1972). The hidden alcoholic. *Quarterly Journal of Studies on Alcohol, 33,* 667–683.
Schneider, J. (1978). Deviant drinking as disease: Alcoholism as a social accomplishment. *Social Problems, 25,* 361–372.
Seeley, J.R. (1962). Alcoholism is a disease: Implications for social policy. In D. Pittman & C. Snyder (Eds.), *Society, culture and drinking patterns* (pp. 586–593). New York: John Wiley and Sons.
Spector, M., & Kitsuse, J. (1977). *Constructing social problems.* Menlo Park, CA: Cummings Publishing.
Straus, R., & Bacon, S.D. (1951). Alcoholism and social stability: A study of occupational integration in 2,023 male clinic patients. *Quarterly Journal of Studies on Alcohol, 12,* 231–260.
Trice, H.M., & Roman, P.M. (1970). Delabeling, relabeling and Alcoholics Anonymous. *Social Problems, 17,* 468–480.
Williams, R.M. (1970). *American society: A sociological interpretation.* New York: Knopf.

CHAPTER 12

The Workplace as Locale for Risks and Interventions in Alcohol Abuse

WILLIAM J. SONNENSTUHL AND HARRISON M. TRICE

Despite 200 years of efforts to prevent alcohol problems, Americans have made little headway (Bacon, 1987). Hilton and Clark (1987), for instance, observe that between 1967 and 1984 the volume of drinks consumed by Americans did not change significantly, that the proportion of respondents experiencing any of nine problem consequences did not change significantly, and that there was a slight increase in the proportion of Americans that reported experiencing one or more alcohol dependency problems. Nor are drinking practices and alcohol problems evenly distributed across class structure or by age and sex: although drinking, sometimes fairly regular and heavy, is more prevalent among those of higher social statuses than those of lower statuses, drinking problems are more prevalent among the poor than the affluent (Cahalan, 1970; Cahalan, Cisin, & Crossley, 1969; Cahalan & Room, 1974; Knupfer, 1967; Mulford, 1964). According to Cahalan:

> Those of lower socioeconomic status and certain sociocultural groups (e.g., Irish-Americans and those of Spanish-speaking origin) tend to have a relatively high rate of problems in relation to the frequency of their drinking and the high amounts they consume. (1987, p. 369)

Although these data probably accurately reflect American drinking practices and problems, social scientists remain uncertain about what the inverse relationship between drinking practices and problems means (Straus, 1975).

Historically, community studies (e.g., Lynd, 1929; Lynd & Lynd, 1937; Hoover, 1990; Warner & Lunt, 1941; Warner & Lunt, 1942) have documented the relationship between social class and a wide range of social phenomena; in his 1944 monograph, Bacon urged social scientists to conduct in-depth community studies in order to understand the functional and dysfunctional value of drinking within American society. Unfortunately, social scientists have ignored his plan and we are, as a consequence, less wise than we might otherwise be about the relationship between drinking practices and problems within the context of social class. In addition, they have tended to focus almost exclusively on the dysfunctional, rarely examining such functional effects as how alcohol contributes to group solidarity (e.g., Mars, 1987), how alcohol promotes "time out" (McAndrew & Edgerton, 1969), and how social drinking "pays off" (Trice & Beyer, 1977). Because alcohol acts as a potent symbol for the shift from work to play (Gusfield, 1987), encouraging people to relax, have fun, and experience human warmth, its prohibition is difficult to imagine.

Preoccupation with the dysfunctional stems from puritanical America's temperance heritage—the symbolic crusades waged around alcohol by the upper classes to keep the lower classes in their place (Gusfield, 1963). In these crusades, alcohol functions as a powerful metaphor for defining what constitutes "right living" and how deviants ought to be managed. Depending upon historical circumstances, deviant drinking has been regarded as being sinful, criminal, or sick, and deviant drinkers have been shunned, punished, or treated (Conrad & Schneider, 1980).

This chapter reviews the research on drinking behavior and alcohol problems in the workplace. During the colonial era, drinking on the job was a common practice, but with the coming of the Industrial Revolution and the arrival of new immigrants to work in the new industries, this practice became the subject of periodic crusades (Ames, 1989; Staudenmeier, 1985, 1987; Trice & Schonbrunn, 1981). For instance, during the temperance era, many companies, concerned about the effects of drinking on productivity and pressured by moral entrepreneurs, adopted either dry-on-the-job policies or policies requiring that employees remain abstinent on and off the job. Between 1910 and 1920, some followers of Fredrick Taylor began to argue that use of alcohol on and off the job adversely effected workers' productivity. In this fashion, temperance was joined with the scientific management principles of Taylor and his followers. After passage of the Volstead Act, some companies such as the Ford Motor Company used prohibition as a means of enforcing paternalism, accusing those sympathetic to employee democracy of drinking and trafficking in alcohol. After repeal, Alcoholics Anonymous (AA) members, industrial physicians, the Yale School of Alcohol Studies (Henderson & Bacon, 1953), and the National Council on Alcoholism (NCA) worked with employers and unions to define alcoholism as a disease and to establish job-based alcoholism programs.

Predictably, some critics (Fillmore, 1989; Fillmore & Caetano, 1982; Wager, 1988) charge that workplace efforts to manage alcohol problems are directed at controlling low-status employees and denying them their rights.[1] For instance, althugh deskilling (Braverman, 1974) and drug screening (Sonnenstuhl, Trice, Staudenmeier, & Steele, 1987) may be more dangerous to employee rights than employee assistance programs (EAPs), Dixon argues without empirical evidence that EAPs

> help arm management with psychotechnologies to mystify and control worker discontent in the same manner that an earlier generation of community psychiatrists and psychologists armed the government so as to control explosive neighborhoods in the 1960s. What may appear as benevolent counselling and therapy initiated and supported at the job site or in the slum-based drop-in counselling center can conceal the more cynical intent to better "adjust" employees and impoverished minority groups to the alienating conditions of work or the structural problems of chronic poverty and racism. (1984, p. 50)

In a similar but less strident vein, Straus (1976a) summarizes industrial programs directed at alcohol as being

> aimed primarily at detection and rehabilitation or at modifying the drinker's behavior through education.... [T]hey do not address the more basic question of identifying what it is about work situations and work experiences that makes it meaningful for employees to drink too much or what programs of preventive intervention might modify these conditions.

Indeed, Levine (1978) argues that we have yet to articulate a working-class perspective of alcohol in modern society because, since the 18th century, policy discussions have been dominated by the world views of capitalists and the middle class. Lurie (1979), on the other hand, writing about Native Americans, characterizes drinking as the world's oldest on-going protest demonstration.

In this chapter, we examine these concerns by first reviewing the research on workplace risk factors that may make it meaningful for employees to drink and then examining the literature on workplace interventions. In doing so, we pay particular attention to the relationship of social class to risk and interventions.

Risk Factors at Work

We do not attribute alcohol problems primarily to causal factors in the workplace; rather, we assume that biological, psychological, sociological, and anthropological factors also play a part in their development. In addition, we assume that, when individuals come to work, they bring these predisposing factors with them and that workplace factors may either exacerbate or dampen them. From a review of the workplace literature on drinking and alcohol problems, we have identified four perspectives used by social scientists to explain the development of alcohol problems.[2] We use the term *perspectives* because these approaches are not significantly refined to justify using the term *theoretical model.*

According to sociological and anthropological literature that emphasizes the role of culture, drinking behavior is learned within the group and can be both functional and dysfunctional (e.g., Bacon, 1962; Bales, 1962; Heath, 1987; Pittman & Snyder, 1962; Snyder, 1978). The group establishes norms indicating how, when, and where one may drink; rationales for drinking; and social controls to ensure that members drink appropriately. Drinking may perform useful functions, such as easing group relations, relieving individual stress, and increasing group solidarity. However, drinking becomes problematic and dysfunctional when individuals exceed approved drinking norms, evade social controls, and disrupt the stability of group life.

Work organizations are cultural entities with their own drinking norms, social controls, and rationales. Our types relate to these cultural properties. The cultural perspective emphasizes the drinking norms that develop within a particular workplace, and we argue that either the occupational subculture, the administrative subculture, or both may promote heavy drinking, which leads to individuals' abusive drinking behavior and creates poor performance. The social control perspective emphasizes the relative absence of workplace sanctions that may create opportunities for heavy drinking and lead to alcohol problems. The last two perspectives—alienation and stress—argue that modern organizations create in individuals a sense of powerlessness or stress that workers learn to relieve through abusive drinking.

Workplace Cultures Perspective

The workplace can give rise to two kinds of drinking cultures: administrative and occupational. The workplace cultures perspective argues that such subcultures contain beliefs about what is and is not appropriate drinking behavior. In some work organizations, temperance is the norm (Staudenmeier, 1987); in others, either the administrative or the occupational subculture may promote heavy drinking, which can lead to alcohol abuse and dependence.[3]

Administrative support for heavy drinking may occur at several levels. In may be organizationwide, where virtually every occasion becomes an opportunity to drink. Conversely, heavy drinking may be confined to specific departments or to specific occasions.

Military organizations provide potent examples of organizationwide support for heavy drinking. Bryant (1974) reports that drinking plays an important ceremonial role in military life, where personnel are required to attend an endless round of parties and receptions. Molloy (1989) describes how specific merchant vessels came to be "alcoholic ships." The entire ship—the captain, supervisory staff, and unlicensed personnel—became locked into a drinking culture. Similarly, Pursch's (1976) lucid description captures the centrality of drinking rituals in naval life:

> In Naval Aviation,... we drink at happy hours, after a good flight, after a bad flight, and after a near midair collision (to calm our nerves).... We drink when we get our wings, when we get promoted (wetting down parties), when we get passed over to (alleviate our depression), at formal dining-ins, change of command ceremonies, chief's initiations, and at "Beef and Burgundy Night." At birthday balls we drink our door prize if we have the lucky ticket.... We "hail and farewell" frequently, and the first liquid that wets the bow of any newborn ship at its christening is champagne. Thus, we drink from enlistment to retirement and from teenhood to old age. (p. 1655)

In other organizations, administrative support for heavy drinking may be confined to specific departments or groups (French & Magee, 1972; Trice, 1965a). For example, Hayhurst (1938, p. 629) reports that industrial workers in certain departments were more prone to drink than those in others because they believed that it was necessary to imbibe alcohol to prevent the effects of poisons and dust to which the workers were exposed. In other instances, support for heavy drinking may be confined to specific occasions such as business lunches, conferences, and office parties. A national survey of 528 business executives, for example, found that 60 percent approved of drinking at a two-day sales conference (Roman, 1982). Tax deductions spur "business meals in which alcohol is equated with food" (Fillmore, 1981, p. 31). At business conferences, alcohol may act as a social lubricant. For example, the second author observed a "managerial retreat" in which drinking became a rite of integration. Concerned about departmental rivalries, a vice president took his feuding managers to a secluded lodge and insisted that they get intoxicated together, believing that their "common hangovers" would promote better communication and cooperation. The vice president eventually institutionalized this drinking rite, reverting to it whenever he felt cooperation became a problem.

Within some occupations, heavy drinking is not viewed as pathological; rather, it is seen as normal, conforming to group standards (Cosper, 1979). Generally, such occupational communities (Salaman, 1974; Van Maanen & Barley, 1984) not only approve of drinking but actively encourage it. For instance, longshoremen (Pilcher, 1972), construction workers (Applebaum, 1984), and railroad engineers (Gamst, 1980; Salaman, 1974) are all close-knit occupational communities with distinctive drinking customs.

Historically, among many occupations, drinking on the job was a common practice (Staudenmeier, 1985; Rorabaugh, 1979). For instance, Sullivan (1906) compared the drinking practices of 19th-century English dockers and coal miners. On the docks, alcohol was readily available and its use was supported by myths about the effectiveness of alcohol as a mechanism for alleviating fatigue. In the mines, however, alcohol was forbidden during the work day and myths supporting alcohol as a palliative did not develop among English miners. In contrast, during the frontier days of the American West, the miners developed a heavy drinking culture because of the "unstructured life the miners led . . . and . . . [because] . . . there was little else to do" (Winkler, 1968, p. 426). Similar drinking traditions grew up among other freewheeling 19th-century occupations, such as stage drivers, lumberjacks, river boatmen, and canal builders (Rorabaugh, 1979).

Today, many occupations continue to support distinctive drinking customs on and off the job. For example, Riemer (1976, p. 6) reports that "in many ways drinking is built into the work culture of the building trades and concomitantly into the entire building process. . . . " Sonnenstuhl and Trice (1987, pp. 233–234) describe the centrality of drinking in the Tunnel and Construction Workers Union—the "Sandhogs." Their occupational culture

> reflects both the Irish drinking heritage of many Sandhogs and the heavy drinking practices generally associated with mining. . . . Bars play an important role in the Sandhog drinking culture, functioning as community bulletin boards and as places to unwind. . . . The easy camaraderie of such drinking groups provides unemployed members with information about work while further cementing gang solidarity. . . .

Similarly, LeMasters (1975, p. 163) observes of construction workers:

> [T]he tavern (or bar) provides a "cover" or "protective coloring" for the man or woman who is already drinking too much and is in danger of becoming an alcoholic.

Epidemiological data also support the main thrust of the cultural model. Trice (1962, p. 509) conducted two studies of the work histories of AA members ($N = 632$). He reported that there were indeed occupational differences in the manner in which alcoholics in higher-status occupations and those in lower-status occupations drank with fellow workers in off-the-job situations. Higher-status workers tended to drink "normally," waiting until they were separated from work associates to drink heavily; lower-status workers, however, drank in an uninhibited fashion, becoming drunk with coworkers. Hitz (1973, p. 503), in a survey of occupational drinking practices and problems, suggested that people predisposed to drinking problems may gravitate or drift into occupations with heavy drinking practices, making for a more pronounced drinking culture than in other occupations. Similarly, Plant (1978, 1981) found that "high-risk occupations" attract recruits who are heavy drinkers and that those who dislike the heavy-drinking culture leave high-risk occupations and return to normal drinking.

Whitehead and Simpkins (1983, p. 477) reviewed 234 studies of 56 occupations, concluding some were at "high risk of alcoholism." High-risk occupations included those labeled "entertainment," Army—based in the United States, Navy—carrier based, pipeline construction, Navy—shore based, Marine Corp, oil rig work, Army—based overseas, alcohol beverage work, salesmen on the road, seafaring occupations, and tavern keepers.

Seeman and Anderson (1983) sought to test the impact of social support on the development of drinking problems. To the researchers' surprise, they found that greater involvement in social support networks was significantly correlated with more drinking and drinking problems, even when work was intrinsically rewarding for workers. These findings suggest that many work-based social networks—especially occupational communities—may encourage the use and abuse of alcohol rather than discourage it.

Social Control Perspective

The social control model predicts that those characteristics that weaken worker's integration into, and regulation by, the work organization are likely to put them at risk of developing alcohol problems. This persepctive is articulated by Trice and Roman (1978, pp. 101–102), who argue that there are two general types of workplace risks that lessen social control: the absence of surpervision and low visibility of job performance. Empirical instances of this model are: (1) work roles with little or no supervision; (2) work roles with little or no interdependency with other roles; (3) work roles with low visibility of performance; and (4) work roles that call for frequent geographical mobility, shift changes, and frequent changes among fellow workers and formal supervision. Mandatory retirement and unemployment, by loosening individual ties to the workplace and its control mechanism, also contribute to alcohol problems.

A number of studies suggest empirical support for the social control model. For instance, Trice (1965a, 1965b) studied a large utility and discovered that employees diagnosed as alcoholic were concentrated in blue-collar jobs engaged in mobile outside work. He also found that employees whose work was closely intertwined with the work of fellow employees were more likely to be diagnosed as alcoholic than those who were free to arrange their own work. That is, where employees were working closely with others or were closely supervised, their deviant drinking was spotted and diagnosed as alcoholism, but where they worked independently and with little supervision, their drinking problems persisted.

Mannello and Seeman (1979) found that railroad workers who had alcohol problems spent more time away from home on the job, and had less contact with one's immediate supervisor than did nonproblem drinkers. Ames and Janes (1987) found that the drinking culture of assembly line workers was able to flourish because supervisors, fearing union grievances, were unwilling and often unable to exercise their authority. Trice and Beyer (1981) also found that geographical mobility was associated with problem-drinking employees in a large manufacturing corporation. Similarly, in a study of federal employees, Roman (1981) tested for the presence of interdependence of tasks, on-the-job mobility, and closeness of supervision as risk factors relative to problem drinking. He found that task interdependence and on-the-job mobility proved to be significantly correlated with the identification of alcohol problems. In this study, however, closeness of supervision was not significantly related to the identification of alcohol problems.

Other occupational studies suggest that both the absence of supervision and low visibility of job performance "give an alcoholism-prone personality a job situation within which his tendency can easily develop" (Hitz, 1973, p. 505). Clinton (1977, p. 42), in an ethnographic study of carpenters, reported that alcoholic-prone members of the union tended to be attracted to "brush jobs," that is, a job carried out more than 40 miles from the hiring hall. And Murray (1975, p. 24) insists that "mobility and consequent estrangement from the stabilizing influence of homelife" characterize such relatively high-abuse occupations as entertainers, commercial travelers, seamen, and persons in the armed forces.

The absence of supervision and low visibility of job performance also play a role in the development of alcohol problems among high-level executives. For instance, ambitious young executives, in order to gain a competitive advantage, may cover up and even encourage heavy drinking by seniors, as was the case in an ethnographic study of two companies with alcoholic presidents (Trice & Belasco, 1970). Similarly, Dubofsky and Van Tine (1977) describe how, when John L. Lewis was vice president of the United Mine Workers Union, his agents encouraged the union president, Frank Hayes, to drink alcoholically so that Lewis could act as president.

Shift work[4] also appears to contribute to the onset of drinking problems. Shift workers follow a schedule that conflicts with most of the working population; they are "personally out of phase with the main community... " (Melbin, 1987, p. 103). A household survey by Smart (1979) conducted in a community near Toronto revealed that the highest rates of alcohol problems were among shift workers and the unemployed.

Mandatory, even voluntary, retirement from work falls within the purview of the social control model because with retirement comes a lessening of group support and reinforcement for what is and is not appropriate use of alcohol (Minnis, 1988). Retirement for many brings not only a severe change in income, but also a sudden and often unplanned loss of structure and purpose in one's daily routine. When compounded with the death of a spouse and the loss of work peers, retirement can bring with it a severe sense of psychological isolation. One type of elderly alcoholic stems from the young problem drinker who as he or she grows older enters into alcoholism. Some researchers (Schuckit & Miller, 1976), however, assert that there is a prominent type of older alcoholic who begins heavy drinking quite late in life. For instance, Rosen and Glatt (1971) found that one-third of the patients under psychiatric and geriatric care were excessive drinkers and concluded that retirement ranked higher than bereavement as a precipitating factor in their abusive use of alcohol.

Occupational obsolescence has many of the same effects as mandatory retirement upon alcohol use. Because of rapid technological change, knowledge and skills acquired during formal education or training may simply not last through a career. Consequently, many workers now become obsolete because they cannot perform new tasks that come to be attached to their work roles. Although the literature does not contain relevant studies that provide statistical estimates, a case study by Trice and Belasco (1970) serves as an example:

> "Engineer D" was superior in his college work and had many job offers upon graduation. During college he drank with fellow students, but reported no role impairments. For approximately seven years after Engineer D was hired, his specialty was central in organizational planning and production. He advanced rapidly in salary and organizational status. During this period, his drinking was normal. Overall company planning, however, soon moved to areas outside his specialty. Engineer D tried, unsuccessfully, to become knowledgeable in the new areas. As his obsolescence became clearer, his drinking became heavier. He was then transferred to an administrative post, a move which subtly constituted a demotion. Subsequently, Engineer D's addiction to alcohol developed.

Telematics, the merging of telecommunications and computers in order to store and manipulate data and information at high speeds with "no respect for physical distance, makes it possible for work ranging from routine clerical tasks to professional writing to occur in the home" (Becker, 1986, p. 200). Obviously, such a technological development would largely reduce the influence of work settings. More likely, with the emergence of "flex-time" in many work places, the structural influences of the workplace will be reduced. Under these arrangements, employees have flexible opportunities to arrange their work schedules as personal needs dictate.

Considerable evidence suggests that unemployment is a definite risk factor in the development of alcohol problems. With unemployment, all direct influences of the workplace are removed. Kasl (1973, p. 512) sees unemployment as "a major life change, most frequently an involuntary 'exit' from the work environment. . . . " Among high-status managerial persons, the laid-off employee tends to feel as if he or she has no future, or has been deeply damaged, describing the experience metaphorically as "losing a leg; dying professionally" (Kaufman, 1982, p. 12).

Unemployment was emphasized in a 1977 study done in Buffalo, New York, and its environs. For males employed full or part time, the percentage of heavy drinking was 40 percent and 43 percent, respectively. But the rate of heavy drinking among laid off males looking for work was 56 percent. Among females the comparable percentages were 10 percent of full-time employees, 10 percent of part-time employees, and 25 percent of those unemployed but looking for work (Barnes & Russell, 1977). Similar findings have been reported by Ojesjo (1980) and Jacobson and Lindsay (1980). As stated by O'Donnell, Voss, Clayton, Slatin, and Room (1976, p. 111) in a study of nonmedical use of psychoactive drugs in the military, "a regular job may serve as a restraining influence on the extent to which men use drugs, or . . . users are less likely to seek or find full-time employment."

Brenner (1973), in a landmark epidemiologic study, clearly demonstrated a distinct relationship between the state of economy and mental illness of all kinds, including alcoholism. Later he reported that national increases in unemployment are "consistently followed within 2 or 3 years by increases in cirrhosis mortality rates. . . " (Brenner, 1975, p. 1282).

Within the past decade, the severe impact of hostile takeovers of companies by "corporate raiders" has drastically underscored the sudden and traumatic removal of employees from the workplace:

> From the available evidence, a conservative conclusion is in just the four years between 1983–1987, well over two million people saw their jobs disappear or deteriorate as a result of takeovers, forced mergers, and other restructuring. During this same time, nearly 10,000 companies traded hands, impacting the millions of workers and managers on the payrolls. (Hirsh, 1987, p. 30)

Panic, depression, and fear are constant results among those who did not possess "golden parachutes." It is not unreasonable to assume that alcohol use and abuse accompanies these severe jolts.

The Alienation Perspective

Ever since the Industrial Revolution disrupted traditional community life, there have been constant criticisms of the secular, highly urbanized and industrialized life that replaced it. Marx argued that, through work, human beings found fulfillment and that the new industrial order alienated workers from their labor, producing in them a sense of meaninglessness, withdrawal, and passivity. Alienated work, then, which is not intrinsically satisfying, prevents workers from achieving their full potential, and is "typical of modern society" (Seeman, 1967, p. 273). Moreover, alienation produces "a consequent decline in physical and mental health, family stabilty, community participation and cohesiveness, and balanced political attitudes, while there is an increase in drug and alcohol addiction, aggression and delinquency" (O'Toole, 1974, p. xvi).

According to the alienation perspective, work roles that lack creativity, variety, and independent judgment and that are not satisfying create in workers a sense of powerlessness that they allegedly learn to relieve through drinking. Powerlessness results because work is assumed to be the central life interest of employees, providing them with meaning and a sense of power. When this central life interest is thwarted, they are said to turn to drink as a way of compensating for their loss. Accordingly, work environments that provide opportunities for independent judgment, thought, and initiative will be beneficial for health and reduce alcohol problems.

Empirical data for the alienation perspective are sparse. Parker and Brody (1982) conducted a household telephone survey of 564 employed women and 484 employed men in the Detroit area in order to study the workplace risk factors related to alcohol problems. They looked at such risk factors as job visibility, freedom from supervision, job complexity, time pressure, availability of alcohol, drinking status of respondent's associates, and job stress; they report that only those factors associated with alienation (i.e., job complexity and job stress) were related to alcohol problems. Seeman and Anderson (1983) argued that most conceptions of alienation are too simplistic and hypothesized that three domains of alienation were related to drinking problems: alienating work experience, a general sense of powerlessness, and general social isolation from supportive networks. They found, on the one hand, that a general sense of powerlessness was consistently associated with heavier drinking and problem drinking. On the other hand, none of the work experiences, such as job satisfaction, job complexity, or level of rewards, were significantly associated with drinking patterns. Moreover, a high degree of general social network experience, including being a part of an occupational community, failed to alleviate the overall sense of powerlessness.

Other researchers have also found a sense of powerlessness to be associated with drinking behavior. McClelland and his colleagues (1972, pp. 334–335), for instance, state, "Men drink primarily to feel stronger; those for whom personalized power is a particular concern drink more heavily." Levy, Reichman, and Herrington (1979) suggested from their personality data that alcoholics perceive themselves as being significantly less powerful than nonalcoholics.

Castor and Parsons (1977) also interpreted their data as indicating a general feeling of powerlessness among male alcoholics.

Researchers have also examined the impact of routine work on drinking patterns. Widick (1976) and Runcie (1980) report on drinking among automobile assembly line workers, suggesting that abusive behavior is related to routine work and boredom. For instance, 40 percent of the workers with whom Runcie worked drank alcohol at lunch time. In this same vein, Hingson, Mangione, and Barrett (1981, p. 735) reported that boring tasks were significantly and positively related to their measures of drinking behavior. Archer (1981, p. 64), however, found little support for the hypothesis that boring work contributes to alcohol problems. She argued that alcoholic workers (as compared to nonalcoholics) experienced themselves as having a greater sense of personal power over their jobs and that they had not lost hope for future promotions or occupational mobility.

Because the alienation perspective emphasizes deficiencies in opportunities for creativity, variety, and individual judgment in job roles as being related to drinking behavior, available studies have operationalized alienation by means of "job complexity" indexes. Unfortunately these studies have produced mixed results. For instance, Seeman and Anderson (1983, p. 72) found such a high positive correlation with socioeconomic status that the variable was confounded. They reached the conclusion that "powerlessness is consistently related to drinking, while substantive complexity is not." Parker and Brody (1982, p. 120) found that job complexity accounted for "some of the variance in alcohol problems," although it actually accounted for less than "never married" among women, and the relationship was more pronounced among men than women in the sample.

Likewise, job satisfaction has been explored within the alienation model, producing mixed results. Mangione and Quinn (1975) examined the relationship between subjective reports of job satisfaction and drug use at work and found a significant association between drug use and low job satisfaction. Unfortunately, these researchers did not distinguish between alcohol and other drugs in their data. Similarly, Herrington, Levy, and Reichman (1980) matched alcoholics and nonalcoholics, and concluded that the latter not only had more job satisfaction, but they also had higher perceptions of success, goal attainment, and job liking. Archer

(1977, p. 1981), however, found no differences on job satisfaction indexes between alcoholics in a treatment center and a matched comparison group of nonalcoholics. Similarly, Seeman and Anderson (1983) found no correlation between job satisfaction and their three measures of drinking behavior—average amount per occasion, how often per month respondents drink, and drinking problems. Also, Newcomb (1988, p. 94) reported on indexes into job satisfaction and drug abuse among young, recent entries into the job market. He concluded that "the relationship between job satisfaction and any type of drug use is not a strong association.... " Earlier, Hardy and Cull (1971), however, had also compared alcoholics and nonalcoholics and reported that nonalcoholics seemed to be better fitted to and satisfied with the job they held than were alcoholics.

In a macro sense, the study by Steele and Hubbard (1985) of management styles and perceived substance abuse also related to the general job satisfaction perspective. Using data from a random sample of employees from seven corporations, they reported that to the extent the workplace values its empoyees, improves working conditions, and appears to be concerned with the welfare and happiness of its workers, to that extent their employees viewed substance abuse to be of less impact than if the opposite climate prevailed, that is, "a punitive reaction to the interests, desires, and contributions of employees... " (p. 283).

The Work Stress Perspective

The work stress perspective focuses upon workplace experiences and events that become translated into life strains. Unlike the alienation perspective, this perspective does not rest on beliefs about the intrinsic nature of work in modern society—such as it being intrinsically dissatisfying. Rather, the work stress perspective addresses those factors that come from the immediate realities of work life. Conceptually, the work stress and the alienation perspectives are distinct; however, in their operationalization the two often become blurred (e.g., Seeman & Anderson, 1983; Parker & Brody, 1982). Barley and Knight (1991) claim that the term *stress* has lost its original research meaning and that in everyday usage it now stands for workers dissatisfied with work. If this is true, use of the concept of stress in research must be reexamined.

The work stress perspective contains a wide range of explicit workplace stressors. These include physical properties of the working environment, changes in job content, machine pacing, monotony, boredom, role conflicts, degrees of control over work processes, work overloads or underloads, extensive overtime, inequality of pay, and job complexity. Such workplace experiences do not necessarily impact on employees directly, but rather they may function to bring together and concentrate earlier emotional distresses, giving them new and more intense meanings (Pearlin, Lieberman, Menaghan, & Mullan, 1981) that workers may attempt to relieve through abusive drinking. Regular and persistent requests to work overtime may be especially stressful. Not only is this risk a pernicious one, but it is attractive because of the almost irresistible attraction of additional, often needed income.[5]

A number of studies suggest the relationship between work stressors and the development of drinking problems. For instance, Trice and Belasco (1970, p. 222), in their study of the work histories of AA members, found that college-trained subjects linked the beginning of their alcoholism symptoms to career commitment. Sherlock (1967, p. 196) reported a similar pattern of onset of drug abuse behavior among physicians and nurses. Margolis, Kroes, and Quinn (1974, p. 660) reported that "escapist drinking" was moderately but significantly associated with role ambiguity, overload, and job insecurity. In a study of Australian workers, Ferguson (1974) found a distinct relationship between the severity of deadlines and variance in workloads and heavy, habitual drinking. Parker and Brody (1982, 120) also report that such job stress items as frequency of feeling tense, worried, or upset about job circumstances accounted for some of the variance reported in loss of control, dependency on alcohol, and alcohol-related job problems. Hingson et al. (1981) reported a positive correlation between stress and measures of drinking. Conway, Vickers, Ward, and Rahe (1981), however, found no association between subjective stress indicators and alcohol use among 34 naval company commanders. Oddly, in this study, alcohol use was significantly higher during low-stress weeks, probably because during high-stress weeks the commanders were less likely to leave their bases.

Fennell, Rodin, and Kantor's (1981) study is a good example of the stress perspective. They used a variety of work stresses as independent variables (i.e., not enough information, can't see results, not enough time, conflicting demands, promotions unfair, no help from supervisor or coworkers, and coworkers are incompetent) and related them to seven reasons for drinking alcohol (e.g., to relax, to forget about the job, to forget worries etc.). They found that a worker in their sample was much more likely to state that a particular reason for drinking is important if he or she experienced any one of the eight different work setting stresses.

Another empirical expression of the stress perspective comes from research that conceives of status inconsistencies as a source of stress. Parker (1979) did a careful, detailed analysis of status inconsistencies using data from a 1974 survey of adult drinkers in metropolitan Boston. He collected data on ethnicity, occupation, and education. The results suggested that status inconsistency does predict drinking abuses and "needs to be seriously considered by analysts of drinking behavior" (p. 93). Somewhat related are the findings of Trice and Roman (1971, p. 195) relative to the "left out syndrome." They found that employees who, for any reason, were not included in the training for the performance of basic changes in office technology were at higher risk of psychiatric disorders of a wide variety, including alcohol abuse.

Social Class and Risk

Our review of risk factor research suggests that administrative and occupational subcultures define for their members how, with whom, and for what reasons one should drink, and either or both subcultures may support norms contributing to alcohol abuse and dependence. The social control perspective postulates that gaps in the organization's social control structure contribute to drinking problems—especially when employees are predisposed to them—by creating opportunities to use alcohol, as when they enable persons to work unseen by or apart from their coworkers and supervisors. The alienation and work stress perspectives view a variety of conditions, including boring work and status inconsistency, as factors in psychological and physiological distress, which employees may seek to alleviate through drinking. Overall, the perspectives suggest that alcohol problems are related to the type of drinking culture in a workplace, the degree to which the organization constrains employees not to abuse alcohol, the feelings of alienation and stress resulting from how work is organized, and the rationales workers devise for drinking.

At first blush, it might seem that all these factors are arrayed against lower-level workers. For instance, occupational cultures appear to be more heavily represented in the research literature than administrative cultures, and lower-status workers appear more likely to feel powerless about their work than higher-status workers. However, a closer look at the research reveals that these factors have been correlated with alcohol problems in both lower- and higher-status workers. Both administrative and occupational subcultures can construct heavy drinking norms, develop alienation and stress rationales for such drinking, and create social controls that encourage rather than discourage such drinking behavior. What these data do not tell us is how such drinking behavior becomes defined as an alcohol problem in the workplace and whether or not lower-level employees in comparison to higher-status ones are disproportionately identified as having alcohol problems. In order to address these questions, we turn to a brief review of the workplace interventions that have been used to identify problem drinkers and change their behavior.

Workplace Interventions

Employee assistance programs, health promotion progams (HPP), and quality of work life (QWL) programs have all been proposed as workplace interventions for the reduction of alcohol problems. Although there is a growing amount of research on employee assistance programs, there is relatively little information on the efficacy of HPP and QWL initiatives (Sonnenstuhl, 1988). Consequently, we shall briefly describe what is known about these two interventions and focus upon EAPs because that is where the data are. For each intervention, we shall briefly describe the way it works and whatever data are available on its use by social class.

Quality of Work Life

Quality of work life advocates believe that alcohol problems are a form of maladaptive coping response to such conditions as job insecurity, hazardous work, role stress, nonsupportive jobs in which workers get little or no feedback, "work addiction," and occupational obsolescence (O'Toole, 1974). Consequently, they advocate restructuring work in order to reduce the effects of these alienating and stressful work conditions and, ultimately, employees' need to drink. Restructuring takes a variety of forms, such as job enlargment, job enrichment, performance appraisals, employee meetings, labor-management committees, quality circles, suggestion boxes, opinion surveys, flexible schedules, team building, semiautonomous teams, action teams, and parallel organizations (e.g., Cummings & Molloy, 1977; Ozley & Ball, 1982; Parker, 1985; Stein, 1983). Generally, these activities are initiated by management, which then solicits union cooperation and encourages the voluntary participation of all employees. Parker (1985) argues that union cooperation in QWL programs amounts to cooptation of the collective bargaining system and that employees' voluntary participation creates divisiveness among union members.

Although the alienation and stress perspectives above suggest that a significant relationship may exist between working conditions and drinking behavior, QWL evaluators have thus far failed to include drinking behavior as a variable in their outcome studies. Perhaps this stems from the stigma associated with the study of alcohol problems, or simply from the predilection of QWL advocates to focus on job satisfaction, productivity, and other global measures of well-being. Whatever the reason, the belief that changing the work organization will alleviate drinking problems remains an untested hypothesis, albeit one grounded in the research on risk factors, and nothing is known about the impact by social class.

Health Promotion

Health promotion or wellness programs seek to change employees' unhealthy living habits, which supposedly put them at risk of contracting chronic diseases, by educating them about the "right" way to live (Parkinson et al, 1982; Conrad, 1987a, 1987b, 1988). Like QWL practitioners, HPP advocates emphasize employees' voluntary participation in the educational programs; however, unlike them, they seek to attract volunteers through fitness and exercise programs, hoping that employees who become involved in these activities will wish to participate in others (Conrad, 1988). Health promoters treat alcohol use as a factor increasing one's risk of death from heart disease, cancer, stroke, accidents, pneumonia, diabetes, cirrhosis, suicide, and homicide (Fielding, 1984a, 1984b), and propose to manage the dangers by educating employees about the "correct" ways to drink. That is, they seek to prescribe for employees, within the context of the scientific and medical literature, the appropriate norms and rationales for drinking.

A review of the literature turned up only one evaluation study of a work site health promotion program's efforts to teach employees to drink. This was the Take Charge program conducted and studied by Shain, Suurvali, and Boutilier (1986, p. 86), which was designed to reinforce "life style supports for moderate (low-risk) consumption of alcohol" and reduce "high risk consumption." Although they report favorable outcomes, the study suffers from serious flaws: (1) a sizable number of employees who chose not to participate, (2) a high dropout rate for those who did participate, and (3) a self-selection bias (that is, those who came already intended to reduce their alcohol consumption when they came to the program).

Overall, HPP's impact on alcohol problems remains highly speculative and experimental. Generally, they attract the conspicuously well—those who already exercise, do not smoke, and use alcohol moderately (Conrad, 1987a, 1987b, 1988; Fielding, 1984b), and those who participate tend to be high-status rather than low-status employees (Conrad, 1987b). In addition, there is little support for the motion that participation in one health promotion activity automatically causes one to participate in others or assume a healthy way of life (Conrad, 1988).

Employee Assistance Programs

In contrast to health promotion and quality of work life programs, there is relatively more information on employee assistance and its impact on alcohol problems (Sonnenstuhl & Trice, 1990). Unlike these other strategies, EAPs are designed to combat alcohol problems and to follow the standards of industrial and labor relations.[6]

EAPs recognize that, because America is a polyglot of cultures, arriving at precise definitions of appropriate drinking and alcohol problems is problematic. Consequently, they take a pragmatic approach to these questions. They do not support temperance. Rather, they prescribe that it is acceptable to drink so long as the drinking does not adversely affect job performance. Conversely, they proscribe drinking that interferes with work. The dimensions of job performance are: (1) quality of work; (2) quantity of work; (3) attendance at work; (4) knowledge of the job; (5) dependability; (6) initiative; (7) cooperation; and (8) need for supervision (Glueck, 1978, pp. 229–300).

EAPs use the job performance standard because it is the standard of industrial jurisprudence upon which both labor and management have been able to agree (Elkouri & Elkouri, 1985). Consequently, drinking behaviors are defined as follows (Trice & Roman, 1978): "Normal drinking" does not exceed community standards, unduly alter behavior, and interfere with effective and efficient work performance. "Abusive drinking" both exceeds the bounds of community definitions of appropriate drinking behavior and impairs job performance. The EAP strategy assumes that with prompt feedback from one's own supervisor and/or peers, the abusive or problem drinker will improve his or her performance and return to normal drinking. "Alcohol dependence" is defined as physiological loss of control over drinking and is generally indicated by the employees' inability to control their drinking and maintain an acceptable level of job performance. Those employees who are unable to return to "normal drinking" when given prompt and consistent feedback from supervisors and peers are assumed to be alcohol dependent, and they are encouraged to seek medical help and/or the support of Alcoholics Anonymous.

Within the context of employee assistance, the definition of drinking behavior as "normal," "abusive," or "addictive" grows out of the interaction between supervisors, coworkers, and the suspected employee. Indeed, the line between prescribed and proscribed drinking behavior is negotiated so that the employee comes to see himself or herself as drinking normally, abusively, or addictively. Moreover, in order to protect the employee from being railroaded into prematurely labeling himself or herself as an alcoholic, the interaction between supervisor and suspected employee follows the industrial jurisprudence principle of progressive discipline and is called "constructive confrontation."

The constructive confrontation strategy requires that the supervisor hold a number of discussions with an employee whose performance is unacceptable. In the confrontational part of the initial discussion, the employee is given the specifics of his or her unacceptable work performance and told that continued unacceptable performance is likely to lead to formal discipline. In the constructive part, employees are reminded that practical assistance is available through an employee assistance program. Subsequent steps in the process depend on the response of the employee. If performance improves, nothing happens; if unacceptable performance continues, several more informal discussions follow.

The constructive part of the informal discussions (1) expresses emotional support and group concern about the employee's welfare; (2) emphasizes that employment can be maintained if the employee performs better in the future; and (3) suggests an alternative course of behaviors that the employee can take to regain satisfactory work performance. The confrontational part of such discussions (1) reiterates expectations of work performance; (2) reminds the employee that he or she is not fulfilling these expectations and that formal sanctions will follow if expectations continue to be violated; and (3) establishes some distance between the employee and other employees who are meeting expectations—thus setting the stage for further sanctions, if needed.

The constructive confrontation strategy is designed to overcome the psychodynamics of denial that characterize alcohol problems (Trice & Sonnenstuhl, 1991). The psychodynamics of denial are: (1) internal feelings of guilt and self-hatred for excessive drinking and associated behaviors; (2) defense mechanisms, especially denial and rationalization, that are activated by the ego from both internal and external criticisms; and (3) interpersonal tactics of manipulation and projection used to counter attempts by others to stop them from drinking (Denzin, 1986). Constructive confrontation overcomes denial by bringing reality to the fore and keeping it in focus. This is done by reiterating that job performance is expected, bringing progressive sanctions to bear upon deviations from job performance, and constructively providing employees with opportunities for treatment.

In attempting to overcome denial, however, some may fear that managers will go too far. Some data suggest that supervisors of lower-status employees use constructive confrontation more often than supervisors of higher-status employees use the strategy. According to Trice and Beyer:

> Apparently, the higher prevalence and visibility of problem drinkers at the lower-status job levels makes for policy salience among supervisors of such employees. At the same time, these managers probably experience more social distance between themselves and their subordinates, facilitating policy use. This relative readiness to use the policy among supervisors of low-status employees is further reinforced by a few perceived risks in its use and by relatively more rewards. (1977, p. 65)

In addition, data from two studies (Beyer & Trice, 1984; Trice & Beyer, 1984) suggest that in well-implemented programs supervisors do not use the strategy indiscriminately against lower-level employees and that punitive forms of discipline, such as suspension and written warnings, are counterproductive. According to these studies, those supervisors who confronted employees at the earliest signs of deteriorating job performance, who used both confrontive and constructive topics in talking with employees, and who did so consistently whenever job performance problems arose were able to improve their employees' job performance and encourage those in need of medical care to seek help. Those supervisors who were merely confrontive and punitive were notably unsuccessful in helping their employees to either improve job performance or seek help.

In addition, labor arbitrators, the final judges of fairness, increasingly are supportive of EAPs and the constructive confrontation process (Denenberg & Denenberg, 1983; Trice & Sonnenstuhl, 1990; Trice & Belasco, 1966). Traditionally, arbitrators who were forced to decide whether employers had treated alcoholic and emotionally disturbed employees fairly, simply asked whether the employees had been disciplined in accordance with management and labor's collective bargaining agreement. If the answer was "yes," the arbitrators let the discipline stand; if "no" they reversed management disciplinary action. In some instances, arbitrators, who considered alcoholism or emotional disorders as illnesses, would reverse management's disciplinary actions with the proviso that the employees accept treatment and resolve their problems. Among today's arbitrators, EAPs are perceived as a middle ground between these two extremes. Arbitrators look favorably upon the fact that employees are disciplined on the basis of performance and that at each step of the progressive discipline process, employees are offered help. If these conditions are met, arbitrators will generally let a disciplinary action stand because the employers, by offering help, have met their obligations to the employee.

Although supervisors' use of the constructive confrontation strategy has received the bulk of attention, the strategy has also been implemented by coworkers, colleagues in occupational associations, and union peers (Bissell & Haberman, 1984; Sonnenstuhl, Staudenmeier, & Trice, 1988; Sonnenstuhl & Trice, 1987). Beyer and Trice (1982, p. 192) have referred to these as "informal pressures from fellow workers that define the lower limits of acceptable behavior." These interventions could be referred to as "peer constructive confrontations," and have in some instances become formalized.

Operation Redblock, an EAP program in the railroad industry, is an example of the use of peer pressure in executing constructive intervention. Eichler, Goldberg, Kier, and Allen (1988, p. iii) studied the program in considerable depth, calling it a "peer referral program [that] is worker driven and worker run, having earned the support of both management and labor." The program was begun because severe disciplinary measures within the railroad industry had failed to deter substance-abusing employees.

Operation Redblock generates norms of acceptable drinking via development of education and peer pressure. It avoids seeking control of workers' off-the-job activities, and does not urge workers to become teetotalers.[7] It does, however, set up prevention committees made up of voluntary worker-based teams that carry out education and, when deemed necessary, constructive confrontation of coworkers. Approximately 90 percent of a committees' efforts goes into promoting the program as an educational device. The other 10 percent is typically spent on confrontations with coworkers who have been referred to the committee by fellow workers.

These confrontations take place away from coworkers and management in a neutral location, in order to protect employees' confidentiality. Members of the committee tell the referred worker that "reporting to work impaired and/or on-the-job substance use cannot be condoned or tolerated" (Eichler et al., 1988, p. 24), while urging the worker to contact the EAP counselor. A second appearance before the committee calls forth a stronger confrontation and the members insist that the offender work with the EAP counselor. According to Eichler et al:

> Co-workers are often the first to know that a fellow employee may have a substance abuse problem. Reporting the troubled individual to the prevention committee affords the co-workers a confidential way of dealing with the potential problem. Committee members, once apprised of the situation, are able to observe and/or approach the reported individual to explore the possibility that such a problem exists. (1988, p. 28)

Similarly, within the Tunnel and Construction Workers Member Assistance Program (Sonnenstuhl & Trice, 1987; Sonnenstuhl, Trice, & Evans, 1988), volunteer union members, who also belong to AA, constructively confront fellow workers with drinking problems. These miners, called "Sandhogs," work in close-knit gangs and face danger constantly. Consequently, gang members who evaluate one another's day-to-day performance are usually the first to know of a coworker's drinking problem. Using a mixture of constructive and confrontive elements, they tell their fellow miner that he is a valued union member, that he suffers from an illness, that recovery and a happy life are possible without alcohol, and that network members are ready to help him. The confrontive elements consist of pointing out how alcohol is destroying his family and work life and emphasizing that continued drinking will ultimately lead to complete loss of work and death.

EAPs construct drinking norms in two ways. First, they establish through their formal policies a clear statement of what constitutes alcohol abuse and dependence—drinking that disrupts normal job performance (Trice, 1962; Warkov, Bacon, & Hawkins, 1965).[8] Whether an employee chooses to drink is his or her own business, but if he or she chooses to drink, alcohol consumption should not interfere with job performance. Second, the constructive confrontation strategy provides a pragmatic technique for identifying problem drinkers and alcoholic employees and motivating them to return to "normal drinking" without jeopardizing their rights. In the give and take, employees learn what is prescribed and proscribed. Constructive confrontation assumes that, when constructively confronted, problem drinkers will stop drinking abusively and return to normal performance. In addition, constructive confrontation, through the use of progressive sanctions, moderates the psychodynamics of denial associated with alcoholism (Denzin, 1986), thus motivating the alcoholic employee to change his or her behavior either alone, with the help of the program's counseling resources, or with the assistance of mutual-aid and self-help groups, such as Alcoholics Anonymous.

Having sung the praises of employee assistance, however, it is appropriate to add a caveat. Many programs labeled as "employee assistance" do not promote the constructive confrontation strategy with its industrial jurisprudence safeguards; they rely instead upon employees' willingness to "self-refer" to the program where they are provided with professional counseling. There is little research available on these programs; consequently, we know nothing about their effectiveness at managing alcohol problems or protecting employee rights. Sonnenstuhl (1986), drawing on ethnographic data, found that one such program tended to diagnose and treat sex discrimination problems as stress-related illnesses rather than refer the women either to the company's internal arbitration or affirmative action programs. Consequently, our remarks do not apply to these programs, which are more appropriately identified as "employee counseling."

Conclusion

In this chapter, we reviewed the research on workplace drinking practices and alcohol problems with an eye toward discovering why lower-status employees (who tend to drink less than higher-status employees) have more alcohol problems than the higher-status employees. This review suggests that the workplace risk factors associated with alcohol problems may be found among both lower- and higher-status employees. That is, both groups may possess drinking cultures that encourage heavy consumption of alcohol and rationalize doing so because of alienating and stressful conditions at work. In addition, individuals from both groups may come to work with predispositions to drink heavily and are likely to get in trouble at work because of their drinking when they are not sufficiently integrated into nonabusive drinking peer networks and are not closely supervised. Relatively more information is available about drinking practices and alcohol problems among lower-status employees than among higher-status employees. This may either represent an accurate reflection of the incidence of alcohol problems or selection biases. We suspect the latter because gaining access for research to higher-status employees generally, and top management in particular, is often an insurmountable problem.

Our review of workplace interventions to mitigate the risk factors associated with alcohol problems likewise produced relatively little data, particularly with reference to how social class effects identification of problem drinkers. Although researchers have suggested that quality of work life programs, which are often targeted toward lower-status employees, might reduce alcohol problems by making work less alienating and stressful, evaluation studies of these projects have not included drinking practices and alcohol problems as variables. Similarly, although health promotion programs, which are generally targeted at higher-status employees, are hypothesized to reduce alcohol problems by teaching employees how to live healthy lives, there are no data from evaluation studies to demonstrate this effect. Indeed, HPPs appear to attract the conspicuously well rather than those who are overweight, do not exercise, smoke, and drink too much. Consequently, we know little about how social class is related to these interventions and how they may contribute to a disproportionate number of alcohol problems in lower-status employees.

In contrast to quality of work life and wellness programs, there is relatively more information about employee assistance programs. Apparently, when these programs are implemented according to good industrial and labor relations practices and supervisors are well trained in the constructive confrontation strategy, lower-status employees are not unjustly accused of having alcohol problems and are not overrepresented in treatment. In these instances, good implementation practices make alcohol problems visible among both higher-status and lower-status employees and arbitration stands as a safeguard to employees' rights. Unfortunately, the data associated with these statements are sparse and more research is required before we can definitively claim that EAPs treat everyone fairly, regardless of social status.

Unfortunately, many of today's EAPS are not built upon the industrial and labor relations practices of job performance, progressive discipline, and arbitration. Rather, these programs encourage employees to self-refer themselves for counseling at the earliest sign of a problem. No evaluation data are available on these programs. Consequently, beyond what these practitioners claim, nothing is actually known about their impact upon alcohol problems, their effect upon employees' rights, or whether they discriminate against lower-status or higher-status employees.

Overall, we still know relatively little about drinking practices and alcohol problems within the workplace. Following Bacon, we should like to urge social scientists to conduct community studies that accurately describe the drinking practices in all social classes and that document their functional and dysfunctional aspects. Such studies are important for workplace research because organizations do not exist in isolation from their communities. Events that occur outside work effect what happens on the job. We thus need studies that link drinking practices in the community with workplace drinking practices and alcohol problems, as well as studies that link wider social efforts for combating alcohol abuse and dependence and efforts that take place within the workplace. For example, what impact do programs addressing intoxicated drivers, public education, advertisements for treatment, and AA activities have upon EAP, HPP, and QWL efforts to reduce problem drinking? Such community studies can have a major impact on our understanding of the role of work site initiatives and their relationship to programs in society at large. At the same time, they would illuminate the relationships between community and workplace drinking practices and between the definitions of alcohol problems at work and within society. In this way, we would begin to generate better insights about the relationship between social class drinking practices and problems.

Within work organizations, we also need to conduct qualitative research so that we may more clearly understand how the workplace culture, social control structures, alienation, and stress relate to drinking behavior, and how these factors interact with social class. Ethnographic studies on a wide range of organizations are necessary for generating thick descriptions (Geertz, 1973) of employees' drinking behavior. Ideally, such studies should compare employees' drinking behavior across many organizations having different administrative and occupational drinking subcultures. This would enable researchers to generate hypotheses they can test as to how workplace culture directly affects drinking behavior and how it indirectly affects it through the rationales employees learn to use for relieving alienation and stress with alcohol. Such studies should also generate hypotheses researchers can test as to how social controls in an organization directly affect drinking behavior and how workplace structures indirectly affect it by generating alienation and stress.

In addition, accurate descriptions of workplace interventions and their relationship to drinking behavior must be developed so that researchers may generate grounded theories for future testing (Glaser & Strauss,

1967). Except for a few rare cases (e.g., Conrad, 1988; Sonnenstuhl & Trice, 1987), few descriptions exist of the actual processes underlying such strategies. Instead, the employee assistance, health promotion, and quality of work life literature is replete with claims of what their practitioners say they do. The gulf between words and deeds is often considerable (Deutscher, 1966), and as long as what actually occurs remains unknown, one cannot say whether a particular program has truly been implemented, whether any changes in alcohol abuse and dependence should be attributed to its efforts, and whether they discriminate against lower-status employees.

Qualitative studies should be undertaken in organizations in which multiple interventions have been introduced. This will help us understand how these interventions interact with one another and the impact of this interaction on alcohol abuse and dependence. For example, we would benefit from knowing the conditions under which employee assistance, health promotion, and quality of work life interventions enhance or detract from one another. Such knowledge is crucial for designing truly comprehensive work site interventions for combating alcohol problems and understanding their impact upon lower-status and higher-status employees.

Notes

The section, "Risk Factors at Work," is adapted from Trice and Sonnenstuhl (1988). The discussion of workplace interventions is based on Sonnenstuhl (1988).

1. Gary (1989) also argues that employee assistance programs discriminate against lower-status workers because these workers are generally employed in marginal enterprises, which have little incentive to adopt programs. Consequently, such workers are likely to be terminated rather than helped.
2. Parker and Brody (1982) proposed three models in their review of the literature: social control, alienation, and availability. Plant (1978, 1981) also suggested in his review of the literature that there were clusters of factors that could act as models for future research. Similarly, Trice and Roman (1978) proposed a number of conceptual models in their earlier work. Our proposed perspectives build upon and extend these earlier insights. Similarly, Ames and Janes (1987), in an ethnographic study of auto assembly line workers, suggested that social control, cultural, stress, and alienation factors were related to drinking behavior.
3. Some studies (Crowley, 1985; Mensch & Kandel, 1988) report finding no relationship between occupation and alcohol and drug use. These studies, however, do not define occupation in cultural terms; rather, they define occupations structurally by lumping workers with similar job titles and characteristics into census categories that fail to consider the cultural essence of occupations (Freidson, 1986). For instance, "professional and technical" included authors,

Baptist ministers, strip teasers, and accountants. "Managers and administrators" may refer to such diverse work roles as political appointees, bank officials, taco vendors, chief executive officers, and tavern operators. This diverse group of workers would not classify themselves as being in the same category because they most likely do not believe in the same things and therefore would be unlikely to associate with one another. Lumping them together confuses the cultural entity, occupation, with the structural features of work and mixes the drinking practices of ministers, strippers, and accountants. Under such circumstances, one would expect to find only spurious relationships between occupational culture and drinking practices. For an additional critique, see Abbott (1988).

4. "Shift work" and "night work" can be used interchangeably.
5. We are indebted to Paul Roman of the University of Georgia for this observation.
6. Bacon, in his research at Yale and through his work with the National Council on Alcoholism, made several significant contributions to the early field of industrial alcoholism programs, the forerunners of employee assistance. His 1951 study with Straus (Straus & Bacon, 1951) demonstrated that the vast majority of alcoholics were employed rather than skid row derelicts. Working with Henderson at the NCA, he popularized the notion of the employed alcoholic as "half-man" and promoted the Yale Plan for business and industry (Henderson & Bacon, 1953).
7. The word *teetotal,* as defined in *Webster's New Collegiate Dictionary,* is formed by reduplication of the initial letter of "total" for the sake of emphasis. Therefore, *teetotal* underscores the total abstinence from alcoholic drinks.
8. Although most discussions focus on the intoxicated employee, considerable evidence suggests that withdrawal distress, commonly called "hangover," can also seriously impair performance (Trice & Roman, 1978).

References

Abbot, A. (1988). *The system of professions.* Chicago: University of Chicago Press.

Ames, G.M. (1989). Alcohol related movements and their effects on drinking policies in the American workplace: An historical overview. *Journal of Drug Issues, 19,* 489–510.

Ames, G.M., & Janes, C.R. (1987). Heavy and problem drinking in an American blue-collar population. *Social Science and Medicine, 25,* 949–960.

Applebaum, H. (1984). Dangerous occupations—Construction work, fire fighting and mining. In H. Applebaum (Ed.), *Work in market and industrial societies* (pp. 146–164). Albany: State University of New York Press.

Archer, J. (1977). Social stability, work force behavior, and job satisfaction of alcoholic and non-alcoholic blue-collar workers. In C. Schramm (Ed.), *Alcoholism and its treatment in industry* (pp. 156–176). Baltimore, MD: Johns Hopkins University Press.

Archer, J. (1981). *Alcoholism and alienation among blue-collar workers: Test of a causal theory.* Unpublished doctoral dissertation, John Hopkins University.

Bacon, S.D. (1944). *Sociology and the problems of alcohol.* Memoirs of the section of studies on alcohol, No. 1, Laboratory of Applied Physiology, Yale University. New Haven, CT: Yale University.

Bacon, S.D. (1962). Alcohol and complex society. In D.J. Pittman & C.R. Snyder (Eds.), *Society, culture and drinking patterns.* New York: John Wiley and Sons.

Bacon, S.D. (1987). Alcohol problem prevention: A "common sense" approach. *Journal of Drug Issues, 17,* 369–393.

Bales, R.F. (1946). Cultural differences in rates of alcoholism. *Quarterly Journal of Studies on Alcoholism, 6,* 480–499.

Bales, R.F. (1962). Attitudes toward drinking in the Irish culture. In D.J. Pittman & C.R. Snyder (Eds.), *Society, culture and drinking patterns* (pp. 572–576). New York: John Wiley and Sons.

Barley, S.R., & Knight, D.B. (1991). Toward a cultural theory of stress. In B. Staw & L.L. Cummings (Eds.), *Research in Organizational Behavior, 14.*

Barnes, G.M., & Russell, M. (1977). *Drinking patterns among adults in New York State: A descriptive analysis of the sociodemographic correlates of drinking.* Buffalo, NY: Research Institute on Alcoholism.

Becker, F.D. (1986). Loosely coupled settings: A strategy for computer-aided work decentralization. In B.M. Staw & L.L. Cummings (Eds.), *Research in organizational behavior* (Vol. 8, pp. 116–142). Greenwich, CT: JAI Press.

Beyer, J.M., & Trice, H.M. (1982). Design and implementation of job-based alcoholism programs: Constructive confrontation strategies and how they work. In D. Godwin (Ed.), *Occupational alcoholism: A review of research issues* (Research Monograph 8, pp. 181–239). Washington, DC: U.S. Department of Health and Human Services.

Beyer, J.M., & Trice, H.M. (1984). A field study of the use and perceived effects of discipline in controlling work performance. *Academy of Management Journal, 27,* 743–764.

Bissell, L., & Haberman, P.W. (1984). *Alcoholism in the professions.* New York: Oxford University Press.

Braverman, H. (1974). *Labor and monopoly capital.* New York: Monthly Review Press.

Brenner, M.H. (1973). *Mental illness and the economy.* Cambridge, MA: Harvard University Press.

Brenner, M.H. (1975). Trends in alcohol consumption and associated illnesses: Some effects of economic changes. *American Journal of Public Health, 65,* 1279–1292.

Bryant, C.D. (1974). Olive-drab drunks and G.I. junkies: Alcohol and narcotic addiction in the U.S. military. In C.D. Bryant (Ed.), *Deviant behavior: Occupational and organizational bases* (pp. 129–145). Chicago: Rand-McNally.

Cahalan, D. (1970). *Problem drinkers: A national survey.* San Francisco: Jossey-Bass.

Cahalan, D. (1987). Studying drinking problems rather than alcoholism. In M. Galanter (Ed.), *Recent developments in alcoholism,* Vol. 5, pp. 363–373.

Cahalan, D., Cisin, I.H., & Crossley, H.M. (1969). *American drinking practices: A national study of drinking behavior and attitudes.* New Brunswick, NJ: Rutgers Center of Alcohol Studies.

Cahalan, D., & Room, R. (1974). *Problem drinking among American men.* New Brunswick, NJ: Rutgers Center of Alcohol Studies.

Castor, D., & Parsons, O. (1977). Four focus of control in alcoholics and treatment outcome. *Journal of Studies on Alcohol, 38,* 2087–2095.

Clinton, C.A. (1977). The use of cultural ecology in an urban occupational group. *Anthropological Quarterly, 50,* 39–44.

Conrad, P. (1987a). Wellness in the workplace: Potentials and pitfalls of worksite health promotion. *Milbank Quarterly, 65,* 255–275.

Conrad, P. (1987b). Who comes to worksite wellness programs? A preliminary review. *Journal of Occupational Medicine, 29,* 317–320.

Conrad, P. (1988). Health and fitness and work: A participant's perspective. *Social Science and Medicine, 26,* 545–550.

Conrad P., & Schneider, J. (1980). *Deviance and medicalization.* St. Louis, MO: C.U. Mosby.

Conway, T.L., Vickers, R.R., Ward, H.W., & Rahe, R.H. (1981). Occupational stress and variation in cigarette, coffee and alcohol consumption. *Journal of Health and Social Behavior, 22,* 155–165.

Cosper, R. (1979). Drinking as conformity: A critique of sociological literature and occupational differences in drinking. *Journal of Studies on Alcohol, 40,* 18–82.

Crowley, J.E. (1985). *The demographics of alcohol use among young Americans: Results from the 1983 National Longitudinal Survey of Labor Market Experience of Youth.* Columbus: Center for Human Resource Research, Ohio State University.

Cummings, T.G., & Molloy E.S. (1977). *Productivity and the quality of work life.* New York: Praeger.

Denenberg, T.S., & Denenberg, R.V. (1983). *Alcohol and drugs: Issues in the workplace.* Washington, D.C.: Bureau of National Affairs.

Denzin, N. (1986). *Treating alcoholism.* Beverly Hills, CA: Sage.

Deutscher, I. (1966). Words and deeds: Social science and social policy. *Social Problems, 13,* 233–254.

Dixon, K. (1984). The coercion of labor by mental health professionals. *Social Policy, 14,* 47–54.

Dubofsky, M., & Van Tine, W. (1977). *John L. Lewis: A biography.* New York: Quadrangle.

Eichler, S., Goldberg, C.M., Kier, L.E., & Allen, J.P. (1988). *Operaton Redblock: Case study of peer prevention substance abuse program for railroad industry personnel.* Washington, DC: U.S. Department of Transportation, Federal Railroad Administration.

Elkouri, F., & Elkouri, E.A. (1985). *How arbitration works* (4th ed.). Washington, DC: Bureau of National Affairs.

Fennell, M.L., Rodin, M.B., & Kantor, G.K. (1981). Problems in the work setting, drinking and reasons for drinking. *Social Forces, 60,* 114–132.

Ferguson, D. (1974). A study of occupational stress and health. In W. Welford (Ed.), *Man under stress* (pp. 33–55). London: Taylor and Francis.

Fielding, J.E. (1984a). *Corporate health management.* Reading, MA: Addison-Wesley.

Fielding, J.E. (1984b). Health promotion and disease prevention at the worksite. *Annual Review of Public Health, 5,* 237–265.

Fillmore, K.M. (1981, April). *Research as a handmaiden of policy: An appraisal of estimates of alcoholism and its cost in the workplace.* Paper presented at the National Alcoholism Forum of the National Council on Alcoholism, New Orleans.

Fillmore, K.M. (1989, November). *Risk factors for alcohol problems: Social and ethical considerations with social attention to the workplace.* Paper presented at the second annual conference of the R. Brinkley Smithers Institute for Alcoholism Prevention and Workplace Problems, Cornell University, Ithaca, NY.

Fillmore, K.M., & Caetano, R. (1982). Epidemiology of alcohol abuse and alcoholism in occupations. *Occupational alcoholism: A review of research issues.* Rockville, MD: U.S. Department of Health and Human Services.

Freidson, E. (1986). *Professional powers: A study of the institutionalization of formal knowledge.* Chicago: University of Chicago Press.
French, E., & Magee, J.J. (1972). The incidence of the emotionally disturbed within the administrative units of a large organization. *Personnel Psychology, 25,* 535–543.
Gamst, F.C. (1980). *The hoghead: An industrial ethnology of the locomotive engineer.* New York: Holt, Rinehart, & Winston.
Gary, J.A., Jr. (1989). *Elements of class structure in employee assistance programs* (Working paper #103). Ithaca, NY: School of Industrial and Labor Relations, Cornell University.
Geertz, C. (1973). *The interpretation of cultures.* New York: Basic Books.
Glaser, B., & Strauss, A. (1967). *Discovery of grounded theory.* Chicago: Aldine.
Glueck, W.F. (1978). *Personnel: A diagnostic approach.* Dallas: Business Publications.
Gusfield, J. (1963). *Symbolic crusade: Status politics and the American temperance movement.* Urbana: University of Illinois Press.
Gusfield, J. (1987). Passage to play: Rituals of drinking time in American society. In M. Douglas (Ed.), *Constructive drinking* (pp. 73–90). New York: Cambridge University Press.
Hardy, R.E., & Cull, J.G. (1971). Vocational satisfaction among alcoholics. *Quarterly Journal of Studies on Alcohol, 32,* 180–182.
Hayhurst, E.R. (1938). Industrial alcoholism. *Industrial Medicine and Surgery, 7,* 629–631.
Heath, D.B. (1987). A decade of development in the anthropological study of alcohol use: 1970–1980. In M. Douglas (Ed.), *Constructive drinking* (pp. 16–69). New York: Cambridge University Press.
Henderson, R., & Bacon S.D. (1953). Problem drinking: The late plan for business and industry. *Quarterly Journal of Studies on Alcohol, 14,* 247–262.
Herrington, S., Levy, M.F., & Reichman, W. (1980). Career development in alcoholics and non-alcoholics. In L. Brill & C. Winick (Eds.), *Yearbook of drug and substance abuse* (pp. 204–212). New York: Human Sciences Press.
Hilton, M.E., & Clark, W.B. (1987). Changes in American drinking patterns and problems, 1967–1984. *Journal of Studies on Alcohol, 48,* 515–523.
Hingson, R., Mangione, T., & Barrett, J. (1981). Job characteristics and drinking practices in the Boston metropolitan area. *Journal of Studies on Alcohol, 42,* 725–738.
Hirsh, P. (1987). *Pack your own parachute: How to survive mergers, takeovers, and other corporate disasters.* New York: Addison-Wesley.
Hitz, D. (1973). Drunken sailors and others: Drinking problems in specific occupations. *Quarterly Journal of Studies on Alcohol, 34,* 496–505.
Hoover, D.W. (1990). *Middletown revisited.* Muncie, IN: Ball State University Press.
Jacobson, G.R., & Lindsay, D. (1980). Screening for alcohol problems among the unemployed. In M. Galanter (Ed.), *Currents in alcoholism* (Recent advances in research & treatment, VIII). New York: Grune & Stratton.
Kasl, S.V. (1973). Mental health and work environment: An examination of the evidence. *Journal of Occupational Medicine, 15,* 509–518.
Kaufman, E. (1982). The relationship of alcoholism and alcohol abuse to the abuse of other drugs. *American Journal of Drug and Alcohol Abuse, 9,* 1–17.
Knupfer, G. (1967). Epidemiologic studies and control programs in alcoholism. *American Journal of Public Health, 57,* 973–986.

LeMasters, E.E. (1975). *Blue-collar aristocrats: Life styles at a working-class tavern.* Madison: University of Wisconsin Press.

Levine, H.G. (1978). *Notes on work and drink in industrializing America.* Paper presented for NIAAA Research Conference on Alcohol Use and the Work Environment, Belmont, MD.

Levy, M.F., Reichman, W., & Herrington, S. (1979). Congruence between personality and job characteristics in alcoholics and non-alcoholics. *Journal of Social Psychology, 107,* 213–217.

Lolli, G. (1953). The use of wine and other alcoholic beverages by a group of Italian extraction. *Quarterly Journal of Studies on Alcohol, 14,* 395–405.

Lurie, N.O. (1979). The world's oldest on-going protest demonstraton: North American Indian drinking patterns. In M. Marshall (Ed.), *Beliefs, behaviors, and alcoholic beverages.* Ann Arbor: University of Michigan Press.

Lynd, R.S., & Lynd, H.M. (1929). *Middletown: A study in contemporary American culture.* New York: Harcourt, Brace and Co.

Lynd, R.S., & Lynd, H.M. (1937). *Middletown in transition: A study in cultural conflicts.* New York: Harcourt, Brace and Co.

Mangione, T.W., & Quinn, R.P. (1975). Job satisfaction, counterproductive behavior, and drug use at work. *Journal of Applied Psychology, 60,* 114–117.

Mannello, T.A., & Seeman, F.J. (1979). *Prevalence, costs, and handling of drinking problems on seven railroads* (Final Report). Washington, DC: University Research Corp.

Margolis, G.L., Kroes, W.H., & Quinn, R.P. (1974). Job stress: An unlisted occupational hazard. *Journal of Occupational Medicine, 16,* 659–661.

Mars, G. (1987). Longshore drinking, economic security, and union politics in Newfoundland. In M. Douglas (Ed.), *Constructive drinking.* Cambridge, England: Cambridge University Press.

McAndrews, C., & Edgerton, R. (1969). *Drunken comportment.* Chicago: Aldine.

McClelland, D.C., Davis, W.N., Kalin, R., & Wanner, E. (1972). *The drinking man.* New York: Free Press.

Melbin, M. (1987). *Night as frontier: Colonizing the world after dark.* New York: Free Press.

Mensch, B.S., & Kandel, D.B. (1988). Do job conditions influence the use of drugs? *Journal of Health and Social Behavior, 29,* 169–184.

Minnis, J.R. (1988). Toward an understanding of alcohol abuse among the elderly: A sociological perspective. *Journal of Alcohol and Drug Education, 33,* 32–40.

Molloy, D. (1989). Peer intervention: An exploratory study. *Journal of Drug Issues, 19,* 319–336.

Mulford, H.A. (1964). Drinking and deviant drinking, U.S.A., 1963. *Quarterly Journal of Studies on Alcohol, 25,* 634–650.

Murray, R.M. (1975). Alcoholism and employment. *Journal of Alcoholism, 10,* 23–26.

Newcomb, M.D. (1988). *Drug use in the workplace.* Dover, MA: Auburn House.

O'Donnell, J.A., Voss, H., Clayton, R., Slatin, G., & Room, R. (1976). *Young men and drugs: A nationwide survey* (NIDA research monograph 5). Rockville, MD: National Institute of Drug Abuse.

Ojesjo, L. (1980). The relationship to alcoholism of occupation, class, and employment. *Journal of Occupational Medicine, 22,* 657–666.

O'Toole, J. (Ed.). (1974). *Work and the quality of life.* Cambridge, MA: MIT Press.

Ozley, L.M., & Ball, J.S. (1982). Quality of worklife: Initiating successful efforts in labor-management organizations. *Personnel Administrator, 27,* 27–33.

Parker, D.A. (1979). Status inconsistency and drinking behavior. *Pacific Sociological Review, 22,* 77–95.

Parker, D.A., & Brody, J.A. (1982). Risk factors for alcoholism and alcohol problems among employed women and men. In D. Godwin (Ed.), *Occupational alcoholism: A review of research issues.* Rockville, MD: U.S. Dept. of Health and Human Services.

Parker, M. (1985). *Inside the circle: A union guide to QWL.* Boston: South End Press.

Parkinson, R.S., & associates (1982). *Managing health promotion in the workplace.* Palo Alto, CA: Mayfield.

Pearlin, L.I., Lieberman, M., Menaghan, E., & Mullan, J. (1981). The stress process. *Journal of Health and Social Behavior, 22,* 337–356.

Pilcher, W.W. (1972). *The Portland longshoremen.* New York: Holt, Rinehart, & Winston.

Pittman, D.J., & Snyder, C.R. (Eds.) (1962). *Society, culture and drinking patterns.* New York: John Wiley and Sons.

Plant, M.A. (1978). Occupation and alcoholism: Cause or effect? A controlled study of recruits to the drink trade. *International Journal of the Addictions, 13,* 605–626.

Plant, M.A. (1981). Risk factors in employment. In B.D. Hore & M.A. Plant (Eds.), *Alcohol problems in employment.* London: Croom Helm London.

Pursch, J.A. (1976). From quonset hut to naval hospital: The story of an alcoholism rehabilitation service. *Journal of Studies on Alcohol, 37,* 1655–1665.

Riemer, J.W. (1976). *Deviance as fun—A case of building construction workers at work.* Paper presented at annual meeting of the American Sociological Association, San Francisco, CA.

Roman, P.M. (1981). Job characteristics and the identification of deviant drinking. *Journal of Drug Issues, 11,* 357–364.

Roman, P.M. (1982). Employee alcoholism programs in major corporations in 1979: Scope, change, and receptivity. In J. DeLuca (Ed.). *Prevention, intervention, and treatment: Concerns and models* (pp. 177–200). Rockville, MD: U.S. Department of Health and Human Services.

Rorabaugh, W.J. (1979). *The alcoholic republic: An American tradition.* New York: Oxford University Press.

Rosen, A.J., & Glatt, M.M. (1971). Alcohol excess in the elderly. *Quarterly Journal of Studies on Alcohol, 32,* 53–59.

Runcie, J.F. (1980). By day I make the cars. *Harvard Business Review, 58,* 106–115.

Salaman, G. (1974). *Community and occupation: An exploration of the work/leisure relationship.* London: Cambridge University Press.

Schuckit, M.A., & Miller, P.L. (1976, May 23). Alcoholism in elderly men: A survey of a general medical ward. *Annals of the New York Academy of Sciences, 273,* 558–571.

Seeman, M. (1967). On the personal consequences of alienation in work. *American Sociological Review, 32,* 273–285.

Seeman, M., & Anderson, C.S. (1983). Alienation and alcohol: The role of work, mastery, and community in drinking behavior. *American Sociological Review, 48,* 60–77.

Shain, M., Suurvali, H., & Boutilier, M. (1986). *Healthier workers: Health promotion and employee assistance programs.* Lexington, MA: Lexington Books.

Sherlock, B.J. (1967). Career problems and narcotics addiction in the health professions: An exploratory study. *The International Journal of Addictions, 2,* 191–205.

Smart, R.G. (1979). Drinking problems among employed, unemployed and shift workers. *Journal of Occupational Medicine, 21,* 731–736.

Snyder, C. (1978). *Alcoholism and the Jews.* Carbondale: Southern Illinois University Press.

Sonnenstuhl, W.J. (1986). *Inside an emotional health program: A field study of workplace assistance for troubled employees.* Ithaca, NY: ILR Press, Cornell University.

Sonnenstuhl, W.J. (1987). The social construction of alcohol problems in a union's peer counselling program. *Journal of Drug Issues, 17,* 223–254.

Sonnenstuhl, W.J. (1988). Contrasting employee assistance, health promotion, and quality of work life programs and their effects on alcohol abuse and dependence. *Journal of Applied Behavioral Science, 24* (4), 347–363.

Sonnenstuhl, W.J., Staudenmeier, W.J., & Trice, H.M. (1988). Ideology and referral categories in employee assistance program research. *Journal of Applied Behavioral Science, 24* (4), 383–396.

Sonnenstuhl, W.J., & Trice, H.M. (1987). The social construction of alcohol problems in a union's peer counseling program. *Journal of Drug Issues, 17,* 223–254.

Sonnenstuhl, W.J., & Trice, H.M. (1990). Strategies for employee assistance programs (2nd ed.). Ithaca, NY: ILR Press.

Sonnenstuhl, W.J., Trice, H.M., & Evans, B. (1988, August). Peers in the tunnels: A union's solution to problem drinking. *ALMACAN, 18,* 13–17.

Sonnenstuhl, W.J., Trice, H.M., Staudenmeier, W.J., Jr., & Steele, P. (1987). Employee assistance and drug testing: Fairness and injustice in the workplace. *Nova Law Review, 11,* 709–731.

Staudenmeier, William J., Jr. (1985). *Alcohol in the workplace: A study of social policy in a changing America.* Unpublished doctoral dissertation, Washington University.

Staudenmeier, William J., Jr. (1987). Context and variation in employer policies on alcohol. *Journal of Drug Issues, 17,* 255–271.

Steele, P.D., & Hubbard, R.L. (1985). Management styles, perceptions of substance abuse, and employee assistance programs in work organizations. *Journal of Applied Behavioral Science, 21,* 271–286.

Stein, B.A. (1983). *Quality of work life in action: Managing for effectiveness.* New York: American Management Association.

Straus, R. (1975). Reconceptualizing social problems in light of scholarly advances: Problem drinking and alcoholism. In N. Dermerath, O. Larsen, & K. Schuessler (Eds.), *Social policy and sociology.* New York: Academic Press.

Straus, R. (1976a). Alcoholism and problem drinking. In R.K. Merton & R. Nisbet (Eds.), *Contemporary social problems* (4th ed.). New York: Harcourt, Brace, Jovanovich.

Straus, R. (1976b). Problem drinking in the perspective of social change, 1940–1973. In W.J. Filstead, J.J. Rossi, & M. Keller (Eds.), *Alcohol and alcohol problems: New thinking and new directions.* Cambridge, MA: Ballinger.

Straus, R., & Bacon, S. (1951). Alcoholism and social stability: A study of occupational integration in 2,023 male clinic patients. *Quarterly Journal of Studies on Alcohol, 12,* 231–260.

Sullivan, W.C. (1906). Industry and alcoholism. *Journal of Mental Science, 52,* 505–514.

Trice, H.M. (1962). The job behaviors of problem drinkers. In D.J. Pittman & C.R. Snyder (Eds.), *Society, culture and drinking patterns* (pp. 493–510). New York: John Wiley and Sons.

Trice, H.M. (1965a). Alcoholic employees: A comparison of psychotic, neurotic, and "normal" personnel. *Journal of Occupational Medicine, 7,* 94–99.
Trice, H.M. (1965b). Reaction of supervisors to emotionally disturbed employees: A study of deviation in a work environment. *Journal of Occupational Medicine, 7,* 177–188.
Trice, H.M., & Belasco, J.A. (1966). Emotional health and employee responsibility (Bulletin No. 57). Ithaca, NY: New York State School of Industrial and Labor Relations, Cornell University.
Trice, H.M., & Belasco, J.A. (1970). The aging collegian: Drinking pathologies among executive and professional alumni. In G. Maddax (Ed.), *The domestic drug.* New Haven: College and University Press.
Trice, H.M., & Beyer, J.M. (1977). Differential use of an alcoholism policy in federal organizations by skill level. In C. Schramm (Ed.), *Alcoholism and its treatment in industry* (pp. 44–68). Baltimore: Johns Hopkins University Press.
Trice, H.M., & Beyer, J.M. (1981). A data-based examination of selection-bias in the evaluation of a job-based alcoholism program. *Alcoholism: Clinical and Experimental Research, 5,* 489–496.
Trice, H.M., & Beyer, J.M. (1984). Work-related outcomes of the constructive confrontation strategy in a job-based alcoholism program. *Journal of Studies on Alcohol, 45,* 393–404.
Trice, H.M., & Pittman, D. (1958). Social organization and alcoholism. *Social Problems, 5,* 294–307.
Trice, H.M., & Roman, P.M. (1971). Occupational risk factors in mental health and the impact of role change experience. In J.J. Leedy (Ed.), *Compensation in psychiatric disability and rehabilitation* (pp. 145–205). Springfield, IL: Charles C Thomas.
Trice, H.M., & Roman, P.M. (1978). *Spirits and demons at work: Alcohol and other drugs on the job* (rev. ed.). Ithaca, NY: New York State School of Industrial and Labor Relations, Cornell University.
Trice, H.M., & Schonbrunn, M. (1981). A history of job-based alcoholism programs: 1900–1955. *Journal of Drug Issues, 11,* 171–198.
Trice, H.M., & Sonnenstuhl, W.J. (1988). Drinking behaviors and risk factors related to the work place: Implications for research and prevention. *Journal of Applied Behavioral Science, 24*(4), 327–346.
Trice, H.M., & Sonnenstuhl, W.J. (1991). Job behaviors and the denial syndrome. In D.J. Pittman and H.R. White (Eds.), *Society, culture, and drinking patterns reexamined.* New Brunswick, NJ: Rutgers Center of Alcohol Studies.
Trice, H.M., & Sonnenstuhl, W.J. (1990). Alcohol and mental health programs in the workplace. *Research in Community Mental Health, 6,* 351–378.
Van Maanen, J., & Barley, S.R. (1984). Occupational communities: Culture and control in organizations. *Research in Organizational Behavior, 6,* 287–365.
Wagner, D. (1988). The new temperance movement and social control at the workplace. *Contemporary Drug Problems, 17,* 539–556.
Warkov, S., Bacon, S.D., & Hawkins, A.C. (1965). Social correlations of industrial problem drinking. *Quarterly Journal of Studies on Alcohol, 26,* 58–71.
Warner, L.W., & Lunt, P.S. (1941). *The Social life of a modern community.* New Haven: Yale University Press.
Warner, L.W., & Lunt, P.S. (1942). *The System of a modern community.* New Haven: Yale University Press.
Whitehead, P.C., & Simpkins, J. (1983). Occupational factors in aloholism. In B. Kissin & H. Begleiter (Eds.), *The pathogenesis of alcoholism: Psychosocial factors.* New York: Plenum.

Widick, B.J. (1976). *Auto work and its discontents.* Baltimore: Johns Hopkins University Press.

Winkler, A.M. (1968). Drinking on the American frontier. *Quarterly Journal of Studies on Alcohol, 29,* 413–445.

CHAPTER 13

Risk Factors for Alcohol Problems: Social and Ethical Considerations, with Special Attention to the Workplace

KAYE MIDDLETON FILLMORE

The goal of this chapter is to examine three themes of research findings from diverse disciplinary perspectives that posit factors which put individuals at risk for alcohol problems. Each of the three has been advanced through longitudinal methodology, a method that closely approximates causal order and is, therefore, taken quite seriously by scientists, policy makers and the interested populace alike (Fillmore, 1988). The three themes are: (1) that antisocial behavior exhibited in youth is an antecedent to adult drinking problems and alcoholism; (2) that there is, in part, a genetic predisposition to alcoholism; and (3) that aggregate-level factors (e.g., price or availability of alcohol) influence change in drinking practices on the individual level.

The strategy for examining each of these themes consists of three steps. First, each is described and critiqued with respect to methodological limitations and the degree to which results can be generalized across diverse cultural groups. Second, using historical and cross-cultural information as tools, each is considered with respect to its application, if and when it is implemented as social policy. Third, each body of research is appraised with respect to a variety of workplace settings should it be implemented as social policy. The rationale for appraising the scientific limitations of these findings in concert with the possible policy consequences of application (considered in social and ethical contexts) is stated below.

First, it is generally recognized that social policy in this and some other industrialized countries is informed or strengthened by advancement in or evaluation by scientific research. The simple fact that the U.S. government (responsible for much social policy in this country) expends monies in support of research in the areas of alcohol, drugs, and mental health serves to justify the claim that contemporary scientific examination is allied with federal interests, certainly economically and sometimes ideologically.

Second, a characterization of the diverse fields of science made more than 40 years ago has even more relevance today: namely, that single research perspectives "have held the others at a safe distance, browsing on [their] own selected pastures and learning more about less and less" (Linton, 1945). Put another way, scientific research tends, for the most part, to rely on assumptions and questions central to only one or another specialty.

Third, the advancement of scientific explanation of human behavior from various disciplinary perspectives often postulates single bodies of explanation. Because there is considerable competition for limited research funds, most in the United States from federal sources, it is not surprising that testing of truly competing hypotheses or truly interdisciplinary interactive hypotheses is atypical in scientific journals. This means that the presentation of results—read by the interested scientific community, translated to the interested lay community, and possibly implemented as policy by the federal or state communities—might be characterized as shedding light on only one part of a larger picture.

Together, the economic and potentially ideological relationships that have grown up between policy makers and research interests, as well as the scientific propensity to yield a crop of findings that is discipline-specific, raise serious questions when these influences are implemented as social policy. In this context, it is important for scientists not only to critically evaluate the evidence from a variety of scientific postulations, but also to consider the possible social and ethical consequences emerging from these findings should they be applied as social policy.

This discussion, which speculates on the relationships between single bodies of research findings and implementation by social policy, necessarily focuses on costs rather than benefits. This is for the simple, but powerful, reason that when social policies are advocated, either by scientists or other members of society, it is their bright rather than their dark side that is typically illuminated. A focus on the dark side of implementation of scientific findings should help balance this equation. Such a focus may also encourage dialogue among both scientists and policy makers in explicating and analyzing the variable social, political, and ethical interpretations of scientific findings as they might be implemented in policy impacting the individual, the community, and at national and international levels. It may also enhance examination of mechanisms that will promote cross-disciplinary research (Fillmore & Sigvardsson, 1988).

Theme 1: An Antecedent to Serious Alcohol Problems Is Youthful Antisocial Behavior (Conduct Disorder)

A tradition of studies from the 1950s sought to locate behaviors in childhood and youth that would ultimately predict alcohol-related adult behaviors (Amundsen, 1982; Jones, 1968, 1971; McCord & McCord, 1960; Monnelly, Hartl, & Elderkin, 1983; Robins, 1966; Vaillant, 1983). These studies include a number of characteristics that reflect their basic underlying assumptions. First, the dependent variable was "alcoholism," reflecting a disease-like conception. Second, the majority of the studies focused on "deviant" youth, reflecting the assumption that "bad behavior" in youth, particularly among males, was somehow related to "bad behavior" or "diseases" in adulthood and that, indeed, the "child is father of the man" (*My Heart Leaps Up,* William Wordsworth, 1770–1850). Third, all but one of the studies were performed in North American cultures.

Taken together, these studies suggested that boys exhibiting antisocial behavior at first measurement tend to become alcoholic more often than boys who do not. Further, they tend to experience other problems in adulthood (see Robins, 1984, for a review of these findings). Antisocial behavior, roughly speaking, is defined as a collectivity of behaviors:

> These predictive behaviors for early childhood were not limited to illegal acts, but involved a set of signs of resistance toward authority, hostility to peers and adults, impulsiveness, and precocious assumption of behaviors reserved for adults: drinking, sexual relations, running away from the parental home, and leaving home. (Robins, 1984, p. 1)

Assessment of some of these studies strongly suggests that this is a social class-linked finding, with the relationship being stronger among lower-status individuals. The majority of the studies used nonrepresentative samples, concentrating on children or adolescents in the most disadvantaged segments of society where the combination of disadvantage and early problems would seem to exacerbate later maladjustment among adults (at least from the white middle-class point of view). The exception to this is Amundsen's Norwegian study of 19-year-old military conscripts (a universe, rather than a sample), which found that lower socioeconomic status and maladjustment in primary school were interacting to predict later alcoholism. Together, an appraisal of these studies strongly indicates that the antisocial personality findings may be largely a function of differences in socioeconomic status.

This socioeconomic qualifier is also found in cross-sectional analyses of the general population. Robins (1984) reports that cross-sectional studies find higher rates of conduct disorder (antisocial behavior) in lower-class urban areas. This suggests that alcohol problems among lower-class men may be linked to the continuation of "bad boy" life styles. If adult alcohol problems are class-linked then by extrapolation it may be suggested that the etiologies of alcohol problems are multicausal and are contingent on group membership. Put simply, early antisocial behavior accounts for alcohol problems among lower-class men but not necessarily among other social groups.

Since these earlier studies were conducted, later studies emerged in the 1960s and 1970s with differing assumptions and designs (Donovan, Jessor, & Jessor, 1983; Hoffman, Loper, & Kammeier, 1974; Jessor & Jessor, 1977; Kandel, Raveis, & Kandel, 1984; Loper, Kammeier, & Hoffman, 1973; Temple & Fillmore, 1985). First, the dependent variable differed. While the earlier studies used a distinct entity or disease, namely "alcoholism," the later studies adopted the concept that disaggregated alcohol problems can be scaled in seriousness along a continuum; these studies also included quantitative measures of consumption. Second, sampling techniques in the later studies were more representative of age-specific portions of the general population. Third, while the focus in the earlier studies was on isolating childhood/adolescent factors that would subsequently explain adult alcohol-related behavior, the later studies used broader models concerned with the processes of development with shorter intervals between measurement. Fourth, and of considerable importance, these developmental processes analyzed in the later studies came to be seen as continuous processes rather than discrete categories, with difficulty distinguishing "transgressive or antisocial behavior" from "the involvement in behaviors that are productive of independence from family and established institutions" (Zucker, 1979).

The findings of these newer studies, all from the United States, cast doubt on the antisocial behavior hypothesis (see Fillmore, 1988, for a detailed discussion of these differences). Furthermore, and of considerable importance, longitudinal studies initiated in the 1960s and 1970s of both young people as well as the adult general population suggest that alcohol-related problems are not necessarily chronic conditions, that is, they could be sporadic (see Fillmore, Hartka, Johnstone, Speiglman, & Temple, 1988, for a recent review of the literature on "spontaneous remission"). These differences in assumptions and findings between the "problem drinking" literature and the "alcoholism" literature, among others, have laid the groundwork for some of the major contemporary controversies in American alcohol research, in particular a challenge to alcoholism as a disease entity.

It has been strongly suggested that the antisocial behavior hypothesis needs reexamination, particularly with an eye to differential operational definition, cohort and/or historical effects and sampling frame, for the quite pragmatic reason that antisocial behavior (now called "conduct disorder") has been accepted in the *Diagnostic and Statistical Manual* (DSM) of the American Psychiatric Association (1987) as a predisposing factor for alcoholism, among other "adult antisocial behaviors." This has rather far-reaching implications for institutions charged with either the prevention or treatment of alcohol problems, among them the possibility that "false positives" may be identified and subjected to intervention measures based on faulty evidence.

There is also concern that this hypothesis is culture bound, primarily because it is derived from limited geographic as well as ideological circumstances. First, it is important to point out that the hypothesis in question specifies that behaviors in youth are associated with behaviors in adulthood. However, the *meanings* of age categories vary greatly across cultures. For instance, unlike the conception of the period of youth in the United States, youth among Lau Islanders ranges from the early teenage years to the late thirties or early forties (Walter, 1982). Second, the age of permission to drink varies as well. In some societies alcohol is accessible for persons of all ages; in others it is socially accessible for only limited ages. Third, the expectations for drinking styles have been found to be age-graded. In Truk, it is noted that drinking is seen as "'young men's work'... drinking and flamboyant drunken comportment is *expected* of young men; in Truk the young man who abstains is 'abnormal,' not the other way around," and adult men are expected to drink less frequently and heavily (Marshall, 1979). Fourth, and closely related to considerations of this hypothesis for the workplace, the age of adult status and the age of entering the labor force differ widely across societies. In France, for instance, 36 percent of those in the workforce had finished their full-time education by age 14 (Rabier & Inglehart, 1980), which would suggest that in this culture the age of greatest prevalence of antisocial behavior, as well as experimentation with alcohol, would occur at the time of first employment.

There are other serious problems in applying this hypothesis to all of human kind. On the one hand is the antecedent factor itself—the concept of the bad child or adolescent. One must ask if all children or adolescents engaging in "drinking, sexual relations, running away from the parental home, and leaving home" (Robins, 1984) are bad, evil, diseased, deviant, or sinful in all societies? And, of course, the concept of alcoholism must be considered in a broader cultural context where questions, for instance, from the DSM-III-R (American Psychiatric Association, 1987) (e.g., Did your hands shake a lot the morning after drinking, did you miss work or school because of drinking or a hangover, did you feel that you should cut down on your drinking or stop altogether?) might connote normative behavior for persons occupying given age, sex, or work roles across different cultural settings.

The applicability of the antisocial hypothesis in particular and individual-level alcoholism hypotheses in general across workplace settings in different cultures is questionable. Illustrative of the culture-bound nature of the relationship of alcohol to work is a report from an Eastern European country:

> It is well known that in almost each business organization [alcohol is consumed] during the break.... In connection with that we shall point out an occurrence domesticated with us which slowly acquires the civil right. Our publicity knows very well that beginning with the first meetings in the morning with public or collaborators, business or private, till the end of the working day, and afterwards, all of them began and end with a cup of hard drink, out of the representative fund, of course. Business relations establishments, business or conversation, contract signing, all other kinds of business cannot be performed without the presence of alcohol. (Stojiljkovic & Vasev, 1973)

This description of the interface between alcohol and the workplace in one Eastern European country makes application of the antisocial behavior theory almost ludicrous when performance in the workplace is liberally bathed with alcohol use. More generally, cross-cultural considerations of definitions of deviance and meanings and definitions of age-related behaviors sharply bring into focus the notion that many of our individual-level hypotheses of the etiology of alcoholism, not to mention its symptoms, are likely to be culture bound.

We now ask how we might specifically conceptualize this hypothesis in relationship to the contemporary American workplace. The dimension that seems to have the greatest relevance for the American workplace is not that of the antecedent factor per se but is that which locates the behaviors under question most squarely among young, lower-class males. It is notable that antisocial behavior (and antisocial personality disorder) and alcohol and drug problems are found predominantly among the young (Myers et al., 1984), and all are more highly prevalent among the lower social classes (Cahalan, 1970; Robins, 1984). Indeed, the high remission rates found for antisocial personality and alcohol problems after the period of youth (in this case after age 29) in the recent Epidemiologic Catchment Area Study stimulated Robins and colleagues to investigate their co-occurrence, finding that "men with alcohol disorders are 11 times as likely to have antisocial personality as are men without alcohol disorders, and for women, the risk is increased 23 times by the presence of alcohol disorder" (Robins, Helzer, Przybeck, & Regier, 1988).

With these empirical facts in mind, the skeptic cannot help but envision, with the application of the antisocial behavior model, the dark side of interventions occurring in American business and industry, particularly among young, lower-class males. Historical evidence from several fronts has identified the negative consequences of drinking among lower-status persons as threatening to the industrial establishment. A review of this historical literature from several nations found a number of such instances:

> The emphasis on a link between the workers' use of alcohol and revolution in France, and the biological deficiencies of the starving industrial workers in Germany, served to justify class differences and measures to control the workers. The connection made between inefficiency and drinking among the workers made the American industrialists active supporters of the Temperance Movement in the name of profit. (Fillmore, 1984)

This analysis has been brought to the present by observations of employee assistance programs (EAPs) in the contemporary American workplace. EAPs utilize medicalized definitions of deviant behaviors—behaviors which, it might be added, appear to be quite "normal" among the young, lower-class persons. Dixon contends that EAPs

> help arm management with psychotechnologies to mystify and control worker discontent in the same manner that an earlier generation of community psychiatrists and psychologists armed the government so as to control explosive neighborhoods in the 1960s. What may appear as benevolent counselling and therapy initiated and supported at the job site or in the slum-based drop-in counselling center can conceal the more cynical intent to better "adjust" employees and impoverished minority groups to the alienating conditions of work or the structural problems of chronic poverty and racism. (Dixon, 1984, p. 50)

Following examination of manuals calling for support of EAPs on the part of both management and labor, Dixon points out that such support "will certainly obscure the repeatedly documented relationship between exploitive social and economic conditions and psychopathology (Brenner, 1969, 1977), thereby ensuring that meaningful demands for structural changes in the workplace will be blunted" (1984, p. 51). While it is clear that unions could act as sentinels for discerning these relationships by directing attention toward the social and economic relationships between workers and management, there is evidence that when unions have become involved in EAPs, they have readily adopted disease model arguments (Roman, 1988), most probably because they depend on recovering alcoholics from Alcoholics Anonymous (AA) to run their programs.

These considerations are perhaps more serious in recent times with increasing preemployment drug screening. Without the protection of collective bargaining from unions, we would hypothesize that large portions of young, lower-class males are effectively being excluded from the workplace for activities that are, indeed, a part of their subcultural baggage (i.e., the subcultures of youth and of the poor) (Roman & Blum, 1990). Recent descriptive evidence from preemployment screening of Postal Service employees (Office of Selection and Evaluation, U.S. Postal Service, 1989) documents that drug-positive applicants are more likely to be black, male and between the ages of 25–35. Similar information from the National Institute on Drug Abuse (1987) finds that 65 percent of those 18–25 years of age have had at least some experience with illicit drugs (44 percent having had experience in the last year). These reports suggest (1) the preemployment screening could unquestionably blunt the objectives of affirmative action programs, and (2) that as the labor force is increasingly dominated by older Americans, young, poor minorities will become even more bifurcated from the larger society (Roman & Blum, 1990).

In sum, one cannot help but envision a hypothesis that cross-sectionally and longitudinally links "bad" behaviors ("bad" from the white middle-class point of view) to membership in given social classes as one that has potential for major social consequences, particularly in the hands of institutions driven by the profit motive. Its darker side includes the potential for control of large groups of people by the captains of industry and, more important, selective exclusion of certain groups from the workplace altogether. In a broader framework, we must ask if it is useful to have a theory that is historically and culturally bound. A positive answer means that this theory may only apply or "fit" the circumstances of modern America or modern Northern Europe. If so, then how do we determine its limits on a pragmatic level?

Theme 2: An Antecedent to Serious Alcohol Problems Is a Genetic Predisposition to Them

Twin studies, half-sibling studies, and, in particular, adoptee studies have laid the groundwork for examining the genetic and environmental contributions to the development of serious alcohol problems. The most powerful design is the adoptee study that I concentrate on herein (from Denmark: Goodwin, Schulsinger, Hermansen, Guze, & Winokur, 1973; Goodwin, Schulsinger, Knop, Mednick, & Guze, 1977; from Sweden: Bohman, 1978; Bohman, Sigvardsson, & Cloninger, 1981; Cloninger, Bohman, & Sigvardsson, 1981; from the United States: Cadoret, Cain, Grove, 1980; Cadoret & Gath, 1978; Cadoret, O'Gorman, Troughton, & Heywood, n.d.). These three studies have supported the premise that a genetic factor is implicated in the development of alcoholism.

Although only a handful of scientists have critically evaluated the adoptee studies (El-Guebaly & Offord, 1977, Fillmore, 1988; Murray, Clifford, & Gurling, 1983; Peale, 1986; Schuckit, 1984; Searles, 1988), these commentaries suggest that there are serious limitations of the generalizability of the findings. Among the criticisms are:

- Persons with psychiatric or behavioral problems are not likely to mate randomly.
- Early deprivation of the parent-child relationship or early environmental stress may exacerbate later alcoholism.
- The ages of follow-up in these studies are not those when the prevalence of chronic alcoholism is highest.
- The sporadic nature of alcohol problems (as reported in epidemiological research of the general population) has not been accounted for.
- There is quite limited information on environmental variables occurring between birth and follow-up.
- The sex of biological parent with alcoholism was not always stated in two of the studies.
- Two of the studies suffered substantial sample loss at follow-up.
- Control groups were inappropriate in two of the studies.
- The criteria for a diagnosis of alcoholism among offspring differed across studies (in one, relying on the "impressionistic" records of the adoption agency for alcohol problems among parents and interviews with adopted parents for alcohol problems among offspring; in another, relying on somewhat "mild " criteria in the evaluation of alcoholism; and in another, relying on Temperance Board and other social records).
- A matching effect may be present in these studies as a part of the adoption process itself where offspring were placed in homes that resembled the social class characteristics of their biological parents.
- In one study, the male probands and controls differed with respect to important variables often associated with alcohol problems (e.g., marital status).
- In two studies, cohort and historical effects were not accounted for.

The potential confounding factors in these studies suggest caution in interpretation and care in reanalysis. Especially noteworthy is the lack of integration in this body of work with other theoretical perspectives and important findings in the contemporary alcohol field. These studies tend to ignore or only pay lip service to the many studies of powerful environmental variables in the development of alcohol problems, for instance, childhood acquisition of "deviant" drinking and subsequent "deviant" drinking (Donovan, Jessor, & Jessor, 1983; Jessor & Jessor, 1977), the relative impact of availability of alcohol on individuals' drinking (Wish, Robins, Hesselbrock, & Helzer, 1979), the relationship between per capita consumption and alcohol problems in societies as a whole (Bruun, Edwards, Lumio et al, 1975) or the effects of economics in reducing alcohol problems on the individual and aggregate levels (Kendell, de Roumanie, & Ritson, 1983). Because some of the advocates of the hereditary argument see alcoholism as an "either/or" concept, the notions of a continuum of alcohol problems and alcohol problems as nonstatic behaviors tend to be dismissed. Data supporting continuum notions include the findings of Cahalan and Room (1974) on the continuum of alcohol problems in the general population; the findings of Fillmore, Bacon, and Hyman (1979) and Roizen, Cahalan, and Shanks (1978) on the sporadic character of drinking problems in the general population; and the findings of Polich, Armor, and Braiker (1981) on the sporadic character of drinking problems among treated alcoholics.

It takes little imagination to envision the implementation of the genetic hypothesis in the American setting. Popular thinking, reflecting the powerful social influence of AA ideology, has already mistranslated the current trend in this research agenda

> away from the search for an inherited mechanism that makes the alcoholic innately incapable of controlling his or her drinking. Rather popular conceptions are marked by the assumption that any discovery of a genetic contribution to the development of alcoholism inevitably supports classic disease-type notions about the malady. (Peele, 1986, p. 69)

Because contemporary American scientific and popular thought on the problems of alcohol are so intermixed historically, it is not surprising that this theory has given popular credibility to the view that "once an alcoholic, always an alcoholic." If, indeed, popular notions endorse this view, a predisposing factor to alcoholism that allegedly resides in genetic makeup has serious implications for social policy.

History, of course, provides significant lessons for us in view of the fact that this is not the first time science has explored this theoretical perspective. At the end of the 19th and the beginning of the 20th centuries, the notion of a hereditary link between alcoholism in parents and problems in their children was popular in medical, psychiatric, and social thinking and was clearly reinforced by temperance advocates (Bynum, 1984). As in contemporary times, there was never a "critical experiment" to demonstrate beyond a shadow of a doubt that a biological basis for alcoholism exists. During this period, recommendations for prevention included inhibiting alcoholics from marrying and having children (Bynum, 1984; Edwards, 1984).

It is not farfetched to imagine that the children of alcoholics might be seen to deserve special consideration such that labeling would take place at an early age and abstinence would be required of them. In fact, the seeds are well sown for this, as evidenced by Adult Children of Alcoholics—a social movement currently sweeping the nation that may be characterized as a new marketing strategy to attract new clientele to treatment in an era of cuts in mental health budgets. In any event, a genetic hypothesis would deflect attention away from consideration of such factors as the availability of alcohol so that, as Edwards (1984) so aptly put it, it "would be a convenient message for those wanting no meddlesome tampering with the liquor supply."

There are also clear historical lessons that are applicable to relating genetic claims in the world of work. Barrows maintains that the association of alcohol abuse with workers in late 19th-century France was used to justify an overarching explanation for the French defeat (Barrows, 1979). Respected scientists published detailed family trees of the debauched ancestry of armed children seized at the end of the uprising. Similarly, in late 19th-century Germany, scientists endorsed the view that "the drunken comportment of starving workers was attributed by their more well-to-do contemporaries to biological deficiencies [drunkenness and alcoholism seen as inherited mental illness] which, in their reasoning, were typical of industrial workers" (Vogt, 1980). This reasoning served to justify control of the superior classes over the degenerated ones.

The place of work is central to life in the Soviet Union. Recent efforts to deal with alcohol problems in this setting include compulsory treatment; appointment to lower remunerated jobs; nomination to a lower post; deprivation of bonus and of the place at a resthouse; postponement of housing conditions improvement; and criticism in factory and wall newspapers, in radio broadcasts, and on satirical posters (Alekseev & Koshkina, 1986). One must naturally wonder how the workplace in such a cultural setting during a crisis economy would deal with alcoholics who were differentiated on a constitutional versus environmental basis, as was the case under national socialism in Germany. Fahrenkrug (1987, p. 20) reports of the era of German national socialism that under the most severe war conditions, "modern alcohol control showed its true face: 'For the registration of alcoholics and dangerous alcoholics, especially early registration,' claimed police chief Major Messer of the Reich Office, 'no dangerous alcoholic, no person who has fallen under the influence of alcohol may . . . remain unknown to the state and party' " (Messer, 1942, as quoted in Fahrenkrug, 1987, p. 20). Among those called upon to report such persons for registration were the health insurance organization, the national insurance office, the German work front, the plant manager, the shop steward, and the factory nurse (Messer, 1942, as quoted in Fahrenkrug, 1987). Under the "Social Inability Law," "socially incapable inebriates obtained the degrading punishment of a complete exclusion from national life and were surrendered to 'annihilation through work' " (Fahrenkrug, 1987, p. 25).

In the vein of this brief historical examination, it is clear that genetic arguments have more often been directed to the lower, most disadvantaged classes, workers included. For instance, sterilization of inebriates under national socialism was concentrated among lower-class males between ages 30 and 40 (Meggendorfer, 1940, as cited in Fahrenkrug, 1987).

Although information is lacking on the social characteristics of those identified within the American workplace as candidates for alcoholism treatment, there is indirect evidence that workers from lower-status jobs are overrepresented (Trice, 1965; Trice & Beyer, 1977; Warkov et al., 1965) and some evidence that higher-status supervisors handle their employees' alcohol problems on a "private, personal basis" compared to lower-status supervisors who tend to use the EAP-type mechanism in a less constructive way (Trice & Beyer, 1977). While from an "objective" criterion (e.g., Calahan, 1970), persons in lower-status jobs and/or in the lower classes are overrepresented as experiencing alcohol problems, it is instructive that practices of intervention (hypothetically motivated on a "humanitarian" motif) in contemporary America are targeting the same class of people as in, for example, fascist "interventions." This gives us pause, when considering the breadth of social circumstances in which a genetic hypothesis might take seed and flower, to consider how and why it comes to pass that the drinking behavior of the most disadvantaged in the society is that most typically targeted for intervention.

Theme 3: Environmental Change on the Aggregate Level Can Govern the Incidence, Chronicity, and Remission of Serious Alcohol Problems on the Individual Level

We now turn to a collection of longitudinal studies that measures changes in the environment hypothesized to affect drinking practices. Some of these studies contribute to a literature sometimes called the public health perspective. The theoretical framework specifically relevant to the alcohol field on which much of this research rests emphasizes situational factors influencing drinking patterns and problems, above and beyond psychological or biological factors, which until the 1960s and 1970s had dominated the field (see Bruun, 1971; Lemert, 1967; Makela, 1978). It is postulated that the social control system and the contexts and situations of drinking may be used to explain much drinking behavior. The complicated questions being addressed, then, are to what degree can drinking patterns and problems change, either on the individual or aggregate level, if the environment is manipulated? Further, if there is a manipulation, to what degree is observed change in drinking the result of the individual(s) being exposed to one or another naturally occurring environment? In contrast to the antisocial behavior and genetic hypotheses, in this instance, individual-level behavior is hypothesized to flow from aggregate-level influences. Furthermore, these studies do not necessarily rely on a disease-like conception of alcoholism but move toward continuums of consumption and problem behaviors.

Nine longitudinal studies were located that assess environmental change. Two studies evaluated the availability of alcohol with respect to consumption and problems. In the first, Robins (1973) studied heroin and alcohol use among U.S. Vietnam veterans before and after leaving Vietnam, with the major finding that the higher prevalence of heroin use in Vietnam was at least partially contingent on its higher availability; alternatively, the higher prevalence of alcohol use in the United States was also partially contingent on its higher availability. In the second, Plant (1979) studied newly recruited manual workers to the brewing industry and compared them to controls recruited to a nonbrewing industry. The exposure to the drink trade did increase alcohol problems among the brewing industry group, interpreted on evidence of greater availability and increased peer pressure.

One study evaluated increased availability of alcohol in the face of rapid industrial change. Caetano, Suzman, Rosen, and Voorhees-Rosen (1983) studied two Shetland Island communities, one more directly affected by the impact of the rapidly developing oil industry. Increases in mean alcohol consumption were noted in both regions but particularly in the target zone and particularly among those subjects under age 30. Availability of alcohol and a rise in income were interpreted to contribute to the increase in consumption (the increase in consumption being attributable to increases in frequency of drinking, rather than quantity per occasion).

One study challenged the single distribution hypothesis (i.e., the relationship between per capita sales and alcohol problems). Fitzgerald and Mulford (1984) examined consumption and problems in the state of Iowa as a function of increases and decreases in per capita consumption on a seasonal basis between 1979 and 1980. They found that heavy drinking and alcohol problems changed little, but usually in the opposite direction of the sales change. It was concluded that the single distribution model is inadequate.

Only one study could be located that evaluated changes in price of alcohol on individual-level drinking practices. Kendell, de Roumanie, and Ritson (1983) evaluated the effects of the increase in the price of alcohol in an area in Scotland. Although this study reported decreases in consumption and adverse consequences among all drinkers, it did not include a control group. The study addresses the criticisms of the per capita consumption argument, the thesis that alcohol problems in the aggregate tend to follow the rise and fall of per capita consumption (Bruun et al., 1975). The criticism has been that even with social controls exerted on a population, those controls (such as price) would only affect "social" drinkers, not heavy or problematic drinkers (Nathan, 1983). This study observed a decrease in consumption and problems even among heavy and problematic drinkers.

Only one study could be located that evaluated a deterrence model in changing consumption on the individual level. Homel (1986) evaluated the impact of roadside random breath tests in one city in New South Wales. While the deterrence model was supported in this research, there was no control community.

Three studies evaluated educational intervention. In the first, Greenfield and Duncan (1984) evaluated an alcohol abuse prevention program using a naturalistic longitudinal design in a U.S. college setting. Levels of exposure to the multicomponent program were confounded by self-selection into various living groups (e.g., fraternities versus dormitories) with widely differing drinking norms, potentially biasing estimates of program impact. These analysts faced up to their complex self-selection bias by modeling the way individual characteristics determine choice of living groups and then using these selection results to adjust the equations for estimating program effects. The analysts regard their findings as most tentative and classify them as a methodological pilot study. However, one finding that has relevance to the debate on the extent of environmental influences is that an agglomeration of like individuals, rather than direct peer influence, accounted for the substantial differences in living-group drinking patterns. In the second study, Giesbrecht, Pranovi, and Wood (1989) evaluated a community education program in a quasi-experimental design study over a two-year period of three Ontario communities for an assessment of the distribution of consumption model (Ledermann, 1956, 1964). Consumption declined in all communities, with a somewhat greater decline in the intervention community. In the third, Bagnall (1987, 1989) evaluated alcohol education programs for 13 year olds in three regions in Britain using experimental and control schools with the finding that there was a modest influence of the education on drinking knowledge and on consumption.

There are major methodological difficulties and limitations with the majority of these studies.

First, there is an absence of control groups or control groups are inadequate due to "self selection" bias. This increases the complexity and difficulty of the questions this research seeks to address. Often the self-selection and control group problems are unavoidable in this kind of research. These studies do not necessarily reflect poor planning. Rather it is because many of them study social forces which are rarely under the researcher's control. However, this presents an inherent bias because not all variations are available for study.

Second, the time between measurement points is inadequate to evaluate the long-term effects of the environmental change. Most of these studies used a three-year period between baseline and follow-up. An equally important question to be addressed is the long-term influence of social change on a population. While accumulation of social and medical statistics over the long term can partially evaluate this issue and while the intervention can get lost in other historical changes, the importance of observing new cohorts "born into" the already existing historical change must be addressed. In other words, while one may conclude that, at least in the short term, heavy drinkers are decreasing or increasing their consumption as a result of changes in price or various interventions, it is quite different to suggest that the social change will influence the same persons over the long term or will influence new generations. In this respect, multiple follow-ups over longer periods of time, as well as incorporation of new birth cohorts into these follow-ups, should yield designs that more specifically address the long-term effects of social controls and social change.

Third, all these studies have been performed in North American or Northern European countries (the exception being one study in New South Wales), limiting the ability to generalize to other cultural contexts.

Much of the groundwork of the studies that seek to evaluate aggregate level influences is based on the public health perspective literature that posits that reduction of alcohol problems might be enhanced by use of a variety of control strategies, most prominent being increases in the price of alcohol and/or decreases in its availability. Bucholz and Robins (1989) point out that this literature makes two assumptions: "first, that governmental policies can affect the general level of alcohol consumption, and second, that a reduction in overall consumption will lead to reduced rates of alcohol problems. These assumptions are independent, and both must be demonstrated to be correct to support the effectiveness of control policies"(pp. 173–74).

It is generally agreed that the first assumption is correct since price and alcohol use are inversely related (Popham, Schmidt, & de Lint, 1976); however, for the purposes here, it might be looked at with greater care. Peele (1987) argues that while this finding may be valid, it is important to take into account repercussions of increasing price or decreasing availability in light of a society's proclivity to replace alcohol with other psychoactive drugs or to manufacture illicit alcohol. Equally important, Peele's review of the relevant literature led him to posit that for such measures to actually "work," the society or relevant group should be homogeneous in its drinking style, the populace must strongly support decreases in availability, and the effort should be conceived of as a major shift in drinking behavior and attitudes. His analysis strongly suggests that, in a country like the United States, the prognosis for effectively utilizing control-of-supply strategies may not be bright.

There are even greater problems with the assumption that a reduction in consumption will be accompanied by a reduction in alcohol problems.

This link is not clear and the studies (e.g., natural experiments and historical studies) used to determine the direction of this relationship are in disagreement (see both Peele, 1987, and Bucholz & Robins, 1989, for discussion of these studies). However, taking into account the range of historical documentation on the issue of alcohol controls and the recent technology that has been applied to use of historical and economic data as related to the consequences of alcohol use, there does seem to be a literature emerging that is better documenting shifts in alcohol problems over long historical periods as a hypothesized result of alcohol controls (Room, 1988).

It is pertinent to address the practicability and applicability of the control argument across diverse cultural settings. It is clear that government policy is intimately related to production, trade, and distribution as they are associated with price, availability, and public attitudes (Farrell, 1985). These relationships differ widely, however. Noteworthy is that some societies are highly dependent on alcohol as a part of their economy. For instance, in Zambia, the brewing industry almost dominates the holding company of that country, employing one-quarter of its employees; in Western Samoa, the revenue from alcohol contributes to 7 percent of all government revenue (Farrell, 1985). Farrell comments:

> The needs of developing countries are compelling and their resources painfully inadequate. So the prospect of employment, production, and tax revenue from a stable, profitable industry such as production and distribution of alcoholic beverages is understandably appealing. (p. 7)

While raising the price or decreasing the availability of beverage alcohol in these cultural contexts would probably, in fact, decrease consumption of legal alcohol, it would also probably increase the consumption of illicit brew and, furthermore, would be at odds with the economic interests of these developing countries.

From a historical and cross-cultural point of view, economic interests in the workplace have been highly implicated in encouraging drinking among workers and, subsequently, using the workers' drinking to institute greater controls. Among the many examples that social historians and anthropologists have made available, Hunt's (1987) description of the mine owners and the Bantu in South Africa illustrates this point:

> Here [in Khoikhoi] the use of alcohol became an important element in the Europeans' increasing need for wage labourers. This need was of paramount importance once gold and diamonds had been discovered. Mineowners encouraged traders to entice the Bantu into debt, and once indebted they could then be signed up for work in the mines. Alcohol was an important element

> within this trading. In addition to assisting the development of a labor force, it was also an important item for sale within the mining compounds. The mineowners erected beerhalls in the compounds and encouraged its sale to the Bantu. This extensive use of alcohol was again not without its problems, and soon the mineowners were complaining of a fall in labour productivity and an increase in labour absenteeism. As a result of these increasing complaints, Prohibition was introduced in 1896, whereby alcohol was banned for the Bantu except in the form of beer sold from government-run beerhalls. (p. 270)

This example illuminates circumstances in which alcohol control measures have emerged as a result of a web of historical circumstances, often motivated by profit and dominance over lower-status social groups.

The control-of-supply argument tends to ignore the rich and varied customs of drinking in diverse social groups and the observation that drinking can represent an expression of group solidarity. In reference to occupational considerations, Cosper (1979) has pointed out that social control theory in general "does not explain why drinking should be valued by, or rewarding to, members of a particular occupational group" (p. 876), and he has alerted us to the importance of drinking in occupational subcultures as "communicative behavior symbolizing social solidarity and the situation, wealth, masculinity, identity or superiority of the group as well as reward or rejection" (p. 886). Our understanding of these meanings of drink and drinking in a wide variety of social groups, including occupational groups, has yet to be elucidated. Noting the observation of Selden Bacon (1976) that in modern times both the public and scientific language of alcohol is almost solely concerned with its problematic aspects, Levine comments that we have yet to articulate a working-class perspective on the role of alcohol in modern society. Rather, "since the end of the 18th century and the beginning of the 19th century discussion... has been dominated by the social outlook or world view of two social classes: larger capitalists and the middle class" (Levine, 1978, p. 24). Despite our ignorance on these matters, one pertinent description of groups subjected to political and economic domination is that drinking constitutes the "world's oldest on-going protest demonstration" (Lurie, 1979).

Like our consideration of the individual-level hypotheses evaluated earlier in this chapter, there is no doubt that in contemporary times, as in the past, it is the lower-status worker who would most likely be the target for and recipient of a wide variety of preventive measures. After all, the "prevention" strategies enthusiastically adopted by the captains of industry at the turn of the century in the United States were directed almost solely toward the lower-status worker and were regarded as successful in removing alcohol from the workplace and the factory door (see Fillmore & Caetano, 1982, for a discussion of this history).

Modest recent findings have suggested that alcohol may still be present in the American workplace, particularly in some industries dominated by blue-collar workers where alcohol and other drugs are consumed on the job lot during breaks and lunch hours (Ames & Janes, 1986). These analysts concluded that "the social organization of the workplace, including the important factors of job alienation, job stress, inconsistent social controls and the evolution of a 'drinking culture,' is implicated... to be a primary vehicle for promoting high levels of alcohol use" (p. 23).

During the American temperance movement, some workplace policies required that workers be totally alcohol free; in contemporary America, at least one industry has required that its employees be entirely cigarette smoke free (Pittman, 1990). As noted above, considerable energy is being devoted to ensuring that potential employees displaying traces of other drugs be excluded from the workplace. These trends in an era of new conservatism with an emphasis on social control raise serious issues of civil liberties.

Recommendations for prevention in the modern worksite seem to follow two general tactics. First, in line with control-of-supply arguments, strategies are proposed to remove alcohol from the workplace. Second, strategies that change the nature of the work role (e.g., creating "a working environment that gives workers greater control over, and thus greater involvement in, the products of their work" [Ames & Janes, 1986] or, from another source, "involving the employees in profit or work sharing schemes whereby personal and co-operative responsibility of the finished job is encouraged," improving recreational facilities and so on [World Health Organization, 1986]) are proposed. Given the inevitable focus of prevention on lower-status workers, and taking into account the profit motive of the American capitalists, it is likely that the former, rather than the latter, would be implemented.

Control strategies, according to Peele (1987), have "neither demonstrated their beneficial impact nor calculated their potential costs, such as increases in police corruption and the criminalization of large numbers of Americans brought on by legal approaches, and the loss of civil rights and diagnostic subtlety resulting from coercive therapy referrals" (p. 74). Moreover, they communicate the message that drinking is hazardous. By doing so, "we actively attack the belief in and the capacity for self-management that remain the strongest prophylactics against substance abuse and addiction" (p. 75).

Conclusion

The goal of this chapter was threefold: first, to examine three "causative" explanations of alcohol problems from the longitudinal literature with respect to their scientific limitations; second, to use history and cross-cultural reports as tools to appraise their relative applicability as social policy; and third, to look specifically at the workplace as a setting for social policy for each of the three explanations with an eye to the "darker side" of policy consequences. The rationale for weaving these three objectives together was based on the observation that a close relationship in contemporary America exists between the source of funding, as sponsored by policy makers, and research interests. Because most scientific explanations in the alcohol field are discipline-specific, often advocating single bodies of explanation, and because social policies are typically advocated with greater emphasis on their benefits, it is important to assess the "dark side" or the possible costs, expressed in social and ethical consequences, of implementing these findings as social policy.

With respect to the objective of a scientific appraisal of the limitations of these research findings, each body of research suffers from methodological weaknesses, self-selection problems, and control group problems. Additionally, each of the explanatory frameworks are geographically biased (i.e., most of the research has been performed in North America and Northern Europe) and there has been almost no attention given to systematically testing findings cross-historically, cross-culturally, or across birth cohorts. Last, each has flown under a disciplinary, even ideological, flag with its own attendant assumptions and with little attention paid to competing hypotheses from other disciplinary perspectives. These biases not only curtail generalization of results but also narrow our vision for assessing diverse social policy implications of research findings.

With respect to the objective of using historical and cross-cultural documents as sources for understanding the conditions and explanations given for various methods of social control, we are alerted to the fact that history and cross-cultural comparison potentially have a great deal to teach us and that caution is warranted when advocating single explanatory frameworks for alcohol-related prevention and intervention. A gaze *across* the three explanatory frameworks from our brief historical and cross-cultural excursion has made one thing clear. Whether the hypothesis has a psychological, biological, or social control origin, we can conclude that the social groups that policies are directed toward and affected by are those that have less political leverage, fewer financial resources, less education, and lower-status jobs (if jobs at all).

This observation raises the controversy between "disease-oriented" and "prevention-oriented" scientists to a higher level if it can be appreciated that policies emerging from both perspectives may be used for gain by the powerful in exploitive economic and social contexts. In contemporary times, the advocacy of health promotion and prevention strategies as well as the humanitarian treatment of alcohol and drug problems, particularly those based on single bodies of evidence, need to be looked at with greater scrutiny and with considerable skepticism with careful appraisal of broader economic and political forces.

With respect to the objective of translating the potential policy implications of these scientific explanations into the workplace setting, the historical and cross-cultural lessons remind us that policies applied to the workplace have an added "dark side." A glimpse behind the doors labeled humanitarian treatment, health promotion, or prevention may reveal nothing more than the profit motive, resulting in coercion and control of lower-status groups, and may incur costs that challenge the boundaries of human rights and even human life.

Relatedly, it is important to point out that policies effected in the workplace, both now and in the past, seem especially prone to a narrowness of vision and that considerable evidence exists to suggest that science has often been the "handmaiden" of policy in these contexts (Fillmore, 1984). While the ideological flags under which these policies have flown can be characterized as "prevention" in America's temperance era to "intervention" in the current era, both have been closely linked to responding to the profit motive as governed by the captains of industry.

All these considerations should give us pause as members of the scientific community. In a period where there is (1) a virtual explosion of scientific information and technological advancement but an implosion of disciplinary examination, in part due to the competition for limited research funds; (2) a cry for effective social policy on a number of fronts, including the workplace, to control or treat alcohol- and drug-related behaviors, and (3) a reciprocal economic, even potentially ideological, relationship between policy makers and the scientific establishment, we, as scientists, must proceed with caution. At the very least, our findings should be discussed in the context of competing explanations; at the most, we should strive to test competing hypotheses, replicate findings in a systematic manner, engage in cross-disciplinary research, and seriously warn the interested reader of the research limitations of findings that may have potential for implementation in social policy. If we do not, our science faces the possible label of ideology.

Acknowledgments

This work was supported by a National Institute on Alcohol Abuse and Alcoholism Research Scientist Development Award (Grant K01 AA–0073). Because this work is an effort to integrate three literatures, portions of this chapter have been published elsewhere or have been adapted from other papers (Fillmore & Caetano, 1982; Fillmore, 1984, 1988). My appreciation is extended to Paul M. Roman for his critical comments and helpful suggestions in the preparation of this manuscript. An earlier version of this chapter was presented as a paper at the Second Annual Conference of the R. Brinkley Smithers Institute for Alcoholism Prevention and Workplace Problems, November 2–3, 1989, Cornell University, New York State School of Industrial and Labor Relations, Ithaca, New York.

References

Alekseev, S.S., & Koshkina, E.A. (1986, June). *Organization aspects of antialcohol activities at industrial enterprises.* Paper prepared for the Informal Consultation on Drug and Alcohol-Related Problems in Employment, WHO and ILO Report, Geneva.

American Psychiatric Association. (1987). *Diagnostic and statistical manual of mental disorders.* Washington, D.C.: Author.

Ames, G.M., & Janes, C.R. (1986). *Heavy and problem drinking in an American blue-collar population: Implications for prevention.* Working paper. Prevention Research Center, Berkeley, CA.

Amundsen, A. (1982). *Who became patients in institutions for alcoholics?* Paper presented at the ICAA Epidemiology Section Meetings, Helsinki.

Anderson, B.G. (1979). How French children learn to drink. In M. Marshall (Ed.), *Beliefs, behaviors, and alcohol beverages: A cross-cultural survey* (pp. 429–432). Ann Arbor: University of Michigan Press.

Bacon, S.D. (1976). Concepts. In W.J. Filstead, J.J. Rossi, & M. Keller, (Eds.), *Alcohol and alcohol problems: New thinking and new directions* (pp. 57–134). Cambridge, MA: Ballinger.

Bagnall, G.M. (1987). Alcohol education and its evaluation—Some key issues. *Health Education Journal, 46,* 162–65.

Bagnall, G.M. (1989, August). *Alcohol education for 13 year olds—A controlled evaluation study.* Paper presented at the Second Meeting of the Collaborative Alcohol-Related Longitudinal Project, Institute for Health and Aging, University of California, San Francisco.

Bohman, M. (1978). Some genetic aspects of alcoholism and criminality. *Archives of General Psychiatry, 35,* 269–276.

Bohman, M., Sigvardsson, S., & Cloninger, C.R. (1981). Maternal inheritance of alcohol abuse: Cross-fostering analysis of adopted women. *Archives of General Psychiatry, 38,* 965–969.

Brenner, M.H. (1969). Patterns of psychiatric hospitalization among different socio-economic groups in response to economic stress. *Journal of Nervous and Mental Disease, 148,* 31–38.

Brenner, M.H. (1977). Health costs and benefits of health policy. *International Journal of Health Services, 7,* 581–623.

Bruun, K. (1971). Implications of legislation relating to alcoholism and drug dependence. In L.G. Kiloh & D.S. Bell (Eds.), *International Congress on Alcoholism and Drug Dependence.* Butterworths, Sydney.

Bruun, K., Edwards, G., Lumio, M., Makela, K., Pan, L., Popham, R.E., Room, R., Schmidt, W., Skog, O.J., Sulkunen, P., & Osterberg, E. (1975). *Alcohol control policies in public health perspective, 25.* Finland: Finnish Foundation for Alcohol Studies.

Bucholz, K.K., & Robins, L.N. (1989). Sociological research on alcohol use, problems and policy. *Annual Review of Sociology, 15,* 163–186.

Bynum, W.F. (1984). Alcohol and degeneration in 19th century European medicine and psychiatry. *British Journal of Addiction, 79,* 59–70.

Cadoret, R.J., Cain, C.A., & Grove, W.M. (1980). Development of alcoholism in adoptees raised apart from their alcoholic biologic relatives. *Archives of General Psychiatry, 37,* 561–563.

Cadoret, R.J., & Gath, A. (1978). Inheritance of alcoholism in adoptees. *British Journal of Psychiatry, 132.*

Cadoret, R.J., O'Gorman, T.W., Troughton, E., & Heywood, E. (n.d.). *Alcoholism and antisocial personality: Inter-relationships, genetic and environmental factors.* University of Iowa College of Medicine, Dept. of Psychiatry, Iowa City, Iowa.

Caetano, R., Suzman, R.M., Rosen, D.H., & Voorhees-Rosen, D.J. (1983). The Shetland Islands: Longitudinal changes in alcohol consumption in a changing environment. *British Journal of Addiction, 78,* 21–36.

Cahalan, D. (1970). *Problem drinkers.* San Francisco: Jossey-Bass.

Cahalan, D., & Room, R. (1974). *Problem drinking among American men.* New Brunswick, NJ: Rutgers Center of Alcohol Studies.

Clausen, J.A. (1978). Longitudinal studies of drug use in the high school: Substantive and theoretical issues. In D.B. Kandel (Ed.), *Longitudinal research on drug use: Empirical findings and methodological issues.* (pp. 235–248). Washington, DC: Hemisphere (Halstead/Wiley).

Cloninger, C.R., Bohman, M., & Sigvardsson, S. (1981). Inheritance of alcohol abuse. *Archives of General Psychiatry, 38,* 861–868.

Cosper, R. (1979). Drinking as conformity: A critique of sociological literature on occupational differences in drinking. *Journal of Studies on Alcohol, 40,* 868–891.

Dixon, K. (1984). The coercion of labor by mental health professionals. *Social Policy, 14,* 47–54.

Donovan, J.E., Jessor, R., & Jessor, S.L. (1983). Problem drinking in adolescence and young adulthood: A follow-up study. *Journal of Studies on Alcohol, 44,* 109–137.

Douglas, M. (1987). A distinctive anthropological perspective. In M. Douglas (Ed.), *Constructive drinking: Perspectives on drink from anthropology* (pp. 3–15). Cambridge: Cambridge University Press.

Edwards, G. (1984). Alcoholism, genetics and society. *British Journal of Addiction, 79,* 353.

El-Guebaly, N., & Offord, D. (1977). The offspring of alcoholics: A critical review. *American Journal of Psychiatry, 134,* 357–383.

Fahrenkrug, W.H. (1987). Conceptualization and management of alcohol-related problems in Nazi-Germany, 1933–45. In S. Barrows, R. Room, & J. Verhey (Eds.), *The social history of alcohol: Drinking and culture in modern society.* Berkeley, CA: Medical Research Institute of San Francisco, Alcohol Research Group.

Farrell, S. (1985). *Review of national policy measures to prevent alcohol related problems.* Report of the World Health Organization.

Fillmore, K.M. (1984, March). Research as a handmaiden of policy: An appraisal of estimates of alcoholism and its cost in the workplace. *Journal of Public Health Policy,* 40–64.

Fillmore, K.M. (1988). *Alcohol use across the life course: A critical review of 70 years of international longitudinal research.* Toronto: Addiction Research Foundation.

Fillmore, K.M., Bacon, S.D., & Hyman, M. (1979). The 27-year longitudinal panel study of drinking by students in college, 1949–1976 (Final report to the National Institute on Alcohol Abuse and Alcoholism under Contract No. ADM281–76–0015). Berkeley, CA: University of California, Social Research Group.

Fillmore, K.M., & Caetano, R. (1982). Epidemiology of alcohol abuse and alcoholism in occupations. In *Occupational alcoholism: A review of research issues* (NIAAA Research Monograph No. 8, DHHS Publication ADM82–1184). Washington, DC: U.S. Government Printing Office.

Fillmore, K.M., Hartka, E., Johnstone, B.M., Speiglman, R., & Temple, M.T. (1988). *Spontaneous remission from alcohol problems: A critical review.* Report commissioned by the U.S. Institute of Medicine.

Fillmore, K.M., & Sigvardsson, S. (1988). "A meeting of the minds": A challenge to biomedical and psychosocial scientists on the ethical implications and social consequences of scientific findings in the alcohol field. *British Journal of Addiction, 83,* 609–611.

Fitzgerald, J.L., & Mulford, H.A. (1984). Seasonal changes in alcohol consumption and related problems in Iowa, 1979–1980. *Journal of Studies on Alcohol, 45,* 363–368.

Giesbrecht, N., Pranovi, P., & Wood, L. (1989). *Research agenda and community interests: Lessons from a prevention project.* Paper presented at the Symposium on Experiences with Community Action Projects for the Prevention of Alcohol and Other Drug Problems, Scarborourgh, Ontario, Canada.

Goodwin, D.W., Schulsinger, F., Hermansen, L., Guze, S.B., & Winokur, G. (1973). Alcohol problems in adoptees raised apart from alcoholic biological parents. *Archives of General Psychiatry, 28,* 238–243.

Goodwin, D.W., Schulsinger, F., Knop, J., Mednick, S., & Guze, S. (1977). Alcoholism and depression in adopted-out daughters of alcoholics. *Archives of General Psychiatry, 34,* 751–755.

Greenfield, T.K., & Duncan, G.M. (1984). *Evaluation of an alcohol abuse prevention program correcting for self-selection.* Paper presented at the 92nd Annual Convention of the American Psychological Association, Toronto, Canada.

Hoffmann, H., Loper, R.G., & Kammeier, M.L. (1974). Identifying future alcoholics with MMPI alcoholism scales. *Quarterly Journal of Studies on Alcohol, 35,* 490–498.

Homel, R. (1986). Policing the drinking driver: Random breath testing and the process of deterrence (Report No. CR42). New South Wales, Australia: Department of Transportation, Federal Office of Road Safety.

Hunt, G. (1987). Spirits of the colonial economy. In S. Barrows, R. Room, & J. Verhey (Eds.), *The social history of alcohol: Drinking and culture in modern society* (pp. 269–270). Berkeley, CA: Medical Research Institute of San Francisco, Alcohol Research Group.

Jessor, R., & Jessor, S.L. (1977). *Problem behavior and psychosocial development: A longitudinal study of youth.* New York: Academic Press.

Jones, M.C. (1968). Personality correlates and antecedents of drinking patterns in adult males. *Journal of Consulting Clinical Psychology, 32,* 2–12.

Jones, M.C. (1971). Personality antecedents and correlates of drinking patterns in women. *Journal of Consulting Clinical Psychology, 36,* 61–69.

Kandel, D.B., Raveis, V.H., & Kandel, P.I. (1984). Continuity in discontinuities: Adjustment in young adulthood of former school absentees. *Youth and Society, 15,* 325–352.

Kendell, R.E., de Roumainie, M., & Ritson, E.B. (1983). Effect of economic change on Scottish drinking habits, 1978–82. *British Journal of Addiction, 78,* 365–379.

Ledermann, S. (1956). *Alcool, alcoolisme, alcoolisation* (Vol 1). Paris: Institut National d'Études Démographiques, Presses Universitaries de France.

Ledermann, S. (1964). *Alcool, alcoolisme, alcoolisation* (Vol. 2). Paris: Institut National d'Études Démographiques, Presses Universitaries de France.

Lemert, E.M. (1967). *Human deviance, social problems and social control.* Englewood Cliffs, NJ: Prentice-Hall.

Levine, H.G. (1978). *Notes on work and drink in industrializing America.* Paper prepared for NIAAA Research Conference on Alcohol Use and the Work Environment, Belmont, MD.

Linton, R. (1945). The scope and aims of anthropology. In R. Linton (Ed.), *The science of man in world crisis* (p. 3). NY: Columbia University Press.

Loper, R.G., Kammeier, M.L., & Hoffmann, H. (1973). MMPI characteristics of college freshman males who later became alcoholics. *Journal of Abnormal Psychology, 82,* 159–162.

Lurie, N.O. (1979). The world's oldest on-going protest demonstration: North American Indian drinking patterns. In M. Marshall (Ed.), *Beliefs, behaviors, and alcoholic beverages* (pp. 127–145). Ann Arbor: University of Michigan Press.

Makela, K. (1978). Levels of consumption and social consequences of drinking. In Y. Israel, F.B. Glaser, H. Kalant, R.E. Popham, W. Schmidt, & R.G. Smart (Eds.), *Research advances in alcohol and drug problems* (Vol. 4, pp. 303–348). New York: Plenum Press.

Marshall, M. (1979). *Weekend warriors: Alcohol in Micronesian culture.* Palo Alto, CA: Mayfield Publishing.

McCord, W., & McCord, J. (1960). *Origins of alcoholism.* Stanford, CA: Stanford University Press.

Meggendorfer, F. (1940). Der schwere Alkoholismus. In *Handbuch der Erbkrankheiten,* hrsg. von A. Gutt, Bd. 3. Berline 1940. S. 278.

Messer, W. (1942). Zur Erfassung Alkoholkranker und Alkoholgefahrdeter, besonders die sogenannte Fruherfassung. In *Die Volksgifte,* Jq. 1942, S. 85f.

Monnelly, E.P., Hartl, E.M., & Elderkin, R. (1983). Constitutional factors predictive of alcoholism in a follow-up of delinquent boys. *Journal of Studies on Alcohol, 44,* 530–537.

Murray, R.M., Clifford, C.A., & Gurling, H.M.D. (1983). Twin and adoption studies: How good is the evidence for a genetic role? In M. Galanter (Ed.), *Recent developments in alcoholism:* Vol. 1. *Genetics, behavioral, treatment, social mediators and prevention, current concepts in diagnosis* (pp. 25–48) New York: Plenum Press.

Myers, J.K., Weissman, M.M., Tischler, G.L., Holzer, C.E., Leaf, P.J., Orvaschel, H., Anthony, J.C., Boyd, J.H., Burke, J.D., Kramer, M., & Stolzman, R. (1984). Six-month prevalence of psychiatric disorders in three communities. *Archives of General Psychiatry, 41,* 959–967.

Nathan, P.E. (1983). Failures in prevention: Why we can't prevent the devastating effects of alcoholism and drug abuse. *American Psychologist,* 139–174.

National Institute on Drug Abuse. (1987, July). *Research on the prevalence, impact and treatment of drug abuse in the workplace.* Announcement No. DA–87–26.

Office of Selection and Evaluation, U.S. Postal Service. (1989). *An empirical evaluation of pre-employment drug testing in the U.S. Postal Services: Interim report of findings.* Unpublished.

Peele, S. (1987). The limitations of control-of-supply models for explaining and preventing alcoholism and drug addiction. *Journal of Studies on Alcohol, 48,* 61–76.

Peele, S. (1986). The implications and limitations of genetic models of alcoholism and other addictions. *Journal of Studies on Alcohol, 47,* 63–73.

Pittman, D.J. (1990). The impact of macro social forces on the distribution of alcohol problems in the workplace. In P. Roman (Ed.), *Alcohol problem intervention in the workplace* (pp. 19–25). Westport, CT: Quorum Books.

Plant, M. (1979). *Drinking careers: Occupation, drinking habits and drinking problems.* London: Tavistock.

Polich, J.M., Armor, D.J., & Braiker, H.B. (1981). *The course of alcoholism: Four years after treatment.* New York: John Wiley and Sons.

Popham, R.E., Schmidt, W., & de Lint, J. (1976). The effects of legal restraint on drinking. In B. Kissin & H. Begleitter (Eds.), *The biology of alcoholism: Vol. 4. Social aspects of alcoholism* (pp. 579–625). New York: Plenum Press.

Rabier, J.E., & Inglehart, R. (1980). *Euro-barometer 9—April 1978, employment and unemployment in Europe.* Ann Arbor, MI: Inter-university Consortium for Political and Social Research.

Robins, L.N. (1966). *Deviant children grown up.* Baltimore: Williams and Wilkins.

Robins, L.N. (1973). The Vietnam drug user returns (Final Report to the U.S. Government Special Action Office for Drug Abuse Prevention. Contract No. HSM–42–72–75).

Robins, L.N. (1984). Changes in conduct disorders over time. In D.C. Farran & J.D. McKinney (Eds.), *Risk in intellectual and psychosocial development.* NY: Academic Press.

Robins, L.N., Helzer, J.E., Przybeck, T.R., & Regier, D.A. (1988). Alcohol disorders in the community: A report from the Epidemiologic Catchment Area. In R.M. Rose & J. Barrett (Eds.), *Alcoholism: origins and outcome* (pp. 15–29). New York: Raven Press.

Roizen, R., Cahalan, D., & Shanks, P. (1978). "Spontaneous remission" among untreated problem drinkers. In D.B. Kandel (Ed.), *Longitudinal research on drug use* (pp. 197–224). Washington, DC: Hemisphere (Halstead/Wiley).

Roman, P.M. (1988). Growth and transformation in workplace alcoholism programming. In M. Galanter (Ed.), *Recent developments in alcoholism* (Vol. 6, pp. 131–158). New York: Plenum.

Roman, P.M., & Blum, T.C. (forthcoming). Employee assistance and drug screening programs. In D. Gerstein & H. Harwood (Eds.), *Treating drug problems* (Vol. 2). Washington, DC: National Academy Press.

Room, R. (1988). *Research on effects of alcohol policy change.* Paper prepared for a meeting on Alcohol Policies: Perspectives from the USSR and Some Other Countries, Balu, USSR.

Schuckit, M. (1984). Prospective markers for alcoholism. In D. Goodwin, K. VanDusen, & S. Mednick (Eds.), *Longitudinal research in alcoholism* (pp. 237–251). Boston: Kluwer-Nijhoff, 1984.

Searles, J.S. (1988). The role of genetics in the pathogenesis of alcoholism. *Journal of abnormal Psychology, 97,* 153–67.

Stojiljkovic, S., & Vasev, C. (1973). *Alcoholism and work.* Paper presented at the International Meeting of Prevention and Treatment of Alcoholism, Belgrade.

Temple, M., & Fillmore, K.M. (1985). The variability of drinking patterns and problems among young men, age 16–31: A longitudinal study. *International Journal of the Addictions, 20.*

Trice, H.M. (1965). Alcoholic employees: A comparison of psychotic, neurotic and "normal" personnel. *Journal of Occupational Medicine, 7,* 94–99.

Trice H.M., & Beyer, J.M. (1977). Differential use of an alcoholism policy in federal organizations by skill level of employees. In C.J. Schramm (Ed.) *Alcoholism and its treatment in industry*. Baltimore: Johns Hopkins University Press.

Vaillant, G. (1983). *The natural history of alcoholism.* Cambridge: Harvard University Press.

Vogt, I. (1980). *Alcohol, casualties and crime: A review of West German studies 1945–1978.* Berkeley: University of California, Social Research Group.

Walter, M.A.H.B. (1982). Drink and be merry for tomorrow we preach: Alcohol and the male menopause in Figi. In M. Marshall (Ed.), *Through a glass darkly: Beer and modernization in Papua New Guinea* (LASER Monograph No. 18, pp. 443–456). Brooks: Papua New Guinea Institute of Applied Social and Economic Research.

Warkov, S., Bacon, S., & Hawkins, A.C. (1965). Social correlates of industrial problem drinking. *Quarterly Journal of Studies on Alcohol, 26,* 58–71.

Wish, E.D., Robins, L.N., Hesselbrock, M., & Helzer, J.E. (1979). The course of alcohol problems in Vietnam veterans. In M. Galanter (Ed.), *Currents in alcoholism* (Vol. VI). New York: Grune and Stratton.

World Health Organization and International Labor Office. (1986, June). *Report of informal consultation on drug and alcohol related problems in employment.* Geneva, Switzerland, June 9–12.

Zucker, R.A. (1979). Developmental aspects of drinking through the young adult years. In H.T. Blane & M.E. Chafetz (Eds.), *Youth, alcohol and social policy.* NY: Plenum Press.

CHAPTER 14

Social Science Research and Alcohol Policy Making

Robin Room

The link between social science research and alcohol policy making reaches back at least 80 years. In the last years of the 19th century, a group of Boston-based academicians and business leaders known as the Committee of Fifty to Investigate the Liquor Problem commissioned a number of studies, extending over several years, of the various aspects of alcohol problems (Billings, 1903; Calkins, 1901; Koren, 1899; Wines & Koren, 1897). A substantial proportion of the studies would now be described as social science, including a survey of drinking patterns among "brain-workers" (Billings, 1903), and a study of alcohol's role in crime that is in some ways still unsurpassed (Koren, 1899). Perhaps the most notable social science effort was the study of *Substitutes for the Saloon* (Calkins, 1901), which involved sending young sociologists and others out to map and report on the place of the saloon and its potential alternatives in 17 big cities of the United States.

At the repeal of Prohibition, as Americans and their legislators suddenly had to face up to designing alcohol control systems, academic social scientists were drawn to alcohol policy issues (Levine & Smith, 1977). Studies were made of the effects of Prohibition (e.g., Bossard and Sellin, 1932; Feldman, 1927; Warburton, 1932), of control systems in other countries (e.g., Thompson, 1935; Wuorinen, 1931), and eventually of the operations of American control systems (e.g., Culver & Thomas, 1940; Harrison & Laine, 1936).

In the 1940s and 1950s, as the alcoholism movement organized itself to influence public policy on the treatment of alcoholism, social scientists were involved in a number of capacities. Selden Bacon played a major role in the organization of what became the national voluntary organization on alcoholism, chaired the first state alcoholism commission, and wrote what were in effect community organizing manuals for the movement (Bacon, 1947, 1949). As state alcoholism commissions became widespread in the 1950s, social scientists played important advisory and sometimes programmatic roles. A social scientist, David Pittman, served for a time as president of the North American Association of Alcoholism Programs. Meanwhile, as urban renewal projects sought to remake the central city and destroy traditional skid rows, social science knowledge about the culture of "urban nomads" was often drawn on as part of the planning (e.g., Bogue, 1963; Dunham, 1954; Rubington, 1958). Social scientists such as Pittman were heavily involved in the eventual adoption in the mid-1960s of the detoxification center as the policy response to the problem of skid row drunkenness (see Room, 1976b).

Social science research also made conceptual contributions to the developing alcohol policies of the 1950s and 1960s. The discovery and description of a population of alcoholics characterized by social stability (Straus & Bacon, 1951; see Straus, 1976) provided evidence for the alcoholism movement's drive to change the social handling of the alcoholic by enhancing the social respectability of alcoholism (Kurtz & Regier, 1975). Sociocultural studies of drinking patterns and problems in different groups (e.g., Bales, 1946; Skolnick, 1958; Snyder, 1958) were seen as pointing to a conclusion that American drinking problems were due to special ambivalence (Ullman, 1958) and lack of consensus (Mizruchi & Perrucci, 1962) about drinking norms—a conclusion that underlay the "responsible drinking" campaigns and policies of succeeding years (Chafetz, 1967, 1971; Plaut, 1967).

In the enormously broadened alcohol arena of the 1960s and 1970s, social scientists have been involved in many aspects of alcohol policy. A social scientist, Robert Straus, chaired the influential Cooperative Commission on the Study of Alcoholism in its formative years, and the commission's reports were all written by social scientists (Cahn, 1970; Plaut, 1967; Wilkinson, 1970). Straus also chaired the National Advisory Committee on Alcoholism in the years of increasing federal involvement in alcohol problems, culminating in the establishment of the National Institute on Alcohol Abuse and Alcoholism (NIAAA) (see National Advisory Committee on Alcoholism, 1968). Harrison Trice and colleagues played an important role in developing and providing a rationale for the strategy of identification and "constructive coercion" or "confrontation" of the alcoholic in the workplace (Roman & Trice, 1967; Trice & Roman, 1972) as industrial alcoholism programs burgeoned in the early 1970s. In recent years, as evaluation studies became a standard instrument of policy, social scientists moved into new and often embattled roles as the bearers of discomforting empirical tidings (e.g., Armor, Polich, & Stambul, 1976; Blane, 1976; Blane & Hewitt, 1977; Blumenthal & Ross, 1975; Gusfield, 1972). In the competition among social and health problems for public funds, social science studies were frequently drawn upon as evidence of an ever-increasing roster and magnitude of alcohol problems (e.g., NIAAA, 1974, 1979; Berry & Boland, 1977).

Social scientists have also been instrumental in recent conceptual developments affecting alcohol policy making. One development is the increasing reconceptualization of alcohol problems in terms of a variety of discrete but overlapping disabilities and problems rather than as a single entity of "alcoholism." This reconceptualization is based in part on general population survey data showing only a modest overlap of drinking problems in the population at large, in contrast to clinical populations (see Clark, 1966; Edwards, Gross, Keller, Moser, & Room, 1977; Straus, 1975, 1976). The conceptual disaggregation of alcohol problems associated with these findings has pointed to new directions for prevention policy making (Bruun, 1970; Gusfield, 1976; Room, 1972). Another development is the reemphasis on the role of alcohol consumption per se—both at the individual level and in the aggregate—in the occurrence of alcohol problems, which has resulted in a renewed focus on the availability dimension of alcohol policies. While Canadian and Scandinavian social scientists have taken the lead in this development (Bruun et al., 1975; deLint & Schmidt, 1968b; Schmidt & Popham, 1980), U.S. researchers have also become involved (Beauchamp, 1976; Medicine in the Public Interest, 1976).

To a large extent, the connections we have sketched above reflect the more official side of social science's relation to alcohol policy making; in some cases the social scientist was acting in an official capacity, in other cases as an adviser to or agent of the authorities. Where new directions of policies were implied or being proposed, there was usually some consensus on the definition of the situation between the social scientist and the policy establishment. Even where the "climate of ideas" at the time may have been hostile to the work (Schmidt & Popham, 1980), there were usually policy makers to sponsor or protect the research work.

However, there are also critical traditions in alcohol social science expressing varying degrees of dissent from a policy consensus. Sometimes the criticism is expressed only in cynical murmurs in convention corridors. Often an unstated divergence can be discerned between social science research and the policy reports based on the research.

Sometimes a divergence can be seen in the writings of an individual social scientist between writings in the policy role and writings outside it. Selden Bacon, writing as an alcoholism movement leader in 1949, criticized dry organizations as "likely to minimize the efforts of a rehabilitation program [and] reluctant to see that prevention emerges from rehabilitation and allied education" (1949, p. 15). Fourteen years later, less centrally involved in the movement, Bacon (1963) was criticizing state commissions on alcoholism for emphasizing only the treatment of alcoholics and treatment-related education, and calling for an equal concern with alcoholism as a public health problem.

Increasingly, critiques of policy-related research are appearing in the social science literature. Light (1976) has challenged the premises of the major economic analysis of the costs of alcoholism (Berry & Boland, 1977), and Marden, Zylman, Fillmore, and Bacon (1976) have criticized as alarmist the reporting of the major federally funded study of adolescent drinking (Rachal et al., 1975). Kurtz and Regier (1975; Regier & Kurtz, 1976) have criticized both the process and the substance of decision making on the Uniform Alcoholism and Intoxication Treatment Act. There has been a considerable controversy involving social scientists over the effects of lowering the drinking age on drunk-driving casualties (see review and references in Whitehead, 1977). The assumptions underlying the "sociocultural" model of drinking problems, emphasizing the role of ambivalence or the absence of moderate drinking norms, have been critiqued (Mäkelä, 1975; Room, 1976a), as have the data and assumptions underlying the "distributionist" model, which emphasizes empirical regularities in the distribution of consumption and relates the overall level of consumption to the rate of alcohol problems (Miller & Agnew, 1974; Parker & Harmon, 1980).

The alcohol social science literature also includes studies that are radically antagonistic to common policy assumptions. A series of skid row studies (Spradley, 1970; Wallace, 1965; Wiseman, 1970) have challenged the official picture of skid row subcultures and their desires. Similarly, Anglo-American theories about Native American drinking have been challenged in the anthropological literature (Levy & Kunitz, 1974). Here also might be mentioned social scientists' involvement in the long and bitter battle over controlled drinking as an outcome of treatment, pitting behavioral psychologists against a fundamental tenet of the alcoholism movement (see review and references in Roizen, 1977a).

Social scientists have thus played a great variety of roles with respect to policy making: policy-making roles, roles in social movements with policy aims, roles as policy advocates and as critics and dissenters, roles as evaluators of policies and programs, and roles as methodological experts. But despite the multiplicity of involvements and roles, there is room for doubt about the effects of social science research on policy. Although policy decisions and social science research findings may often coincide, it would be rash to conclude that the policy decisions were dictated by the research findings.

The experience of Scandinavia, Finland in particular, is instructive in this regard. For an American social scientist with policy interests, Finland appears as a kind of utopia. Finnish alcohol social science has a strong research tradition and continuing institutions dating back more than 25 years, funded out of state alcohol monopoly funds, with regular advisory roles to the national legislature and to the treatment system authorities.

Alcohol social science has had a more systemic viewpoint and a more cumulative research tradition in Finland than in the United States. Over the years, a number of pioneer studies with policy implications have been carried out: for instance, Kuusi's study (1957) of the effects of opening alcohol monopoly stores in rural villages, Lanu's study (1956) of the operation of the buyer surveillance system that limited eligibility to purchase alcohol, Törnudd's study (1968) of the preventive effect of fines and jail terms in cases of public drunkenness, and a cooperative study led by Mäkelä (1974) of the effects of a liquor store strike. In many cases, policy changes appear to have been brought about by the research findings. But as a researcher who has been intimately involved in this history, Kettil Bruun is skeptical that the social science findings played a decisive role:

> A dissertation by Lanu (1956) is often said to have led to the extinction of the buyer surveillance system in Finland—yet the investigation started when the system had already been heavily criticized. Similarly, the decriminalization of public drunkenness was supported by Törnudd's sophisticated study (1968). Nevertheless the explanation for the acceptance of his findings may be that prisons were crowded and the police had more important duties to attend to—in fact, in its proposals to Parliament, the Cabinet [overinterpreted] the implications of the research findings. (Bruun, 1973)

In the American experience of recent decades, also, there is reason to question the apparent influence of social science research on policy. As Bruun (1973) puts it, "social research produces logical arguments rather than logical conclusions regarding policy and action." Thus, social science research tends to be taken into policy discussions in circumstances where it fits a preexisting structure of argument.

The Influence of Social Science Research on Policy

Let us examine several of the historical instances where the strongest case could be made for a social science influence on policy.

1. Straus and Bacon's 1951 study of "Alcoholism and Social Stability" is often cited as instrumental in changing the picture of the alcoholic from the skid row bum to a more middle-class and respectable image. And, indeed, this study found that clients showing up for outpatient treatment at the Yale Plan Clinics did not resemble the classic jobless population that had been the original wartime justification of the clinics. But from the point of view of policy, the study did not create a new direction, but rather provided a new argument for an existing policy direction. The ideology of the nascent alcoholism movement had for at least the preceding five years centered on the idea that the alcoholic was a sick person who

could be helped and was worth helping, and had emphasized that the alcoholic was just like other folk except for a predisposition to alcoholism. The two popular movies made in the 1940s under movement influence—*The Lost Weekend* and *Smash Up*—portrayed middle-class alcoholism (Johnson, 1973, pp. 282–283, 387–388). Differentiation of alcoholism from skid row associations has long been an article of faith for the alcoholism movement (Johnson, 1973, pp. 370–373). The Straus and Bacon study did not create a new argument, but rather provided useful and dramatizable support for an existing line of argument.

2. Social scientists are cited as providing a crucial argument in policy decisions over the decriminalization of public drunkenness: "experts say that the vast majority of chronic alcoholics... would voluntarily join in an effective, comprehensive treatment program" (President's Commission on Crime, 1967, p. 79). This argument served to bridge the gap between civil liberties lawyers' insistence on no compulsion beyond the detoxification process and the need to see decriminalization as a public health measure and as slowing down the "revolving door" of drunkenness arrests (Room, 1976b). Policy imperatives created a need for a research position, rather than vice versa. And, according to Lemert, the research position was falsifiable at the time: "the irrelevance of a treatment model for Skid Row alcoholics could easily have been discovered from available research had codifiers, legislators and judiciary been oriented and organized to do so" (Lemert, 1976). Certainly the research position was falsified by later experience (Room, 1976b).

3. For the first five years of NIAAA's existence, the major federal policy on preventing alcohol problems was the promotion of norms of "responsible drinking." Behind this policy was the assumption that America had a high rate of drinking problems because of a lack of agreed-upon drinking norms:

> The rate of alcoholism... has been shown to be low in groups whose drinking-related customs, values, and sanctions are widely known, established, and congruent with other cultural values. On the other hand, alcoholism rates are higher in those populations where ambivalence is marked.... Ours is a Nation that is ambivalent about its alcohol use. This confusion has deterred us from creating a National climate that encourages responsible attitudes toward drinking for those who choose to drink. (Chafetz, 1971)

This argument, the essence of which also appeared in the work of the 1960s Cooperative Commission on the Study of Alcoholism (Plaut, 1967; Wilkinson, 1970), was directly derived from a social science analysis (Ullman, 1958) drawing on a series of social science studies of ethnoreligious factors in drinking patterns and problems—a comparative analysis in which the comparison with Anglo-Protestant U.S. patterns had no empirical study as a base. The argument has in fact been criticized in recent years on both empirical and theoretical grounds (Mäkelä, 1975; Room, 1976a; Whitehead & Harvey, 1974). But the argument can be seen as arising as the solution to a specific policy dilemma. Johnson (1973, pp. 327–349) has traced the emergence, starting as early as 1953, of a group of clergy, state administrators, and social scientists interested in alcohol problems who were uneasy with the single-minded focus of the orthodox alcoholism movement on the disease concept of alcoholism and providing treatment for the alcoholic. The group contended that "the social uses of alcohol as well as the prevention of problem drinking were both subjects that should be given greater attention" (Johnson, 1973, pp. 327–328). After several years of discussions, the group secured the grant that became the Cooperative Commission on the Study of Alcoholism.

The group, and after it the Cooperative Commission and eventually NIAAA in its formative years, thus wished to refocus attention away from an exclusive concern with the treatment of alcoholics and toward alcohol problems in the society at large. Such an emphasis was, however, the classical territory of the temperance movement, and there was a desire for several reasons to avoid temperance arguments or identification with the temperance movement. In the wake of the debacle of Prohibition, any association with temperance would be impolitic. It would also be thought unwise to alienate or directly confront the alcoholic beverage industries. Furthermore, the group and the commission staff were generally liberal and libertarian in politics and averse to preventive strategies implying control or constraint.

As I have argued elsewhere (Room, 1976a), the ambivalence argument provided an attractive solution to this dilemma: it tended to place the onus for American alcohol problems on conflict over drinking norms and thus implicitly on the temperance movement, it deemphasized amount of drinking as a potential issue, and it pointed to an optimistic program of reducing drinking problems by teaching new rules of drinking. Ostensibly, the argument derived from a small number of social science studies, which became perhaps the most frequently cited studies in alcohol social science. But most of the studies were of one culture rather than explicitly comparative; those that were comparative were to some extent misinterpreted (Mäkelä, 1975). None of the studies were of cultural change over time (Room, 1971), although the policy argument concerned how to change rates of drinking problems. The social science analysis did not produce the policy argument; rather, it served as plausible evidence to buttress policy predilections.

4. The finding that alcohol problems in general populations showed less overlap between problems and less continuity through time than in the

classical picture of alcoholism based on clinical populations has been presented by several authors, including myself (Room, 1972), as holding implications for policy. In fact, the approach of disaggregating alcohol problems and planning for their prevention and treatment differentially by problem area has entered policy discussions and has influenced recent policy-oriented definitions of alcohol problems (Edwards et al., 1977). But, rather than viewing this line of research and findings as resulting in policy changes, the research may be viewed as an incidental artefact of a shift in ideas and policy orientations in a segment of the alcohol policy world that considerably predated the research program. As noted above, as early as 1953 a group interest can be discerned in a wider view of alcohol problems in the society at large, rather than just in treatment of the alcoholic. This group allied itself in the late 1950s with the newly emerging federal interest in alcohol studies in the National Institute of Mental Health (NIMH), securing federal funding for the Cooperative Commission on the Study of Alcoholism. Part of the provisions of this grant was that its research director be someone with strong academic credentials, but "who had not previously been associated with the field of alcoholism" (Johnson, 1973, p. 338). This provision reflected feelings that alcohol research had been too ingrown and too dominated by the orthodox alcoholism movement. By indirection, Straus referred to these feelings in 1960 in characterizing a nascent "second generation of alcoholism researchers":

> You want the freedom to think. You insist on tightness and rigor in methodology. You are seeking methods to involve the best scientific minds in this field... you have been free, remarkably free, from contamination of an emotional involvement with the alcohol problem.... You are rejecting for the researcher the multiple roles of justifier, rejustifier, composer of progress reports, passer of tin cups. Above all, you are rejecting those who would irresponsibly popularize your research and your thinking. (Straus, 1960)

Contemporaneously with the Cooperative Commission grant, NIMH made a grant for the California Drinking Practices Study to make surveys of drinking practices and eventually of drinking problems in the general population. The approach was to be inductive and empirical; again, the principals involved in the actual study—Ira Cisin, Ray Fink, Genevieve Knupfer—were new to the alcohol field. To the extent possible, the study was thus insulated from the presumptions about alcohol problems of the alcoholism movement, and began in a period when the conceptual disaggregation of alcohol problems was already in the air (the 1967 report of the Cooperative Commission is titled *Alcohol Problems* rather than alcoholism, and includes a specific discussion of the "range of alcohol problems"). To a considerable extent, the emphases in the study's eventual reports on the diversity and separability of drinking problems (Clark, 1966; Knupfer, 1967) and on avoidance of the global term *alcoholism* (Knupfer, 1967; Cahalan, 1970) were foreordained in the conditions of the study's inception.

5. As noted above, in recent years there has been a renewed emphasis on the importance of the amount of alcohol consumption in the occurrence of alcohol problems, and consequently on the availability dimension—and particularly on the price of alcohol relative to the cost of living—as a potential means of reducing alcohol problems. The argument has been stated succinctly by an international group predominantly composed of social scientists, in a widely circulated report:

> changes in the overall consumption of alcoholic beverages have a bearing on the health of the people in any society. Alcohol control measures can be used to limit consumption: thus, control of alcohol availability becomes a public health issue. (Bruun et al., 1975, pp. 12–13)

In line with this argument, NIAAA's 1977 draft "National Plan to Combat Alcohol Abuse and Alcoholism" stated that "there is strong evidence that as consumption rises, so do primary and secondary problems related to the use of alcohol. Since the trend in this country during the last decade has been towards ever increasing total consumption, a major effort is required to stabilize this increase" (*Alcoholism Report,* August 26, 1977, p. 1). Under strong pressure from alcoholism movement groups, NIAAA soon backed down from this goal (*Alcoholism Report,* September 23, 1977, p. 7; October 14, 1977, pp. 2–3; October 28, 1977, pp. 2–3; December 9, 1977, pp. 1–2).

A large part of the alcoholism movement's objections to the goal of stabilizing alcohol consumption was a desire to avoid antagonizing the alcohol beverage industries, which have long maintained links with the movement. It had been an article of faith in the alcoholism movement since its inception to avoid "political" entanglements, and particularly involvement in wet/dry struggles, which were seen as potentially detracting from the primary goal of securing humane treatment for the alcoholic. The issue of the relevance of population consumption levels to an "alcohol abuse and alcoholism" agency thus squarely poses the choice between the alcoholism treatment policy frame of the alcoholism movement and the alcohol-problems-in-society policy frame of the Cooperative Commission and its initiators—although ironically with an approach diametrically opposed to the Cooperative Commission's, an approach labeled by some commentators as "neo-Prohibitionist."

The conventional scientific history of the modern consumption control argument attributes the primary work to a group of Ontario Addiction

Research Foundation (ARF) social scientists, Jan deLint, Wolfgang Schmidt, and Robert Popham, who, beginning in 1968 (deLint & Schmidt, 1968a, 1968b) published a long and cumulative series of analyses, drawing on earlier work by the French demographic analyst Sully Ledermann, on the invariant qualities of the distribution of alcohol consumption in the population. DeLint and Schmidt's original data was drawn from a study of wine and liquor purchases in Ontario provincial stores that had been initiated in 1961 and that, judging by the descriptions of it in the ARF annual reports of the early 1960s, originally had broader aims than establishing the distribution of consumption.

In deLint and Schmidt's original report (1968b), submitted for publication at the end of 1967, analysis and discussion is limited to a presentation of the logarithmic normal nature of the empirical curve of distribution of consumption, a cautious description of Ledermann's speculations about the reasons for the shape of the curve, and a remark on the lack of bimodality in the curve, so that "a definition of alcoholism based solely on quantity of drinking must ipso facto be arbitrary." No mention is made of policy implications of the data.

In the ARF *Annual Report* for 1967, transmitted in March 1968, a discussion of the study by the policy-making director of the ARF, David Archibald, included some conservative policy-oriented speculations:

> It is interesting... to speculate on what our chart would look like if total consumption increased.... Given the same shape of curve,... any such upward development... seems likely... to increase the number of persons who, while perhaps they would not appear alcoholic in behavior, would nevertheless be prone to liver damage and other physical diseases that are usually associated with high-volume drinking. (ARF, 1968, p. 21)

In a paper presented in September 1968, the researchers venture a modest policy comment:

> It should be noted that an increase in the per drinker consumption invariably leads to an increase in the consumption of amounts dangerous to health.... The Cooperative Commission on the Study of Alcoholism [Plaut, 1967] has suggested that it does not matter whether its proposals... will lead to an increase in alcohol consumption or not.
>
> It would appear that we are faced here with a dilemma. To reduce the more traditional alcohol problems on this continent, namely the problem of intoxication and the problem of alcoholism in the psychiatric sense (gamma alcoholism) we might want to liberalize liquor legislation.... At the same time... an increase in overall levels of alcohol consumption (will lead) to an increase in the prevalence of organic diseases attributable to excessive alcohol use. (deLint & Schmidt, 1968a)

In Archibald's review of progress in the 1968 ARF report (transmitted April 1969), caution is thrown to the winds, in a blunt statement of policy advocacy:

> It has been found that rates of alcohol consumption hazardous to health are inextricably linked to the general level of alcohol consumption. This implies that the only feasible approach to the prevention of this problem is to effect a decrease in the average level of consumption within the drinking population as a whole. (ARF, 1969, p. 24)

This documentary history suggests strongly that modestly presented research analysis was seized upon by the policy-making level of ARF as an organizing tool for a coherent alcohol policy position. In any event, the researchers involved subsequently rose to the occasion, publishing a series of reports substantially less cautious in their argument than the initial reports. Instead of being presented as an empirical regularity, the form of the distribution of consumption was presented as "for all practical purposes... unalterable" (Popham, Schmidt, & deLint, 1971), and the problems linked closely to consumption level were assumed to extend beyond cirrhosis to "alcoholism" in general (deLint & Schmidt, 1971).

In the 1970s, the work of Ledermann and of deLint, Schmidt, and Popham was subjected to a barrage of critical examination (see reviews in Miller & Agnew, 1974; Parker & Harmon, 1980; Smith, 1976). In the wake of this, Schmidt and Popham (1980) conceded that in some of their work "we may be justly accused of some overstatement and oversimplification," acknowledging that "to a degree this was due to a deliberate strategy to secure a hearing for a point of view which ran counter to the prevailing sentiment." On the other hand, the fact of an empirical regularity in the form of distribution of alcohol consumption in different populations remains relatively unshaken (Guttorp & Song, 1977; Skog, 1977).

In view of the considerable and controversial literature that surrounds the issue of the distribution of consumption, it is ironic that the exact shape of the distribution is in fact not important in the argument for price manipulations or other controls as a prevention strategy. So long as a strong temporal relationship between the overall consumption level and cirrhosis mortality can be shown, whether the consumption is distributed lognormally or otherwise is largely immaterial to policy considerations. From this perspective, the whole argument over the Ledermann curve has been a diversion from the policy issues of the interrelations of prices and other controls, consumption levels, and cirrhosis mortality. On these issues, evidence was available well before 1968. In 1960, John Seeley, then research director of ARF, published an article showing strong relations over time between alcohol prices relative to disposable income, consumption level, and cirrhosis mortality. The article was a refinement of an earlier analysis by a newspaper writer (Erratum, 1961). Seeley's article did not shrink from discussing the policy implications:

> It appears that deaths from liver cirrhosis, though small in number, are increasing rapidly, and rise and fall with average alcohol consumption. It also appears that alcohol consumption rises and falls inversely with alcohol price. It is sufficiently credible to justify a social experiment to determine whether an alcohol price increase would reduce liver cirrhosis mortality, while simultaneously furnishing a sizable increase in government revenue.... (Seeley, 1960)

In the ARF *Annual Report* for 1960, the report of the Research Department pursues the same line, remarking indeed that "perhaps a large part of the problem [of cirrhosis deaths] could be dealt with by government tax policy" (ARF, 1961, p. 51). Archibald's "Annual Review of Operations" also discusses the study and its implications, but with considerably more caution and several caveats (ARF, 1961, pp. 10–11). In succeeding annual reports, discussion of the issue, and indeed in general all policy issues except those concerning treatment, disappeared.

Thus, it seems that Seeley's initial research report fell on stony ground in terms of its influence on policy: the researcher's enthusiasm for its policy implications was not shared by those in policy-oriented positions. Conversely, in 1967 and 1968 the implications of the research were eagerly picked up; the research reports served as a catalyst for policy initiatives actually more directly related to Seeley's earlier work. The policy climate was changing, at least in Canada, probably partly in response to the general increase in consumption levels in the 1960s.

But in the United States the climate of attitudes toward alcohol controls has shifted more slowly than in Canada or Scandinavia. Much of the technical criticism of the "ARF position" has come from south of the border, and is linked to a strong disposition among many social scientists as well as policy makers against recognizing any possibility of relationship between alcohol controls and alcohol problems. This was perhaps most clearly expressed on the record in the reaction of social scientists and the public health establishment to an independent analysis, presented at the American Public Health Association meetings of 1966, of temporal relations between consumption level and cirrhosis mortality. The paper concluded by arguing in guarded terms that "governmental fiscal and regulatory measures can be effective in reducing alcohol consumption and lowering mortality from cirrhosis of the liver" (Terris, 1967). All papers presented at the same session *except* this one were printed in the June 1967 *American Journal of Public Health.* This resulted in the curious situation of the prepared discussion of the session papers, more than half of which focused on the Terris paper, appearing without a paper on which it was commenting (Elinson, 1967). This discussion, by a social scientist, manifested considerable unease over the paper's "provocative analysis." In addition to citing various pieces of counterevidence, the discussant joked about "having a drink or two" at lunchtime "before some impulsive local government is led by Dr. Terris' skillful presentation" to alter control laws, and suggested that, as with a possible association of cervical cancer with frequency of intercourse, there might be knowledge better left unknown: "the implications for prevention—if this were a factor—[might be] just too horrible to endure. I think most of us have a similar feeling about alcohol" (Elinson, 1967).

An extraordinary editorial footnote to this discussion, in giving the reference to Terris's presentation, further dissociated the official organs of public health from any policy implications of Terris's paper:

> A summary of Terris' paper appeared in the APHA 1966 conference report issue of "Public Health Reports," March, 1967, vol. 32, No. 3. The summary in "Public Health Reports" carries the headline, "Restrict Alcohol Availability to Reduce Liver Cirrhosis," and refers to a paragraph toward the end of Terris' mimeographed paper—a paragraph which was *not* read at the meeting, although the full mimeographed paper, which included this paragraph, was distributed to the press. (footnote to Elinson, 1967)

Terris's paper, apparently including the offending paragraph, was finally printed in the *American Journal of Public Health* six issues later (Terris, 1967). Its publication may have been aided by the fact that Terris was by then president of the American Public Health Association. In more recent years, public health journals have shown a greater receptivity to papers in the tradition of Terris's (Brenner, 1975)—including a social scientist's challenge that public health's disinterest in alcohol controls has resulted from a philosophy of "accommodation with the prevailing ethical paradigm" of "market justice" (Beauchamp, 1975).

6. A recent well-publicized interplay of social science research and policy making has occurred over the issue of whether alcoholics can ever return to controlled drinking, touched off by a report written by three social scientists at the Rand Corporation (Armor, Polich, & Stambul, 1976). Abstinence as the goal of alcoholism treatment has long been a fundamental credo of the alcoholism movement, and clinical reports of return to moderate drinking (e.g., Davies, 1962) have been seen as a substantial threat to the movement's therapeutic paradigm, and have not been allowed to pass unchallenged (Roizen, 1977a, 1977b). However, the battle over the Rand Report surpassed previous skirmishes in intensity and duration.

The Rand Report was based on data from NIAAA's monitoring and evaluation system for federally funded alcoholism treatment centers, which routinely collected intake and follow-up data on all clients, and on a special follow-up study performed under NIAAA contract by the Stanford Research Institute (Ruggels, Armor, Polich, Mothershead, & Stephen et al., 1975). The monitoring and evaluation system was regarded by the NIAAA leadership as a major element in projecting an image of NIAAA as a dynamic, forward-looking agency (Chafetz, 1974).

As control of the treatment system slipped from the grasp of the voluntary-action-based alcoholism movement into the hands of government bureaucracies, the goal of abstinence was increasingly eroded. To justify public expenditures, the treatment system needed to show strong and indeed startling improvements in those treated, and abstinence was too harsh an outcome criterion to serve this purpose. Well before the Rand analysis, state and federal evaluation systems had moved away from the criterion of abstinence as their measure of success. This development was parallel to but differently motivated from a similar shift by behavioral psychologists interested in inculcating controlled drinking.

The Rand social scientists, relatively new to the alcohol field, were oriented to the world of governmental agencies and not the world of the alcoholism movement, and may not have expected the ferocity with which their findings on the issue of moderate drinking among those treated for alcoholism were greeted. However, they joined the battle with gusto and some success. In the ensuing outpouring of printed matter, it is notable that only a few social scientists who had close movement ties lined up with the movement position (e.g., David Pittman, NCA press conference materials, July 1, 1976).

But the Rand Report did not have much effect on policy, at least in its immediate aftermath. Roizen (1977b) has noted that the controversy over controlled drinking obscured a number of other important findings with policy implications in the report. Under pressure from the alcoholism movement, NIAAA did not disown the report or its findings, but simply asserted their irrelevance to policy: in a "HEW News" press release of June 23, 1976, Ernest Noble, the new director of NIAAA, stated his feelings that "abstinence must continue as the appropriate goal in the treatment of alcoholism."

Conclusion

It is clear from these six vignettes of interactions between alcohol social science research and policy making that the relationship does not fit a rationalistic paradigm where scientists autonomously discover "new knowledge," which is then carried into action in policy decisions. In some cases, in fact, it could be argued that the policy produced the knowledge rather than vice versa. Nor is it realistic to regard social scientists as disinterested scholars following wherever the facts may lead. Sometimes, indeed, social scientists seem to have had difficulty distinguishing between the world as they found it and the world as they wanted it to be.

Often the apparent linkage between social science research and policy making reflects that both are manifestations of general social trends. Noting that "in discussions of research and policy, research findings are often regarded as causes of policy and action," Bruun (1973) cautions that, if we recognize "that science is valuebound, at least when social science research problems are formulated, then research, policy and action could all be influenced by the same overall forces in society.... Research could be seen as a modern instrument of debate on research policy, primarily on alternate means derived from the same basic values, rather than on alternative goals."

The evidence suggests, then, a considerable skepticism about social science's independent influence on policy making. Nevertheless, it would also be a mistake to conclude that social science research has no independent influence at all. As Bruun (1973) puts it, "the big decisions will always be taken primarily on the basis of values—the small, but still important ones might, however, be improved by social research." Gusfield (1975) notes that as knowledge accumulates and is disseminated, it does set limits to the parameters of public debate. Thus, the long-run effects of research may differ from and be more important than the short-run effects. As Max Planck sadly remarked about physical science, "a new scientific truth does not triumph by convincing its opponents and making them see the light, but rather because its opponents eventually die, and a new generation grows up that is familiar with it" (quoted in Kuhn, 1962, p. 150). Of course, this is comfort to the scientist concerned with policy only to the extent one can be sure that one is riding the wave of history.

That social science knowledge is potentially "subversive" (Gusfield, 1975) of received definitions of the situation is recognized by both social scientists and policy makers, and is a continuing source of strain between them. The social scientist tends to be defined and appreciated by those in the policy arena as a bearer of technical skills—as a survey researcher, an evaluator, a population estimator, a multivariate analyst. The social scientist prefers a different definition of his or her contribution:

> The primary contribution which we have to make is conceptual. In a field which concerns social interaction and involves a complex of human behavior, the greatest need... is for theoretical orientation or explanatory concepts to provide a focus for research design and for hypothesis testing.... Ranking far down in importance, but nonetheless significant, are our methodological and

> technical skills. The particular tactics employed by a sociologist are often, unfortunately, stressed as his sole contribution and I, for one, resent being regarded as sample-minded. (Pearl, 1960)

In the era of the alcoholism movement, alcohol social science research has suffered particularly from operating with assumptions at odds with the prevailing model. If alcoholism is a disease, then the exciting research advances are to be expected from biomedical and not social science research. Marty Mann commented frankly on this in the early days of the movement, in a letter to Howard Haggard, then director of the Yale Center of Alcohol Studies. Commenting on the reaction of a federal official to the physiological research at the center, she noted:

> I would venture to say he was most impressed with that phase of our work.... Many businessmen to whom I talk of this work seem mainly interested in the same thing. Perhaps it gives more hope. And I mean physiological, not social research. (quoted in Johnson, 1973, p. 290)

The common assumptions of social science research have also conflicted with the assumptions of the alcoholism movement (see Room, 1979). Perhaps as a result, the proportion of federal research expenditures supporting social science research tended to drop as the research moved from NIMH to NIAAA jurisdiction (Room, 1979).

While alcohol social science research has a long history of interrelation with policy making, our analysis has underlined the ambiguity of the relationship. Nevertheless, there is clearly no choice but to continue the relationship, and we may expect both social scientists and policy makers to keep on trying. And we may expect them both to keep on finding it trying.

Afterword, December 1989

Apart from a few changes of tense and updated references, the text above is as it was written in April 1978. There is little in it that I would change. In my view, Kettil Bruun's conclusions are still a reliable guide to the relation between social science research and policy making, not only in the alcohol field but also more generally.

Are there any new lessons on the subject from the 1980s? There is certainly more material on some of the case studies mentioned above. In the early 1980s, discussions of alcohol control issues began to enter the public arena and to be taken up within the public health constituency (Moore & Gerstein, 1981; Room, 1984; World Health Organization, 1980). The diversionary debate about the exact shape of the distribution of alcohol consumption simmered down, and research on the effects of alcohol controls and other alcohol policies took on new vigor. The United States may now have overtaken Canada and Scandinavia in "new temperance" thinking and debate, though innovations in actual policy have often been primarily symbolic (e.g., adding warning labels to alcoholic beverage containers). In this case, social science thinking proved to be at least the harbinger of more general cultural shifts in thought about alcohol.

The issue of controlled drinking for alcoholics has remained contentious (Roizen, 1987), with the flash points in the controversy usually involving behavioral psychological thought; in trying to subsume the alcoholism disease concept into a broader analysis, rather than directly opposing it, sociologists have played a quieter subversive role (Room, 1983). There was a second round to the Rand study, but with a more muted public controversy; this time, NIAAA staff simply reinterpreted the findings into a more publicly acceptable form (Room, 1980).

For most of the 1980s, U.S. alcohol social science research lost ground in relative terms to other alcohol research traditions, and the links between researchers and bureaucratic policy formation became attenuated. In part, this trend reflected changes that came with the advent of the Reagan administration. In its first flush, that administration had a specific aversion to anything called or smacking of "social research" (choosing not to go down fighting over a name, the Social Research Group in Berkeley became the Alcohol Research Group). But the relative decline in the influence of social science research in the alcohol field was mostly a byproduct of policies not specifically directed against it. The Reagan administration's shift of treatment and prevention services money into block grants destroyed NIAAA's Prevention Division, which had been the main nexus between alcohol researchers and alcohol policy discussions. Stripping NIAAA of its treatment and prevention grant functions also resulted in an agency oriented to survival in terms of research prestige, in an organizational environment that allocated prestige primarily to microbiological research.

In the later 1980s, the balance began to shift back. Homelessness and AIDS emerged as issues which, in alcohol studies as elsewhere, directed attention back to the social dimensions of health and social problems. The promises of quick breakthroughs from microbiology or genetic studies, with significance for management of or policy on alcohol problems, have begun to come into question. Prevention came back into NIAAA's organizational chart, so far in the form of a branch (one step down from a division). The rhetoric and legislative activity on illicit drugs, which reached an extraordinary pitch in the late 1980s, have in the long run tended to sweep alcohol issues up into the debate, and have brought new resources to the measurement of alcohol-related burdens in the society.

In the longer run—in terms of the "long waves" of alcohol consumption (Mäkelä et al., 1981), and of societal response to alcohol problems (Blocker, 1989)—the 1980s appear to have been a time when a new trend in the cultural position of alcohol, to which the Reagan administration was largely irrelevant, became consolidated. Per capita alcohol consumption in the United States has been dropping steadily since 1981, while the level of societal concerns about heavy drinking and its consequences has risen (Room, 1989). The concerns have been manifested in several popular movements, most notably in Mothers Against Drunk Driving, Remove Intoxicated Drivers, and other grassroots antidrinking-and-driving movements; and in the Adult Children of Alcoholics and other "codependence" movements, which focused attention on the drinker's effects on those around him or her. The concerns have also been manifested in the growth of a professional and paraprofessional alcohol treatment capacity and establishment, and in shifts in this constituency's thinking and practice toward a more proactive and interventionist stance. Generally speaking, sociologists and other social scientists have played little role in the leadership of these movements; for that matter, there has been too little in the way of critical social science analysis of these movements.

Social scientists such as Dan Beauchamp have played more of a role in another manifestation of the new concerns—a strand of activism on alcohol controls and policy, operating under a public health rubric, which has developed out of the perspectives noted in the fourth and fifth case studies presented above. To the great displeasure of the alcoholic beverage industry, this strand formed part of the official policy mix as the surgeon-general turned to drinking-driving issues in the waning days of the Reagan administration (Koop, 1989). But as the groups promoting a "public health approach" to alcohol issues have become more caught up in political action, they have tended to diverge away from the research concerns and institutions of social alcohol research.

Let me add a few last words on Selden Bacon's place in this longer historical perspective. Each of us is to some extent a child of his generation, and Bacon, I am sure, does not find congenial many of the trends I have just described for the 1980s. In common with other scholars from the "wet generations" whose work stretched from Odegard (1928) to Pittman (1980), Bacon has long had a fundamental disbelief in alcohol controls in any form (e.g., Bacon & Jones, 1963) and an aversion to punitive approaches to dealing with alcohol problems. He played a leading role in the first years of the movement to establish an alternative societal vision, one which focused on providing humane treatment for the alcoholic. But as the alcoholism movement developed and became institutionalized, he became critical of it (Bacon, 1963); eventually, indeed, the institutions of the movement have circled back to an openness to coercive approaches and indeed to alcohol control policies. From the perspective of a practical politician, Bacon's career in alcohol politics suggests that sociologists, like poets and artists, are useful in making a revolution, but unreliable, even subversive, when it comes to consolidating it.

From the perspective of alcohol research, it is a different story. Bacon was the first and founding alcohol sociologist of modern times. Not only in his own work, but through the diverse contributions of his students, and in turn of their students, Bacon's legacy in alcohol research continues and increases. The tradition of critical thought that Bacon has epitomized and established will serve society and scholarship well as we seek to transcend the dialectic dynamic of the "long waves."

Acknowledgments

This chapter was originally prepared for a conference on the Utilization of Social Research in Drug Policy Making, sponsored by Columbia University's Center for Socio-Cultural Research on Drug Use, Washington, DC, May 3–5, 1978. Its preparation was partly supported by a national research center grant from the National Institute on Alcohol Abuse and Alcoholism to the Social Research Group (now the Alcohol Research Group). It has benefited from discussions with Robert Straus, Don Cahalan, Ron Roizen, Harry Levine, and Walter Clark.

References

Addiction Research Foundation. (1961). *Tenth annual report for the year ending December 31, 1960.* Toronto: Alcoholism Research Foundation.

Addiction Research Foundation. (1968). *Seventeenth annual report: 1967.* Toronto: Alcoholism and Drug Addiction Research Foundation.

Addiction Research Foundation. (1969). *Eighteenth annual report: 1968.* Toronto: Alcoholism and Drug Addiction Research Foundation.

Armor, D., Polich, J.M., & Stambul, H.B. (1976). *Alcoholism and treatment* [The Rand Report]. Santa Monica, CA: Rand Corp.

Bacon, S. (1947). The mobilization of community resources for the attack on alcoholism. *Quarterly Journal of Studies on Alcohol, 8,* 473–497.

Bacon, S. (1949). The administration of alcoholism rehabilitation programs. *Quarterly Journal of Studies on Alcohol,* 10, 1–47.

Bacon, S. (1963, October). State programs on alcoholism. A critical review. In *Selected papers delivered at the 14th annual meeting* (pp. 1–18). North American Association of Alcoholism Programs, Miami Beach, Florida.

Bacon, S., & Jones, R.W. (1963). *The relationship of the Alcoholic Beverage Control Law and the problems of alcohol* (Study Paper No. 1). New York: New York State Moreland Commission on the Alcoholic Beverage Control Law.

Bales, R.F. (1946). Cultural differences in rates of alcoholism. *Quarterly Journal of Studies on Alcohol, 6,* 480–499.

Beauchamp, D. (1975). Public health: Alien ethic in a strange land? *American Journal of Public Health, 65,* 338–339.

Beauchamp, D. (1976). Exploring new ethics for public health: Developing a fair alcohol policy. *Journal of Health Politics, Policy and Law, 1,* 338–354.
Berry, R.E., Jr., & Boland, J.P. (1977). *The economic cost of alcohol abuse.* New York: Free Press.
Billings, J.S. (Ed.). (1903). *The liquor problem: A summary of investigations conducted by the Committee of Fifty, 1893–1903.* Boston: Houghton, Mifflin.
Blane, H.T. (1976). Education and mass persuasion as preventive strategies. In R. Room & S. Sheffield (Eds.), *The prevention of alcohol problems: Report of a conference* (pp. 255–288). Berkeley, CA: University of California, Social Research Group.
Blane, H.T., & Hewitt, L.E. (1977). *Mass media public education and alcohol: A state of the art review.* Report prepared for NIAAA, Rockville, MD.
Blocker, J.S. (1989). *American temperance movements: Cycles of reform.* Boston: Twayne Publishers.
Blumenthal, M., & Ross, H.L. (1975). Judicial discretion in drinking driving cases: An empirical study of influence and consequences. In S. Israelstam & S. Lambert, *Alcohol, drugs and traffic safety* (pp. 755–762). Proceedings of the Sixth International Conference on Alcohol, Drugs and Traffic Safety, Addiction Research Foundation, Toronto.
Bogue, D.J. (1963). *Skid row in American cities.* Chicago: University of Chicago Community and Family Study Center.
Bossard, J.H.S., & Sellin, T. (Eds.). (1932). Prohibition: A national experiment. *Annals of the American Academy of Political and Social Science, 163,* 1–233.
Brenner, H. (1975). Trends in alcohol consumption and associated illnesses. *American Journal of Public Health, 65,* 1273–1292.
Bruun, K. (1970). Alkoholihatat mahdollisiman vahaiaisksi [The minimization of alcohol damage]. *Alkoholipolitiikka, 35,* 185–191. Abstracted in *Drinking and Drug Practices Surveyor, 8,* 15, 47 (1973).
Bruun, K. (1973). Social research, social policy and action. In *The epidemiology of drug dependence: Report on a conference* (pp. 115–119). Copenhagen: World Health Organization.
Bruun, K., et al. (1975). *Alcohol control policies in public health perspective* (Vol. 25). Helsinki: Finnish Foundation for Alcohol Studies.
Cahalan, D. (1970). *Problem drinkers.* San Francisco: Jossey-Bass.
Cahn, S. (1970). *The treatment of alcoholics.* New York: Oxford University Press.
Calkins, R. (1901). *Substitutes for the saloon: An investigation originally made for the Committee of Fifty.* Boston: Houghton, Mifflin.
Chafetz, M. (1967). Alcoholism prevention and reality. *Quarterly Journal of Studies on Alcohol, 28,* 345–348.
Chafetz, M. (1971). Introduction. In *First special report to Congress on alcohol and health* (pp. 1–4). Washington, DC: National Institute on Alcohol Abuse and Alcoholism.
Chafetz, M. (1974). Monitoring and evaluation at NIAAA. *Evaluation, 2*(1), 49–52.
Clark, W. (1966). Operational definitions of drinking problems and associated prevalence rates. *Quarterly Journal of Studies on Alcohol, 27,* 648–668.
Culver, D.C., & Thomas, J.E. (1940). *State Liquor Control Administration: A statutory analysis.* Berkeley, CA: University of California Bureau of Public Administration.
Davies, D.L. (1962). Normal drinking in recovered alcoholics. *Quarterly Journal of Studies on Alcohol, 23,* 94–104.

deLint, J., & Schmidt, W. (1968a, September). *The distribution of alcohol consumption.* Paper presented at the International Congress on Alcohol and Alcoholism, Washington, DC.

deLint, J., & Schmidt, W. (1968b). The distribution of alcohol consumption in Ontario. *Quarterly Journal of Studies on Alcohol, 29,* 968–973.

deLint, J., & Schmidt, W. (1971). Consumption averages and alcoholism prevalence: A brief review of epidemiological investigations. *British Journal of Addiction, 66,* 97–107.

Dunham, H.W. (1954). *Homeless men and their habitats: A research planning report.* Detroit, MI: Wayne University, Dept. of Sociology and Anthropology.

Edwards, G., Gross, M.M., Keller, M., Moser, J., & Room, R. (Eds.) (1977). *Alcohol-related disabilities.* Geneva: World Health Organization.

Elinson, J. (1967). Epidemiologic studies and control programs in alcoholism: Discussion. *American Journal of Public Health, 57,* 991–996.

Erratum, (1961). *Canadian Medical Association Journal, 84,* 34.

Feldman, H. (1927). *Prohibition: Its economic and industrial aspects.* New York: D. Appelton.

Gusfield, J.R. (1972). A study of drinking drivers in San Diego County. Report prepared for the Department of Public Health, San Diego County, San Diego, CA

Gusfield, J.R. (1975). The (f)utility of knowledge? The relation of social science to public policy toward drugs. *Annals of the American Academy of Political Science, 417,* 1–15.

Gusfield, J.R. (1976). The prevention of drinking problems. In W.J. Filstead, J.J. Rossi, & M. Keller (Eds.), *Alcohol and alcohol problems: New thinking and new directions* (pp. 267–292). Cambridge MA: Ballinger.

Guttorp, P., & Song, H. (1977). A note on the distribution of alcohol consumption. *Drinking and Drug Practices Surveyor, 13,* 7–8.

Harrison, L.V., & Laine, E. (1936). *After repeal: A study of liquor control administration.* New York: Harper and Brothers.

Johnson, B. (1973). *The alcoholism movement in America: A study in cultural innovation.* Unpublished doctoral dissertation, University of Illinois, Urbana-Champaign.

Knupfer, G. (1967). The epidemiology of problem drinking. *American Journal of Public Health, 57,* 973–986.

Koop, C.E. (Ed.). (1989). *Proceedings, Surgeon General's Workshop on Drunk Driving.* Washington: U.S. Government Printing Office.

Koren, J. (1899). *Economic aspects of the liquor problem: An investigation made under the direction of a sub-committee of the Committee of Fifty.* Boston: Houghton, Mifflin.

Kuhn, T. (1962). *The structure of scientific revolutions.* Chicago: University of Chicago Press.

Kurtz, N.R., & Regier, M. (1975). The Uniform Alcoholism and Intoxication Treatment Act: The compromising process of social policy formulation. *Journal of Studies on Alcohol, 36,* 1421–1441.

Kuusi, P. (1957). *Alcohol sales experiment in rural Finland* (Vol. 3a). Helsinki: Finnish Foundation for Alcohol Studies.

Lanu, K.E. (1956). *Poikkeavan alkoholikayttaytymisen kontrolli* [Control of deviating drinking behavior]. Helsinki: Finnish Foundation for Alcohol Studies.

Law and Contemporary Problems (1940). *Alcoholic Beverage Control,* 7, 4, pp. 543–751 (a special issue).

Lemert, E.M. (1976). Comment on the Uniform Alcoholism and Intoxication Treatment Act: The compromising process of social policy formulation. *Journal of Studies on Alcohol, 37,* 102–103.

Levine, H., & Smith, D. (1977). A selected bibliography on alcohol control, particularly before and at repeal (Working Paper F71). Berkeley, CA: University of California, Social Research Group.

Levy, J.E., & Kunitz, S. (1974). *Indian drinking: Navajo practices and Anglo-American theories.* New York: John Wiley and Sons.

Light, D. (1976, September/October). Costs and benefits of alcohol consumption. *Society, 11,* 13, 18–22, 24.

Mäkelä, K. (1974). Types of alcohol restrictions, types of drinkers and types of alcohol damages: The case of personnel strike in the stores of the Finnish Alcohol Monopoly. In *Papers presented at the 20th International Institute on the Prevention and Treatment of Alcoholism* (pp. 16–26). (Manchester, England) Lausanne, Switzerland: ICAA.

Mäkelä, K. (1975). Consumption level and cultural drinking patterns as determinants of alcohol problems. *Journal of Drug Issues, 5,* 344–357.

Mäkelä, K., Room, R., Single, E., Sulkunen P., & Walsh, B., with 13 others. (1981). *Alcohol, society and the state: I. A comparative study of alcohol control.* Toronto: Addiction Research Foundation.

Marden, P., Zylman R., Fillmore, K., & Bacon, S. (1976). Comment on "A national study of adolescent drinking behavior, attitudes and correlates." *Journal of Studies on Alcohol, 37,* 1346–1358.

Medicine in the Public Interest. (1976). *A study in the actual effects of alcoholic beverage laws* (2 Vols.) Report to NIAAA under contract ADM 281–75–0002.

Miller, G.H., & Agnew, N. (1974). The Ledermann model of alcohol consumption; description, implications and assessment. *Quarterly Journal of Studies on Alcohol, 35,* 877–898.

Mizruchi, E.H., & Perrucci, R. (1962). Norm qualities and differential effects of deviant behavior: An exploratory analysis. *American Sociological Review, 27,* 391–399.

Moore, M., & Gerstein, D. (Eds.). (1981). *Alcohol and public policy: Beyond the shadow of Prohibition.* Washington, DC: National Academy Press.

National Advisory Committee on Alcoholism. (1968, December). Interim report to the secretary of the Department of Health, Education and Welfare.

National Institute on Alcohol Abuse and Alcoholism. (1974). *Second special report to the U.S. Congress on alcohol and health from the secretary of Health, Education and Welfare.* Washington, DC: U.S. Government Printing Office.

National Institute on Alcohol Abuse and Alcoholism. (1979). *Third special report to the U.S. congress on alcohol and health: Technical support document* (DHEW Publication No. ADM 379–832). Washington DC: U.S. Government Printing Office.

Odegard, P. (1928). *Pressure politics: The story of the Anti-Saloon League.* New York: Columbia University Press.

Parker, D., & Harman, M. (1980). A critique of the distribution of consumption model of prevention. In Thomas Harford, Douglas Parker, & William Light (Eds.), *Normative approaches to the prevention of alcohol abuse and alcoholism* (NIAAA Research Monograph No. 3, pp. 67–88). Washington: U.S. Government Printing Office.

Pearl, A. (1960). Report of sociologists and social psychologists. In *Multidisciplinary programming in alcoholism investigation* (pp. 24–25). Division of Alcoholic Rehabilitation, California Department of Public Health, Berkeley.

Pittman, D. (1980). *Primary prevention of alcohol abuse and alcoholism: An evaluation of the control of consumption policy.* St. Louis: Washington University, Social Science Institute.
Plaut, T. (1967). *Alcohol problems: A report to the nation.* New York: Oxford University Press.
Popham, R., Schmidt, W., & deLint, J. (1971). Epidemiological research bearing on legislative attempts to control alcohol consumption and alcohol problems. In J. Ewing & B. Rouse (Eds.), *Law and drinking behavior* (pp. 4–16). Chapel Hill: University of North Carolina, Center of Alcohol Studies.
President's Commission on Crime in the District of Columbia. (1967). The drunkenness offender. Reprinted from the 1966 report as Appendix E of President's Commission on Law Enforcement and Administration of Justice, *Task Force Report: Drunkenness.* Washington, DC: U.S. Government Printing Office.
Rachal, J.V., Williams, J., Brehm, M., Cavanaugh, E., Moore, R., & Eckerman, W. (1975). *Adolescent drinking behavior, attitudes and correlates. A national study* (Final Report prepared for NIAAA). North Carolina: Research Triangle Institute.
Regier, M., & Kurtz, N. (1976). Policy lessons of the Uniform Act: A response to comments. *Journal of Studies on Alcohol, 37,* 382–392.
Roizen, R. (1977a). *Alcoholism treatment's goals and outcome measures: Conceptual, pragmatic and structural sources of controversy in the outcome debate* (Working Paper F61). Berkeley: University of California, Social Research Group.
Roizen, R. (1977b). Comment on the "RAND Report," *Journal of Studies on Alcohol, 38,* 170–178.
Roizen, R. (1987). The great controlled drinking controversy. In Marc Galanter (Ed.), *Recent developments in alcoholism* (Vol. 5, p. 245–279). New York and London: Plenum.
Roman, P., & Trice, H. (1967, August). *Alcoholism and problem drinking as social roles: The effects of constructive coercion.* Paper presented at the Society for the Study of Social Problems Meetings, San Francisco, CA.
Room, R. (1971, August). *The effects of drinking laws on drinking behavior.* Paper presented at the Society for the Study of Social Problems annual meetings, Denver, Colorado.
Room, R. (1972). Notes on alcohol policies in the light of general-population studies. *Drinking and Drug Practices Surveyor, 6,* 10–12, 15.
Room, R. (1976a). Ambivalence as a sociological explanation: The case of cultural explanation of alcohol problems. *American Sociological Review, 41,* 1047–1065.
Room, R. (1976b). Comment on the Uniform Alcoholism and Intoxication Treatment Act; The compromising process of social policy formulation. *Journal of Studies on Alcohol, 37,* 113–143.
Room, R. (1979). Priorities in social science research on alcohol. *Journal of Studies on Alcohol* (Supplement no. 8), pp. 248–268.
Room, R. (1980). New curves in the old course: A comment on Polich, Armor and Braiker, "The Course of Alcoholism." *British Journal of Addiction, 75,* 351–360.
Room, R. (1983). Sociological aspects of the disease concept of alcoholism. In *Research advances in alcohol and drug problems* (Vol. 7, pp. 47–91). New York and London: Plenum.
Room, R. (1984). Alcohol control and public health. *Annual Review of Public Health, 5,* 293–317.
Room, R. (1989). Cultural changes in drinking and trends in alcohol problem indicators: Recent U.S. experience. *Alcologia* (Bologna), *1,* 83–89.

Rubington, E. (1958). *What to do before skid row is demolished.* Philadelphia: Greater Philadelphia Movement.

Ruggels, W., Armor, D., Polich, J., Mothershead, A., & Stephen, M. (1975). *A follow-up study of clients at selected alcoholism treatment centers funded by NIAAA.* Menlo Park, CA: Stanford Research Institute.

Schmidt, W., & Popham, R. (1980). Discussion of paper by Parker and Harman. In T. Harford, D. Parker, & W. Light (Eds.), *Normative approaches to the prevention of alcohol abuse and alcoholism* (pp. 89–105, NIAAA Research Monograph No. 3). Washington: U.S. Government Printing Office.

Seeley, J. (1960). Death by liver cirrhosis and the price of beverage alcohol. *Canadian Medical Association Journal, 83,* 1361–1366.

Skog, O.J. (1977). On the distribution of alcohol consumption (Monograph No. 4). Oslo: National Institute for Alcohol Research.

Skolnick, J.H. (1958). Religious affiliation and drinking behavior. *Quarterly Journal of Studies on Alcohol, 19,* 452–470.

Smith, N.M.H. (1976). Research note on the Ledermann formula and its recent applications. *Drinking and Drug Practices Surveyor, 12,* 15–22.

Snyder, C.R. (1958). *Alcohol and the Jews: A cultural study of drinking and sobriety* (Monograph No. 1). New Haven: Yale Center of Alcohol Studies.

Spradley, J.P. (1970). *You owe yourself a drunk.* Boston: Little, Brown.

Straus, R. (1960). Research in the problems of alcohol—A twenty-year perspective. In *Multidisciplinary programming in alcohol investigation* (pp. 28–31). Division of Alcoholic Rehabilitation, California Department of Public Health, Berkeley.

Straus, R. (1975). Reconceptualizing social problems in light of scholarly advances: Problem drinking and alcoholism. In N.J. Demerath, O. Larsen, & K.F. Schuessler (Eds.), *Social policy and sociology* (pp. 123–134). New York: Academic Press.

Straus, R. (1976). Problem drinking in the perspective of social changes, 1940–1973. In W.J. Filstead, J.J. Rossi, & M. Keller, (Eds.), *Alcohol and alcohol problems: New thinking and new directions* (pp. 29–56). Cambridge, MA: Ballinger.

Straus, R., & Bacon, S. (1951). Alcoholism and social stability: A study of occupational integration in 2,023 male clinic patients. *Quarterly Journal of Studies on Alcohol, 12,* 231–260.

Terris, M. (1967). Epidemiology of cirrhosis of the liver: National mortality data. *American Journal of Public Health, 57,* 2076–2088.

Thompson, W. (1935). *The control of liquor in Sweden.* New York: Columbia University Press.

Törnudd, P. (1968). The preventive effect of fines for drunkenness: A controlled experiment. *Scandinavian Studies in Criminology, 2,* 109–124.

Trice, H., & Roman, P. (1972). *Spirits and demons at work: Alcohol and other drugs on the job.* Ithaca, NY: Cornell University.

Ullman, A.D. (1958). Sociocultural backgrounds of alcoholism. *Annals of the American Academy of Political and Social Science, 315,* 48–54.

Wallace, S.E. (1965). *Skid row as a way of life.* New York: Harper and Row.

Warburton, C. (1932). *The economic results of Prohibition.* New York: Columbia University Press.

Whitehead, P. (1977). *Alcohol and young drivers: Impact and implications of lowering the drinking age* (Monograph No. 1). Ottawa: Non-Medical Use of Drug Directorate.

Whitehead, P.C., & Harvey, C. (1974). Explaining alcoholism: An empirical test and reformulation. *Journal of Health and Social Behavior, 15,* 57–65.
Wilkinson, R. (1970). *The prevention of drinking problems.* New York: Oxford University Press.
Wines, F.H., & Koren, J. (1897). *The liquor problem in its legislative aspects: An investigation made under the direction of a Subcommittee of Fifty.* Boston: Houghton, Mifflin.
Wiseman, J. (1970). *Stations of the lost: The treatment of skid row alcoholics.* Englewood Cliffs, NJ: Prentice-Hall.
World Health Organization. (1980). *Problems related to alcohol consumption* (Technical Report Series 650). Geneva: World Health Organization.
Wuorinen, J.H. (1931). *The prohibition experiment in Finland.* New York: Columbia University Press.

CHAPTER 15

Sociology and Public Advocacy: The Applications of Selden D. Bacon's Work to Alcohol Problem Policies

MELVIN H. TREMPER

Some scientists leave their mark upon a field through one or more landmark discoveries. Others leave their mark through a lifetime of continuous leadership and striving for fuller understanding within a chosen field. Selden Bacon affected the field of alcohol studies in both these ways.

For much of his career Bacon treated problems in the study of alcohol not only as research problems, but also as social and political problems. He recognized the impact of "everyday" conceptions of alcohol, and the human behaviors surrounding its use, on "scientific" conceptions of the same phenomena. He also recognized the importance of "scientific" concepts for altering the "everyday" concepts. He was part of a movement that has brought fundamental changes to the way alcoholism is viewed in a broad array of social institutions, including law, medicine, and the workplace.

This chapter will briefly sketch Bacon's main ideas about the study of alcohol and relate these ideas to current trends and events in the alcohol and drug abuse field. This field is loosely defined to encompass government policy makers, members of the alcohol and drug abuse treatment and prevention sector, and researchers pursuing answers to questions related to the manufacture, distribution, and human use of alcohol and drugs.

My experience with Bacon was primarily in his role as teacher. He was director of the Rutgers Center of Alcohol Studies at which I spent four enjoyable and productive years as a graduate student. During this time I had the opportunity to hear him in many informal discussions among center staff. Toward the end of my career at the center I began my dissertation. When the original chair of my committee left Rutgers, Bacon took over the position.

In this role he expanded my vision of where my research fit into a much larger framework of sociological and even Western philosophical thought. Some of his comments were longer than the chapters he was reviewing. Yet, they were never harsh, merely insistent that I at least consider all the possible implications of my findings.

All the while he was guiding my dissertation, he was also managing the many affairs and issues that required resolution in order to keep the center functioning, writing articles, and globe-trotting to carry the torch of the study of alcohol issues at various seminars and conventions. Amidst all this, he still found time to comment so copiously on a graduate student's work.

Contrary to the formal concern for discipline Bacon often called for, I will now embark on a more or less informal examination of his impact on the alcohol and drug field from a framework loosely constructed from the speeches and writings of Bacon himself. My perspectives on these issues are formed both by my years at Rutgers and my years as an official in a state agency charged with coordinating and administering the state's alcohol and drug abuse program.

Bacon as Leader

Bacon's career in the field of alcohol studies has spanned more than 40 years. His contributions were hardly confined to the academic pursuit of knowledge about alcohol. In addition to his scholarly work, he devoted a large amount of time working to transform the image of alcoholism and alcoholics in the minds of other professionals and the general public.

In 1935, while still a graduate student, he joined the newly formed Section of Studies on Alcohol at Yale University (Who's Who, 1973). The section was formed by Howard Haggard within the Laboratory of Applied Physiology. Haggard and others at the laboratory had devoted several years to the study of the metabolism and other physiological aspects of alcohol. They learned much about the physiology of alcohol ingestion. However, they concluded that such studies "contribute only indirectly to the understanding of how inebriety arises. Nor have such studies yielded any tools which could be applied to the prevention of inebriety" (Jellinek, 1945). Thus, the alcohol studies section was created to examine the phenomenon of alcohol from a variety of perspectives.

The section soon became the Yale Center of Alcohol Studies. The center was devoted not only to a multidisciplinary scholarly inquiry into virtually every question related to alcohol, but was also dedicated to disseminating what was already known, and what they were to discover, to the rest of the world. The center also played a central role in creating the modern clinical approach to alcoholism treatment, and in legitimizing alcoholism as a treatable illness.

In the early years of the Yale center, there was little "infrastructure" in the field of alcohol studies. There were few research findings, no body of methods, and there was no theoretical framework within which to place these items. From the beginning, Bacon played a role in defining the fundamental questions of this new field, and in defining sociology's place within it.

These contributions to the development of the content of this field of study are discussed below. First, I present a brief discussion of the creation of the support system required to collect and disseminate the knowledge generated by the researchers at the center. I also mention some of Bacon's efforts to promote an awareness of alcohol issues among professional groups and the general public.

The center's resolve to disseminate information led to the development of a variety of vehicles. The most scholarly of these was the *Quarterly Journal of Studies on Alcohol.* Many of the early works of Bacon and others at the center appeared in specially published monographs, or in the pages of the new journal. Bacon served on its editorial board for many years (Who's Who, 1973).

The journal was directed at the academic community. In addition to this effort the Yale center also supported the creation of mechanisms to educate the general public. This included the National Committee for Education on Alcoholism (NCEA). The goal of this committee was "to bring to all America the message that alcoholism is an illness which can be banished by definite therapeutic means" (Chafetz & Demone, 1962). The committee was cited as one of the few private organizations devoted to dispensing information to the public about alcoholism and as the only such group that did not preach a "dry" or prohibitionist philosophy as the solution for alcohol problems (McCarthy & Douglas, 1949). Bacon was a charter board member and officer of the NCEA (Milgram, 1986). This committee evolved into the National Council on Alcoholism and ultimately into the National Council on Alcoholism and Drug Dependence, which continues its role as a national source of information on alcoholism as a disease and has established itself as a lobbying organization on alcoholism issues before Congress.

A third channel for communication was the Yale Summer School of Alcohol Studies. This was the first school designed to bring information about alcohol and alcoholism to a wide range of practicing professionals: temperance workers, social workers, physicians, clergymen, law enforcement personnel, and eventually the new occupation of alcoholism counselors. This information was presented from the integrated perspective of a wide variety of disciplines. This integrated perspective was a hallmark of the center's approach and one of Bacon's enduring themes.

Finally, to facilitate the transfer of knowledge, the center's publications and documentation section, under Mark Keller, began to collect and classify extant literature on the subject of alcohol. This collection led to the development of the Classified Archives of Abstracts of Alcohol Literature, which for years was the world's premier source of published works on alcohol.

Bacon was a faculty member at the summer school's first session in 1943. He became director of the summer school in 1951 and remained director until 1962 (Milgram, 1986). That year the Center of Alcohol Studies left Yale for Rutgers and Bacon was appointed the new center's director. He held this post until his retirement in 1975 (Milgram, 1986).

In addition to his involvement in the activities of the two centers, Bacon contributed to a wide variety of policy making and advisory groups. He was a member of the Connecticut Prison Association from 1944 to 1954, a member of the board of directors of the North American Association of Alcoholism Programs from 1964 to 1974, including seven years as vice president, and a member of the Subcommittee on Alcohol and Drugs of the National Safety Council from 1961 to 1974 (Who's Who, 1973). During this period he continued to lecture at the Rutgers Summer School of Alcohol Studies, and made numerous presentations before professional groups.

Bacon's contributions to these organizations have had a notable impact on the field. However, they represent only a fraction of Bacon's overall contribution. In addition to these activities, Bacon engaged in a number of research projects and contributed many publications on alcohol-related research, theory, and public policy. These writings and their impact on the field are the subject of the rest of this chapter.

Bacon as Scholar

Many of Bacon's early writings deal with the epistemology of the study of alcoholism. He insisted that the study of alcohol must not be limited to the study of "outrageous events" or problem areas. To borrow an example he used, studying alcoholism had the same relationship to the study of alcohol as studying volcanoes had to the study of geology. Both alcoholism and volcanoes were fascinating extremes, but displayed limited aspects of the more general phenomenon being investigated. Also, one who studied only alcoholism or only volcanoes would derive an equally limited understanding of not only the broader phenomenon but also the very topic in which he was so vitally interested. Thus, alcoholism can only be understood in a sociological sense if one has an understanding of the cultural factors associated with the drinking of alcohol within a particular society. Further, alcoholism can only be fully understood if one comprehends the sociological, psychological, biological, economic, and other aspects of alcohol.

In an early work, Bacon staked a claim for the validity of the sociological study of alcohol problems. He described the sociological areas of interest: "the sociologist is interested in the customs of drinking, the relationship between these customs and other customs, the way in which drinking habits are learned, the social controls of this sort of behavior, and those institutions of society through which such control issues" (Bacon, 1954, 14). He was confident that analysis of such social factors would prove fruitful; since physiological approaches were inadequate, "there seems little reason to believe that there is a physiological need for alcohol, or that drinking stems from an inherited craving... certainly no virus carries alcoholic behavior to the human system" (Bacon, 1954, 15). He urged that "drinking behavior" be studied, not just "problem drinking." He argued that this was necessary because it would not be possible to fully understand pathological drinking unless it could be seen in the context of drinking in general. "By observing the drinking behavior, its concomitants and later consequences, a keener understanding can be obtained, resulting in a more discriminating and more widely oriented perception of why people behave in this fashion. This, in turn, can lead to the more efficient control of drinking which is held to be dysfunctional" (Bacon, 1954, 22).

This work embodied a theme that was to pervade Bacon's writings: in order to understand problem drinking and alcoholism, we first have to understand the total structure of the culture as it relates to alcohol. The concept that alcohol problems and alcoholism cannot be understood in isolation from the cultural matrix of which they are a part has come to play an important role in the alcohol field. It has gone beyond academic acceptance and has found practical applications in the areas of treatment and prevention. It is now recognized that alcoholism cannot be successfully treated or prevented by focusing on the individual alone.

Later in this 1954 work, Bacon foreshadowed decades of concern about the effects of "peer group pressure" on initiating drug use by indicating that learning about drinking typically occurred in the teen years "when parental authority is low." Further, this learning usually occurred in places "out of reach of normal society which may be colored by an aggressive, even an anti-social atmosphere" (Bacon, 1954, 24). The prevention literature of recent years has recognized the importance of Bacon's concern with the context in which alcohol use is learned, who it is learned from, and the general set of linkages of the individual to "prosocial activities," in determining the probability of the abusive use of alcohol by an individual.

Of course, treatment practitioners have long known that it is not wise to encourage alcoholics to return to their barroom buddies if they want to remain sober after treatment. But the impact of other social environmental factors has taken longer to become recognized. A third of a century ago, Joan Jackson wrote about the response of the family system to the successful treatment of the alcoholic, and the impact this response had on the alcoholic's recovery (Jackson, 1954). A decade ago, "family treatment" began to be implemented in alcoholism treatment facilities. Now, alcoholism is often referred to as a "family disease."

In California, this has been taken several steps further. Pioneering programs in that state focus on creating entire communities that are conducive to recovery. These "social model" programs display an explicit recognition that the attitudes toward, and opportunities for, drinking will play a major role in determining whether or not drinking occurs. They then use this recognition to attempt to create communities where these factors are as favorable as possible to supporting sobriety. While effecting community changes is a recent development in the treatment field, it has a somewhat longer history in the prevention arena. Two approaches that reflect this awareness are: (1) efforts to change public policy about the price and availability of alcohol and the consequences for its use; and (2) efforts to change "community attitudes" toward drinking and drug-taking behavior.

In addition to creating theoretical constructs that led to practical application, Bacon also produced quantitative research. Typically, however, he connected the results of such research to a set of larger issues. For example, Straus and Bacon's (1951) study of occupational integration in 2,023 male clinic patients was a pioneering report of the characteristics of the alcoholic population. While the study was limited to individuals in treatment, it did not utilize the then-common approach of studying homeless males or incarcerated public inebriates.

Several seminal points were made in this study. One was that the characteristics of alcoholics as reported by the current literature were a function of the type of setting in which the alcoholics were studied rather than the "true" characteristics of alcoholics. For example, they found a much higher degree of social integration in their population as represented by marital status and employment than was typically thought to be the case with alcoholics. They explained the difference by pointing out that their cases were drawn from the clients of a new type of treatment setting, not from the back wards of mental hospitals or the local drunk tanks. As was typical of Bacon's work, the authors did not say that either portrait of the alcoholic was "right," but that both represented certain elements of the total picture. The finding that alcoholics are much like everyone else has formed the core of many campaigns to develop and improve prevention, treatment, and intervention programs. It has formed a major part of the effort to remove the stigma of alcoholism. The concept that alcoholism is not a condition unique to the dregs of society, but a disease that can strike anyone, is a critical component of the field's public educational messages. While Bacon did not originate this concept, the research evidence he supplied did much to bolster the argument.

In this same work Straus and Bacon (1951) put forth another concept that has become the keystone of many public policy approaches to alcohol problems: the "career" of alcoholism marked by steady deterioration of the drinker's social position with continued alcoholic drinking. They suggested that earlier studies had focused on the last stages of deterioration in the alcoholic career and did not provide a complete picture of the career path.

This depiction of a progressively declining career path was prominent in 19th-century temperance literature, well before the publication of Jellinek's work on the typical progression of alcoholism. This work, based on interviews with members of Alcoholics Anonymous (AA) about their drinking histories, has been immortalized as the "Jellinek Curve," often depicted in wall charts and popular literature designed to educate both the general public and clients in treatment about the inevitability of the course of untreated alcoholism. In this format and others, the concept of an alcoholic "career" or progression of ever-worsening symptoms has been the basis for the strategies of state and federal governments for dealing with alcohol problems. Treatment systems have been devised to focus on the "early stage alcoholic." Concomitantly, methods and strategies have been devised to channel alcoholics into treatment at an ever-earlier stage in the progression. These strategies of "early intervention" have themselves become the domain of specialized professionals.

If alcoholics can be successful targets of intervention at an early stage of this downward spiral, they will be spared the future pain and trauma of continued alcoholism. The creation of publicly funded treatment systems is justified by reference to the savings, in both human and economic terms, inherent in providing treatment to "early stage" alcoholics.

Straus and Bacon (1951) also raised several issues that have become central to many early intervention strategies. They discussed the issue of losses to business due to the alcohol-induced lapses of employees, and presented information showing that often the alcoholic was "protected" from discovery by well-meaning coworkers. These concepts have formed the core of promotional campaigns for employee assistance programs (EAPs) for decades.

The National Institute on Alcohol Abuse and Alcoholism (NIAAA) recognized the importance of EAPs in the early 1970s and created a special funding program that resulted in each state hiring two EAP specialist consultants, one for private sector work organizations and the other for public sector work organizations. The "Thundering Hundred," as these consultants were known, met with varying degrees of success. More recently, the federal government has again stepped up its efforts in this area. Federal legislation now requires federal contractors receiving $25,000 or more in federal funds to certify that they are maintaining a "drug-free workplace." There must be an educational program to inform employees about the policy and about the dangers of drug abuse. In addition, an EAP, while not required, is recommended as a desirable mechanism for maintaining a drug-free workplace.

The government has initiated a toll-free telephone information line, developed a training course, produced videos, and published several booklets and pamphlets to provide employers with information of EAPs. To some extent these contemporary developments are reflections of Bacon's influence.

The final contribution of "Alcoholism and Social Stability" to the field was based on the observation that few referrals to treatment came from physicians or employers. Bacon and Straus attributed this to a lack of awareness of alcoholism among these groups, and argued that there was a need for increasing such awareness. This awareness would prompt others to suggest treatment for alcoholics before they could descend to the final depths. Programs promoting community awareness have also become an important component of most public strategies to combat alcohol and drug abuse. These strategies focus on increasing the awareness of community members of the nature of alcohol and drug abuse so that they might more readily refer those who display the symptoms of such abuse. Treatment "gate-keepers" such as social service workers are also the foci of such information diffusion.

This work illustrates the nature of Bacon's approach to the use of research. He not only presented the data, but also outlined the implications of the data for the way society thought about and reacted to alcoholism. The concepts that many, if not most, alcoholics were employed, "socially integrated" individuals whose full productive capacities were impaired by alcohol but who could be restored through treatment remain cornerstones of the philosophy of the alcohol treatment field to this day. The work also contains the seeds of the concepts of "early intervention" and "community awareness" that drive the design of many of today's social response programs.

But Bacon's efforts were not limited to theoretical frameworks and specific data collection. The need for clarity and precision in the definition of terms and the construction of taxonomies was an enduring theme in his work. He called for precision in the fundamental concepts of the field of alcohol studies: *drink, drinker,* and *drinking.* Until these could be imbued with some semblance of common meaning, there would be no way to compare these phenomena across cultures, or across time within the same culture.

Although many studies of drinking and drinkers are conducted using either no explicit definition or idiosyncratic definitions of these terms, there is at least some effort to adopt a common standard for this purpose. The National Institute on Drug Abuse's (NIDA) High School Senior Survey, conducted by the Institute for Survey Research at the University of Michigan (Johnson, O'Malley, & Bachman, 1989) has made a conscious effort to utilize standard questions and definitions since its inception in 1972. Consequently, it has provided reliable and ongoing baselines of self-reported drug use that are of great importance to planners and policy makers. Similar efforts have been made in NIDA's Household Survey on Drug Abuse conducted since 1974.

Although there is still no way to scientifically assign social disapproval to a particular act of ingesting a particular amount of a particular drug in a particular setting with particular results, at least we are now moving closer to the time when we can actually see what these particulars are, how they are distributed throughout society, and how this distribution actually changes over time.

Nevertheless, some progress has been made. One example is in assessing drinking behavior via self-report surveys. For the most part researchers recognize the need to employ consistent definitions of quantities ingested, as well as the frequency of ingesting episodes, when they survey drinking habits within selected populations. While there are some commonly used measures, there is no universal standard, even among researchers. Of course, when it comes to the users of this information there is even less precision. There is, however, much commonality. All are convinced that there is a level of drinking that is clearly definable as "problem drinking" and that there is another level that is clearly definable as "alcoholic drinking." What is still not clear is what the limits of these levels are. It is also not clear whether "alcoholic drinking" can be defined solely in terms of drinking behavior, or whether other elements, such as the consequences of the drinking, must be considered.

Debate over these issues stems in part from the differing physiological and psychological responses of different individuals to equivalent doses of alcohol, and to the different social responses to both the drinking and the consequences of drinking. These "scientifically" relevant elements of the debate are complicated by the moral, political, and policy implications attached to the various definitions. For example, in the late 1970s the concept of "responsible drinking" was widely advocated as a potential goal of prevention programs. A decade later it is rarely used and, in fact, is officially designated as less-than-desirable terminology (Office of Substance Abuse Prevention, 1989). The rise and demise of the term *responsible drinking* and its potential companion, *responsible drug use,* merit a brief discussion here to illustrate the impact of politics and policy on the development of the field. In part, the adoption of this term was a result of the children of the 1960s coming into positions of authority and putting forth their value positions that not all drug use was necessarily "bad" and therefore problematic or abusive. It also coincided with the stewardship of Morris Chafetz as director of the NIAAA. The "soft on drugs" approach of the drug professionals has been largely discredited as a public policy position, in part due to the efforts of "grassroots" organizations such as the National Federation of Parents, which used very effectively their political connections to directly reach Congress and the Office of the President to affect federal policy. Chafetz has been excoriated and discredited in the alcohol field because (1) he is a psychiatrist and "everyone knows" how psychiatrists treat alcoholics, and (2) he accepted financial support for a foundation that he heads from the liquor industry, allegedly causing him to present the industry's "party line" instead of the truth.[1]

Another of Bacon's concepts that has gained currency in recent years is his concept of "addictionology." In this concept, addiction and its ancillary subject matter are viewed as a specialized field of inquiry in and of itself, like geology or economics. Bacon advocated that studying alcohol from the particular perspectives of many different fields of study should give way to a single unified discipline. There has been some movement toward creation of a special field of study in the years since Bacon first proposed this concept:

1. The federal government, through NIAAA funding, now designates 12 institutions as National Alcohol Research Centers. Although each one has a slightly different focus, it is clear that studying issues related to alcohol comprises their central mission.
2. An array of specialized journals and other forms of documentation has arisen. The *Journal of Studies on Alcohol* is now but one of many scientific journals devoted to documenting the findings of alcohol-related studies.
3. Specialized indexes and computerized data bases such as CORL and ETOH have been created specifically to provide access to alcohol-related literature. There is even a specialized Society of Alcohol Librarians and Information Specialists dedicated to the acquisition and dissemination of information in the area.
4. A coherent cross-disciplinary approach is also being established in clinical settings, utilizing teams consisting of substance abuse counselors, specially trained physicians, social workers, nurses, and others. Of course, the very term *substance abuse counselor* is by now a taken-for-granted designation of a professional specialty. In many ways it

> epitomizes the concept of a specially focused approach to alcohol problems. In other ways, since its knowledge base is limited to clinical treatment skills, this profession does not represent the "compleat" addictionologist envisioned by Bacon.

As a reflection of the changing times, Bacon later advocated a "meta-unification" of addictionology that would combine alcohol and other drugs into a single field of study (Bacon, 1969). While for many years social and cultural differences in the characteristics of user populations promoted separate consideration of alcohol and other drug addiction, blurring of the differences between user populations has led to the creation of a single framework for studying the processes of addiction of both alcohol and other drugs. This framework has led to the merging, not only of scholarly research, but also of a myriad of practical institutional approaches to the problems surrounding these various agents.

One example of this merging can be seen in the organizational structure of state government agencies dealing with the problems of alcohol and drug abuse. Forty-nine of the 50 states have combined alcohol and other drug abuse agencies. It can also be seen in the treatment field. A recent report on the characteristics of alcohol and drug abuse treatment centers disclosed that in 1982 one-third of the centers identified themselves as treating both alcohol and drug abuse (National Association of State Alcohol and Drug Abuse Directors, 1988). In 1987, this proportion had increased to two-thirds. The remaining programs professed to treat alcoholics only or drug addicts only.

Various professional organizations are combining previously separate alcohol-and-drug-focused organizations into one combined entity, or are changing their names to eliminate a single-focus identity. For example, the Association of Labor-Management Administrators and Consultants on Alcoholism recently changed its name to the Employee Assistance Professionals Association. Most states that have requirements for professional certification or licensure for workers in the field have combined standards and issue a single certificate of competency called a chemical dependency or substance abuse counselor certification instead of issuing two separate certifications for alcohol and drug abuse treatment.

The federal Office of Substance Abuse Prevention and the federally supported "prevention research centers" focus on the combined problems of alcohol and drug abuse in one concerted effort. Much of this consolidation has come about not through any desire to achieve some conceptual unity, but rather in recognition of the fact that few, if any, substance abusers today "specialize" in one substance. Alcohol is consumed by almost all those who enter drug treatment, and a variety of drugs is often used by those seeking alcohol treatment. Clinicians have recognized certain similarities in the dynamics of the development of the process of dependency and in the methods that appear to reverse the process.

In some areas, particularly law enforcement efforts, a sharp distinction continues to be made between "illegal drugs" and alcohol. Many community members eager to wage the war on drugs in their neighborhoods are puzzled when professionals advise them to focus energies on alcohol as the number one drug of addiction. Thus, the world is still not entirely rational, nor in agreement about many of the issues that Bacon raised in his four decades of prolific writings.

Conclusion

In retrieving material to write this chapter I was once again reminded that many examples may be found of very insightful works about alcohol in society, ranging from Benjamin Rush in the 1700s to today's latest journal article. Yet many of the early works soon faded from general awareness and the stark facts of the physical effects and social costs of alcohol use were relegated to a dusty library to be "discovered" anew by some later "pioneer."

I often wonder if this is the time when permanent progress will be made. Issues will be resolved, cultural attitudes will be changed, and society will learn to deal with alcohol's volcanoes since it has now mastered the science of its geology. Or will these years of laboring to raise awareness again fade into complacency or mythology?

Bacon and others "at the center" had a commitment to applying fruits of their research to practical issues. As shown above, disseminating their findings was an integral part of the center's total functioning. This aspect was deemed so important that when means to inform a given audience did not exist, they were created. Given this dedication and the contributions many of the early research findings made to enhancing the quality of treatment and improving the general social image of alcoholics, it is ironic that today "research" is often a dirty word in the field. There is a barrier between those who engage in clinical practice and those who engage in research into the causes and treatment of substance abuse.

Part of the reason for this barrier may be due to the different emphasis of today's research. While early research findings emphasized the similarity between alcoholics and other members of society, and demonstrated that treatment could be effective, much current research is focused on the efficacy of treatment itself in search of more sophisticated, cost-effective methods of delivery treatment services. This type of research is perceived as threatening to many practitioners. Not only does it question their current assumptions and practices, but it also sometimes results in conclusions that are inconsistent with the "common knowledge" of the field. In addition, clinical research is often conducted by psychiatrists and clinical psychologists. Both of these groups have for years been viewed as not sympathetic to, or knowledgeable about, alcoholism.

In the last five years there has been a marked increase in the sophistication of evaluation and other research directed at issues in the substance abuse field. The field has wholeheartedly embraced some of the products of this research. For example, advances in biochemical and genetic research have provided additional support for the field's long-held belief that alcoholism is the result of a biochemical "trigger." Results of this research have been eagerly shared within the field. However, many practitioners have pervasive mistrust of research techniques and researchers. This results in a highly selective use of research findings and a general resistance to applying them until they have been accepted by the field's political process.

Results of epidemiological study, evaluation, and other research are adopted or adapted and selectively used to bolster positions and support conclusions already made.[2] Results that do not fit the field's needs or preconceptions are ignored or, worse yet, attacked and discredited. The longest and most bitterly fought battle has been whether an alcoholic can, after treatment, return to drinking without once again falling into the uncontrolled, steadily worsening pattern of alcoholic drinking that existed prior to treatment. While some in the treatment field will say privately that "perhaps" there are some "problem drinkers" who can "sometimes" return to social drinking with little risk, they still condemn public dissemination of information that supports this position on the grounds that it will lead many alcoholics who cannot safely return to drinking into what could be fatal experimentation. An early study of the effectiveness of the outcome of the federally supported NIAAA alcohol treatment system, the Rand Report, provided evidence that some of those treated had returned to a stable pattern of nonproblem drinking 6 to 18 months after treatment. This one finding was the focus of a furious attack on the entire report. Alcoholism practitioners made extensive use of the media to challenge the report's credibility, and also turned to Congress to suppress NIAAA's dissemination of the report's findings.

Whatever the accuracy of the report's findings on patients returning to "social drinking," the furor around this single element obscured all the other findings in the report and ultimately led to the resignation of Morris Chafetz, who was NIAAA director when the report was released. Suppression of this report slowed, but did not stop the process of assessing the effectiveness of alcoholism treatment. It took almost a full decade before some of the conclusions concerning the relative cost effectiveness of different levels of treatment and the benefits of brief interventions were brought to light. By that time an entire industry of inpatient treatment units had been firmly established.

Now, that industry is facing the same "equal treatment" as the rest of the health care sector into which it had so long campaigned to be integrated. However, this equal treatment is not in the form of increased resources and prestige as the field had hoped; rather, it is in the form of the same type of third-party payer review of and attempt to control procedures that the rest of the health care industry is experiencing. If the conclusions of the Rand Report and other studies had been more carefully attended to at the time, the alcohol and drug abuse treatment field might have already been far ahead of the rest of the health care industry in the provision of specialized, intensive, and cost-effective outpatient services, and the more flexible, shorter-term inpatient services that are only now being developed in an attempt to respond to the demands of payment sources.

Although much could be added about the impact of Bacon's research on the field, the above examples are sufficient to illustrate the nature of his contributions of this type. The concern with public education and social change expressed at Rutgers contributed to the center of alcohol studies leaving Yale. Ultimately these activities came to be viewed as having no place within Yale's academic setting. Decades later, federal policy makers have come to realize the need to combine academic research with its dissemination, if not to the general public, then at least to clinical practitioners and public policy makers. This has led to the funding of a network of centers on alcohol and drug abuse issues. The centers vary from those primarily immersed in advanced research to others that perform significant functions in the area of disseminating knowledge to policy makers.

The federal Office of Substance Abuse Prevention (OSAP) has supported the growth of an organization called Substance Abuse Librarians and Information Specialists. The purpose of this group is to share information about both substance abuse and about how to access, collate, store, and retrieve information for those who may need it. OSAP has also supported the development of a specialized electronic data base of alcohol information and the creation of a "thesaurus" of standardized terms by which information may be categorized and accessed. The National Clearinghouse of Alcohol and Drug Information also continues to be supported. This organization responds to requests from the general public, as well as policy makers and researchers, on alcohol- and drug-related issues. OSAP is also structuring a Regional Alcohol and Drug Resource Network as a distribution system.

Dissemination of information within the field remains something of an art form. Often similar ideas are met with resistance or acceptance depending on their source and their manner of presentation. For example, in the past year or so the concept of "relapse prevention" has been gaining wider and wider currency. As part of its continuing education program, the New England School of Addiction Studies held a three-day workshop on special topics in substance abuse in the Spring of 1989. One of the best attended workshops was titled "Relapse Prevention." It was presented by Tammi Bell, a colleague of Terry Gorski. Both, in turn, are associated with Father Martin, a priest who has been active in the field for many years. The workshop was so well received that some states decided to invite Bell to come and present her workshop at local sites.

The general thrust of the Gorski-Bell-Martin message is not very different from that of G. Alan Marlatt, the alcoholism researcher. Each says that, rather than being an exception, return to the use of alcohol or drugs is the norm for those completing treatment. They both advise that there are ways to reduce this return to the use of chemicals. They both say that these methods require the therapist to recognize "danger signs" indicating that a relapse is beginning and that the client can also be trained to change his or her behavior in response to these danger signs so as to substantially reduce the probability of a return to the use of chemicals. Both advocate changes in the delivery of treatment, and especially in the design of aftercare and the treatment of "slips." Typically those who relapse are seen as "unmotivated" or otherwise as "failures." If they are permitted to reenter treatment they are exposed to the same therapeutic regimen they experienced the first time. Both say labeling relapsers as failures is counterproductive and that reapplying the same mode of treatment is not the best way to treat a relapse.

Yet, Gorski and his model have become widely popular, while Marlatt is either unknown or actively repudiated.[3] The key difference to the acceptance of the message of these two men is that Gorski is associated with a well-known figure who supports the "AA model" and he attributes the tendency to relapse to a "Post Acute Withdrawal Syndrome." This syndrome has a substantial biochemical base, which is completely consistent with the "allergic reaction" or "chemical trigger" concept that is so popular with recovering persons working in the field. Marlatt is a behaviorally oriented psychologist who stresses that relapse is a function of learned behavior, and that by learning different behavior, a person can significantly reduce the probability of relapse. Marlatt's message is lost in the "static" created by the field's associations of "learning theory" with "teaching alcoholics how to drink." Thus, much of Marlatt's message is being communicated, albeit in a different form. The moral of this anecdote is that the linkage of research to practice via communication is an art form, one that Selden Bacon mastered well.

Notes

1. Mistrust of psychiatrists and mental health clinicians in general is endemic in the field even today. For example, Mel Shulstad, cofounder and first president of the National Association of Alcoholism Counselors, was reported as saying: "he looked askance at [an organization] because it appears that its leaders are trained primarily in the field of mental health" (*Addiction Letter,* 1989).
2. A special study group under legislative mandate to determine the nature and extent of alcohol problems in a New England state consciously constructed a report based on a carefully selected set of references. Low-point outliers were explicitly rejected in estimating ranges and averages of the percentage of alcohol-related or "caused" cases among a wide range of medical, social, and economic problems. To an extent, this technique helped "maximize" the reported problems.
3. Personal observation of audience reaction to a presentation by Marlatt sponsored by a professional education consortium.

References

Addiction Letter, 5, 6, (1989, June). Washington, D.C.: Manisses Publications.

Bacon, S.D. (1954). Sociology of the problems of alcohol: Foundations for a sociologic study of drinking behavior. *Quarterly Journal of Studies on Alcohol, 4,* 402–445.

Bacon, S.D. (1969) Relevance of social problems of alcohol for coping with problems of drugs. In J. Wittenborn (Ed.), *Proceedings of the Rutgers symposium on drug abuse.* Springfield, IL: Charles C Thomas.

Bacon, S.D. (1976). Defining adolescent alcohol use: Implications for a definition of teenage alcoholism. *Journal of Studies on Alcohol, 37,* 1014–1019.

Bacon, S.D. (1977). On the prevention of alcohol problems and alcoholism. *Journal of Studies on Alcohol, 39,* 1125–1147.

Chafetz, M., & Demone, H.W. (1962). *Alcoholism and society.* New York: Oxford University Press.

Jackson, J.K. (1954). The adjustment of the family to the crisis of alcoholism. *Quarterly Journal of Studies on Alcohol, 15,* 562–586.

Jellinek, E.M. (1945). Introduction in Yale Center of Alcohol Studies: Alcohol, Science and Society: Twenty-nine Lectures with Discussions as given at the Yale Summer School of Alcohol Studies. New Haven: *Quarterly Journal of Studies on Alcohol.*

Johnston, L.D., O'Malley, P.M., & Bachman, J.G. (1989). Drug use, drinking, and smoking: National survey results from high school, college, and young adult populations. (Publication number ADM 89-1638). Washington, DC: National Institute on Drug Abuse, U.S. Department of Health and Human Services.

Marlatt, A. (1985). Chapters 1–5 in A. Marlatt & J.R. Gordon (Eds.), *Relapse prevention: Maintenance strategies in the treatment of addictive disorders.*

McCarthy, R.G., & Douglas, E. (1949). *Alcohol and social responsibility: An educational approach.* New York: Thomas Y. Crowell Company and Yale Plan Clinic.

Milgram, G.G. (1986). The summer school of alcohol studies: An historical and interpretive review. In D.L. Strug, S. Priyadarsini, & M. Hyman (Eds.), *Alcohol interventions: Historical and sociocultural approaches* (pp. 59–74). New York: Haworth.

National Association of State Alcohol and Drug Abuse Directors. (1988). *National drug and alcoholism treatment unit survey, 1987.* Washington, D.C.

National Institute on Drug Abuse. (1989). *National household survey on drug abuse, 1988.* Rockville, MD.

Office of Substance Abuse Prevention. (1989). *Prevention Pipeline, 2, 6* (inside cover, no page). Rockville, MD: U.S. Alcohol Drug Abuse and Mental Health Administration.

Straus, R., & Bacon, S.D. (1951). Alcoholism and social stability: A study of occupational integration in 2,023 male clinic patients. *Quarterly Journal of Studies on Alcohol, 12,* 231–260.

U.S. Department of Health and Human Services. (1987). Drug abuse and drug abuse research: The second triennial report to the Congress from the secretary. Washington, D.C.: U.S. Government Printing Office.

Who's Who in America, 1972–73. (1973). (37th ed., Vol. 1.). Chicago, Il: Marquis Who's Who, Inc.

CHAPTER 16

An Old Warrior Looks at the New

SELDEN D. BACON

It has often been noted that one can learn from failures and mistakes, even learn more than from victories and smooth performances. Of course, no two events are ever identical. Although alcohol can in many ways be called a drug, it differs from heroin, cocaine, or other materials that are currently called drugs in this country. Even so, movements to "do something" about alcohol can be compared to movements to "do something" about "drugs." This is especially appropriate if the same society and about the same time period are involved. It is also appropriate because the current attack on "drugs" often specifies the inclusion of "alcohol."

Programs to do something about alcohol refer to whatever it is about alcohol beverages and their uses that is widely held to be a "problem." In this country, movements to do something about alcohol problems go back at least to the beginning of the 19th century and in many ways have been constantly present ever since. By contrast, while drug problems received societywide attention in the middle of the 19th century, what may be called national movements to "do something" about drugs never showed the persistence, widespread concern, or national popular strength of the alcohol problem movements. Perhaps the current (1985–1990) concern about drugs marks the high point of this sort of public reaction in American history.

It is suggested here that those who wish to "do something" about "drug problems" in this country in the closing years of this century might gain some relevant insights, some worthwhile orientation, and perhaps even some detailed knowledge from consideration of the many attempts to alleviate, control, or even eliminate "alcohol problems" in their society over the past 150 to 175 years, attempts that are being continued at this very time.

As the last comment suggests, the chances of "learning something" may be enhanced for the antidrug group because of the all-too-obvious existence of many mistakes and many partial or greater failures exhibited by the alcohol problem movements. However one may define or describe "alcohol problems," it is clear that they have not been eliminated. And it is very difficult to find *any* convincing evidence that these "problems" have been reduced or controlled. The usual labels for those problems over the past 60 years have been "public drunkenness," "youthful misuse and abuse," "chronic inebriety and/or alcoholism," and "highway accidents (of a major sort) significantly related to alcohol." There is little compelling evidence that these have shown any meaningful reduction in incidence or prevalence over the past 30 to 40 years. Indeed, the most discussed of such possible changes occurred in the late 1960s and early 1970s when it was believed that a real reduction in youthful drinking (or, perhaps, youthful "excessive" drinking) took place. The chief explanation for this change, unfortunately, was that over these few years the use of other drugs had been substituted extensively for alcohol in this age category.

The clear lack of success in reducing, controlling, or eliminating the "problems of alcohol" over these many decades does not provide one with adequate information (to say nothing of understanding) about either the problems or the programs. There have been changes in the nature of the alcohol problems. There have been changes in the customs of drinking. There have been changes in what various publics in the society felt to be "problems." There have been changes in many relevant fields of knowledge for the study of both problems and programs.

The clear lack of success in reduction, control, or elimination gives us little in the way of overall evaluation of the persons and groups involved in those programs, however. The motivations and the organizational abilities, the imagination and technical accomplishments, the leadership and dedicated memberships exhibited over these almost 200 years are hard to match. Such a conclusion is not affected by the fact that mistakes were made and that there were fanatics and even corrupt persons in such organizations, as well as in the countermovement organizations.

But for those attempting to attack effectively drug problems, study of the programs, leadership, techniques, goals, communications, organization, institutional alignments and alliances, observational and recording techniques, major assumptions, and studies of the alcohol-problem movements could hardly be irrelevant. Indeed, major modes of attack adopted by the current "war on drugs" show something like identity with the major modes of attack adopted by the "anti" or "control" alcohol movements.

These modes of attack, announced by President George Bush in September 1989, were (a) upgrading law enforcement, including special support for police, courts, and prisons, as well as developing far-reaching legal testing devices; (b) increasing education about the dangers of drugs above all for children and teenagers; (c) upgrading treatment facilities for drug addicts; (d) providing a central, overall, and powerful agency for the federal government to provide responsible, coherent, and clearly observable federal leadership in the conduct of this war; and (e) reevaluating international relationships in the light of drug problems and providing significant support for international efforts.

The major modes of attack by those fighting the war on alcohol (or alcohol problems) were (a) educating *everyone* about the evils of alcohol; (b) above all, educating children and teenagers; (c), increasing law enforcement capacities through supporting enforcement agencies and courts and, after 1940, through developing legal testing devices; and (d) upgrading treatment facilities for alcohol addicts, starting late in the 19th century and only becoming a major attack after 1950. Although the antialcohol movement was involved, especially in the 1920s, with border patrol and antismuggling activities, this was never a major element in such programs. A major effort on their part that is not germane to the current drug "war" dealt with control of sales under a system of legalized purchase of alcohol beverages, establishing agencies for this purpose at the federal, state, and local government levels. Although control through legalization has been discussed and even strongly backed by some, the current war on drugs has not yet adopted this strategy.

The parallels between major modes of attack by movements against alcohol problems and by movements against drug problems are starkly obvious. Indeed, among important critics of the president's proposed program, there is little if any argument with the statement of major modes of attack as such. Rather, the criticisms are of two sorts: first, that the proposed amount of money awarded to the three major branches of attack (i.e., education, enforcement, treatment) is inadequate, even grossly inadequate; and second, that the proportionate shares to the three divisions should be reordered. But the three major lines of attack themselves are not questioned at all. A minor exception (suggested both by the president and the critics) might be noted: the call for neighborhood education and leadership. This was also an important and, at times, a major policy for the antialcohol movements.

The two movements can also be compared in terms of "who will carry out the program." Personnel of the criminal justice system (judges, prosecuting attorneys, juries, probation and parole officers, police and various departments of inspection) will be the major actors in the enforcement mode of attack. School teachers and to some extent social agency personnel, the clergy, and volunteer groups (including recovered and recovering addicts), as well as school administrators from various levels, would be the major actors in the educational mode. Physicians, nurses, clinical psychologists, social workers, lay counselors, and other therapeutic personnel (again including volunteer recovered and recovering addicts), as well as health administrators from various levels, would be the actors in the treatment mode. The current war on drugs would include some in international diplomacy—actors rarely important in antialcohol circles. Alcohol distribution control programs would include administrators of price and sale conditions, actors not considered appropriate in the dominant circle of those attacking drugs.

As one looks at this descriptive listing of those who are to carry out the programs, two characterizations are obvious: first, the listing is almost (not entirely) identical for the two categories of movements; second, in all instances the program is to be carried out by "somebody else." This suggests that a legislative body or a citizen committee (made up of those enthusiastically supporting or even authoring a program design) has authorized, provided an organizational blueprint, and made available financial resources for carrying out a specified policy. The legislation probably is "not exactly" what the citizen groups were calling for. Similarly, an industrial corporation or a church or a military organization may adopt such a program and authorize organization for action with a specified budget.

In the alcohol field in the past 50 years, delegation of authority and responsibility to "carry out the program" through various "other" groups has frequently been alleged as a major cause for the inefficiency or even failure of many programs. Those in executive and high administrative positions in the currently developing war on drugs might well note some of these experiences.

The first category of "others" to be considered are teachers. If any relatively powerful group decides to change American attitudes and behaviors (and feels there is some great ethical principle involved), American teachers are a fairly immediate target for "carrying out" the message. To almost any neutral outsider it would be no surprise that the teachers asked to undertake this additional task would frequently object. Further, they would object even more if they did not happen to agree with the message or did not "like" the sponsoring group. To fuel such objections further, program sponsors might demand that the teachers prepare themselves by taking special training. Granted enough power by the interested movement, the teachers will really not be asked, they will be told. Not surprisingly, the teaching group with the least prestige in the educational world and the least political power within that world will be given the undesirable job. Those reviewing the past 100 years of experience in dealing with alcohol problems should not be surprised to find that schools commonly "farmed the job out" to the local temperance group. They should also not be surprised to learn that students frequently reported that "alcohol education" was the poorest course in the school.

In movements to control, reduce, or eliminate problems of alcohol, the same difficulties emerged in the world of law enforcement. Many police are not overly enthusiastic about "rassling drunks." Control over or responsibility for the "drunk tank" is not, to put it mildly, the most prestigious police job. Many policemen are not too favorably inclined toward large segments of alcohol control legislation. Special alcohol control police (as in Alcoholic Beverage Control boards or commissions) are frequently among the lowest paid and least professional of police authorities. Whether true or not, there is widespread belief that varying levels of graft are particularly related to enforcement of alcohol-control programs. It is worth noting that other than traffic violations, drunk and disorderly arrests numerically form the single largest category of all police arrests (sometimes other of these cases are "hidden" under different labels). This author once commented on this fact to the then-dean of police administration, O.W. Wilson, and added that it might be considered odd that his two-volume work in that field never mentioned the topic. Actually, it is not at all odd; drunk and disorderly arrests are the nadir of police work.

Alcoholism, hardly recognized as a professionally appropriate condition for attention by physicians or nurses as late as 1945 or 1950, gradually became labeled as a "disease"—a label at least formally accepted by medical and nursing professional associations. Questionnaire studies showed that in the 1940s, about 35 or 40 percent of physicians answered affirmatively to the question, "Is alcoholism a disease?" This number became almost 75 percent or more after 1960. How many of those answering in the affirmative would themselves accept alcoholics (as such) for treatment? If the proportion were as high as 50 percent, this author would be surprised. Some felt that it was very fortunate that most physicians who would accept alcoholics would quickly turn them over to Alcoholics Anonymous, but this is far removed from any medical regimen.

During the 1940s public health officers ridiculed the notion that alcoholism was "a public health problem." Leaders of the mental hygiene movement advised those in the emerging alcoholism treatment movement that the mental health movement could not adopt alcoholism as an appropriate target for their programs. They informally encouraged these newcomers concerned with alcoholism, but at the same time pointed out that they had their own problems. The mental hygiene movement could not take on the odium and almost certain warfare that would follow upon their incorporating alcoholism problems into their own programs. At the time it seemed an understandable position, even to those in the new alcoholism movement.

The conclusion suggested is that the health, treatment, and related groups did not like drunks, did not consider them proper patients, and did not feel alcoholism to be a public health or a medical responsibility. It might surprise some of those professionals to learn that many alcoholics detested physicians and other health professionals, and widely publicized their views. The comparison would relate to the attitudes and practices of health professionals toward those addicted to "drugs." "Turning the drug problem over to the medical world" is not at all a simple matter, not if even vaguely comparable to the experiences with alcoholism.

In each of these three modes the roles of administrators (as distinguished from clinical practitioners, from most sorts of law enforcement agents, and from those teachers primarily working with students) come to be of major importance, largely determining policies, tactics, budgets, record keeping, training, and public relations. Administrators' perceptions of goals, of criteria of success, and of other overriding issues were often not only different but sometimes sharply different from perceptions about the same items held by those on the "front lines" responsible for program implementation. This was painfully clear in some of the treatment programs adopted by penal, military, and industrial groups. It was blatantly manifest that a pre-1920 philosophy of prohibition-favoring temperance groups was influential in some of the 1950 alcohol education programs (whose sponsors publicly attacked these views). As many in the post-1950 alcohol problem "attack groups" were to find out, their need to survive required organization and administration. The consequent bureaucratic structures, whether of educational, law enforcement, or commercial or political health types, took over more and more control. It may well be that these weaknesses will not be so significant for the antidrug movement, but they may be "just as" or "even more" significant. It is important for the "new warriors" to recognize these earlier sources of failure.

A related matter concerns "reports of success." Bureaucratic regimes frequently seem to enjoy or even to require public statements that claim or suggest this or that sort of victory, that declare that their "new" program is really, even wonderfully, efficient, claiming new breakthroughs and "records." The success rates claimed by certain treatment modalities in both alcohol and drug intervention provide occasionally ridiculous examples, reminding one of the 1850 claims by mental hospitals of 100 percent success. The drug enforcement claims of this or that new record in pounds or tons of drugs confiscated also exemplify this practice. In both fields claims by educational programs of success, efficiency, and the like (despite no evidence of any reduction in the problem according to scientific measurement) are all too frequent. Such claims often seem to be believed to be "true" by their authors as well as by some publics.

Yet another matter incorporates the selective labeling of either the "problem actors" or of that segment of problem actors chosen as targets, whether for law enforcement, education, or treatment. In alcohol problem fields, various categories were singled out as the core or major segment of the problem or as targets for this or that program action. For many years it was "generally believed," for example, that skid row "bums" composed the alcoholism or chronic inebriate problem. Some may have expressed the notion that the alcoholic population was chiefly found among the American Indians or those of Irish descent. On the converse side, certain industrial and military alcoholism treatment programs clearly excluded the "upper ranks" from consideration.

In certain current drug programs, particular segments of some school populations are selected for "testing." It may even be recognized that certain other students are omitted; however, the fact that school staff (of any sort) are omitted from testing is ignored. Just who composes the target population of the current War on Drugs varies, depending on the major mode of attack: in Colombia, bankers, gangsters, lawyers, and even farmers form one target; American high school students and grade school students (two rather different groups) form another; chronic users of illegal drugs or addicts form a third; American criminal "dealers" in illegal drugs, preferably "big-time" rather than "small fry," form a fourth. People involved in the newly declared war might remember lack of consensus about "enemies" in the old war. President Franklin Roosevelt expressed a very popular opinion in the late 1920s about opponents and supporters of Prohibition, the so-called wets and drys: "a plague on both your houses."

A final area of comparison concerns national (and also regional and local) organization. Throughout both wars, and in the different focal areas, the lack of any single or coherent structure for attack, whether at national, state, or even village levels, has been a typical characteristic. During the 1960s there was an attempt by a federal government agency to list and categorize federal agencies, departments, or bureaus, with responsibilities and budget for dealing with alcohol and related problems. After two years the attempt was given up because of its scope and complexity.

In the 1960s, the emerging federal organization for dealing with alcoholism called a conference that was to have representatives of all governmental and religious and secular groups concerned with the problems. To this writer it was amazing that although distillers and brewers, AAs and professional therapists, city police, dry organizations, health educators, religious groups and alcoholism researchers were all there, not a single representative of state alcohol and beverage control commissions or boards was present. Indeed, they had not been invited. It seemed, however, that representatives of this group did not feel snubbed and indeed they did not even seem curious about the meeting. At that same meeting a representative of the federal Bureau of Alcohol, Firearms, and Tobacco asked this writer for some explanation of why his agency had even been invited. These examples describe the lack of coherent policy on the federal level and the consequent lack of coherent governmental structure in relation to problems of alcohol.

This structural and policy weakness was even more evident in governmental groups concerned with enforcement, treatment, or education. Not only did much older wet-dry philosophies and controversies persist within such groups, but they were exacerbated by more recent disagreements. These revolved about social, psychological, and biological elements in origins and development of alcoholism, about modes of treatment, about the appropriate roles of arrest and confinement, about the utilization of spiritual understanding and therapy, and about different educational approaches. Whatever conflicts existed, they flourished mightily in the alcohol problems area. Politically satisfactory resolutions to these disagreements do not appear to have resulted in coherent, let alone effective, attacks on problems, at least in the alcoholism field. That such disagreements obtain in the field of "other" drug problems is hardly deniable.

The preceding points of comparison are all very broad. Each could be elaborated around technical matters, almost smothering the reader with the lengthy and always argumentative characteristics of matters of medication and diagnosis, of arrest policy and "victim rights," of teaching techniques and teacher training; for this presentation it merely will be asserted that similarities are multitudinous and striking across programs directed either toward alcohol or toward drugs.

Quite a different mode of comparing alcohol and drug approaches focuses on the matter of research and communication. A great deal of information, "data," and communication about alcohol and alcohol "problems" are available. The largest part of the communication is noticeably remote from scientifically oriented research. Rather, studies are usually directed toward proving or disproving this or that particular "answer," gaining an ally, condemning an enemy, and so on. Students of the history of programs about "other" drugs will recognize the orientation and characterization.

It is suggested that ridicule or careless disregard of such research strategies represents an inappropriate response. These modes of communication form an important part of such movements. They are functional. They need to be analyzed. No one needs to accept the modes of reasoning, the logic, or even the deliberate falsification of these messages. However, their power, location, character, and relation to other belief systems are worth serious consideration. One may even query how a national movement in this country to "do something" about alcohol or other drugs could emerge, let alone survive, without such approaches.

What may be termed more serious study of alcohol problems is open to significant criticism. These criticisms, together with suggestions of how more scholarly and scientific approaches could be developed, have been discussed at some length in other presentations (Bacon, 1984, 1987), and are briefly summarized here.

First, a clear weakness of most of the alcohol problem movements is the manifest poverty of definition of the problem or problems under attack. This was less of a problem for the prohibitionists, but has been a major shortcoming of almost every other group. The post-1940 movements dealing with what is labeled alcoholism present an obvious example. The definitions provided by various groups result in estimates of the size of this problem varying by a factor of ten, for example, from 3,000,000 to 30,000,000 persons afflicted with the "problem." The groups that use these different definitions frequently and publicly deride each other.

Second, an equally obvious shortcoming may be found in definitions of success (in enforcement, education, treatment) in the various programs. Indeed, claims of "success" in the range of 80 to 100 percent are not unknown, and hinge upon questionable definitions of such success.

A third weakness has to do with the lack of historical perspective in studies of either problems or programs, not that this is an unusual shortcoming of studies of any American "problem." Whether anything proposed in the last 75 years is "new" is at least doubtful.

A fourth criticism centers on the pathologic orientation of studies. They are restricted to consideration of the awful, the strange, the evil, the frightening, the "sick," or the "problematic." In every so-called scientific field (e.g., astronomy, physics, chemistry, geology, biology, physiological psychology, anthropology), study of the bizarre or frightening (the teretologic) has frequently initiated study and has always been useful for testing both representativeness of data and generalization. However, study of the phenomenon per se (rather than focusing only on what are labeled the strange, "bad," or painful aspects of that phenomenon) has proven to be the effective and potentially fruitful orientation, even allowing greater understanding of the "pathologic" segments of the whole. In the alcohol field even modern studies of "drinking" and "drinkers" (which might consider the "normal" dimensions of such behaviors) are centered around symptoms of alcoholism.

A fifth weakness relates to scientific study that is dominated by the search for answers. This domination is heavily reinforced by research funding groups with manifest and powerful needs for particular answers, such as industries and government. Indeed, research that might lead to questioning of major assumptions of such policy groups can expect a rather cold reception.

A sixth criticism concerns the adoption of prestigious and professionally "correct" targets and modes of study, quite apart from their relevance to accepted descriptions of "problems." To the amazement of some, a large proportion of physiologic, toxicologic, and other biologic studies, heavily financed and pursued over the past 50 or 60 years, has had little contribution to make to the understanding of alcoholism or other problems related to alcohol, let alone adding to understanding or development of treatment. Much of this research has been of high quality and may well have contributed to an understanding of the liver, brain, or metabolism, but its relation to alcoholism or its treatment is hard to discern. On the other hand, the motivations for pursuit of such prestigious research are not difficult to discern.[1]

A seventh weakness represents yet a further violation of scientific criteria, namely the typical absence of observation in studies of both alcohol "problems" and of alcohol beverage consumption or consumers. Those authoring such studies rely almost entirely on the records and individual assertions of others, most of whom are not researchers by any stretch of the imagination. Reports of tax receipts or sales are accepted as evidence of consumption in a given population. Records of arrests by police for drunkenness are accepted as evidence of the numbers of persons involved. In most scientific fields such alleged scientific data would be a subject for joking. That such so-called data and statistical correlations of combinations of such numbers are taken seriously is striking evidence of the low quality of research in some areas of alcohol studies.

This listing of weaknesses is brief and incomplete but sufficient for present purposes. Two questions are immediately pertinent:

1. Do parallel criticisms obtain in relation to knowledge, understanding, and communication in the field of "other drugs"?
2. Are the weaknesses subject to corrective action?

At least to this author, answers to both questions are in the affirmative. This is not to suggest either swift or simple correctives, but the "new warriors" possess certain potential advantages over past and current groups attacking the problems of alcohol.

Those attacking the "other drug" problems have a "model" to use as a guide. The "old warriors" had nothing remotely resembling such a model.

Those attacking alcohol problems have developed a variety of deeply held beliefs, traditions, and purposes. They have emerged with a variety of organizations, some of which have persisted for 40 years or more. This "variety" includes conflicting as well as tactically-different-but-not-necessarily-hostile viewpoints, goals, and procedures. Not only do some of these groups dislike and distrust each other, but they also have generated and promoted contrasting images of themselves and others with general publics. Comparisons of these images may not only be negative but even characterized by active hostility. Clearly, there are differences between various categories or groups in the "other drugs" field (e.g., disagreements about methadone treatment, about legalizing drugs), but there is nothing even mildly comparable to the cleavages within the alcohol problems field or to the combative attempts to influence public attitudes about that field.

In the alcohol problems field there has almost always been an active "pro-alcohol beverage" group—small, practically limited to the beverage industry, but well organized (despite internal battles). There has also been a much larger, though usually not as well organized, anti-dry category. There appear to be no parallels for these "countergroups" in the drug problems field. There just is not a "home industry" for heroin, for cocaine, or, in any major sense, for marijuana. There is no recognized retail industry, nor are there "tie-in" businesses and industries comparable to restaurants, bottlers, or advertisers. Nor in association with other drugs is there a centuries-old network of behaviors, songs, symbols, and argot, mostly of a most positive and favorable nature, that clearly characterizes the alcohol world. The movement to battle the other drugs does not face this combination of organized and unorganized opponents who have, without exception and without cessation, harassed and at times defeated the antialcohol and antialcohol problem movements. These anti-dry (if not always pro-wet) forces often were strengthened by alliances, more or less open, with religious, ethnic, occupational, and political groups. Clearly, there is objection to some of the techniques, the assumptions, and the attitudes of those fighting the War on Drugs. However, this opposition has in no way approached the level of organization and bonds with significant allies that have characterized opponents faced by the alcohol problem programs.

A more general point might be labeled the "soft" thinking so often characteristic of anti- or control- or alleviate-alcohol problem movements. This is indicated by behavior and processes such as:

1. Describing the whole in terms of only one of the many parts of the whole.
2. Allowing metaphors to become descriptions of reality.
3. Confusing the ideal or the wish with reality.
4. Utilizing one's own movement's propaganda as factual evidence.
5. Accepting the proposition that because one thing preceded another, it necessarily was the cause of that other.
6. Accepting the assertions of Mr. or Ms. X as an expert in *all* areas of knowledge because he or she was an authority in one such area.
7. Asserting that recognizably complex phenomena may be satisfactorily described and fully understood in the terms of one field of expertise.
8. Relying upon ignorance, deliberate falsification, and manifestly ethnocentric (or other group-centric) thinking.

Many of these modes of thinking and explaining have long been used in the alcohol movement field. Sometimes the results have become something like "revealed truth," lasting for decades. Whether the relative youth of the drug field will make the lot of the "new warriors" somewhat less difficult from that of those hoping to do something about alcohol problems is far from certain. The misnaming of the field's targets as "drugs" is an obvious and basic problem for the new group. They seek to end traffic in some drugs and favor greater control of other drugs. Overall their targets are actually only a small percentage of all drugs, surely less than 10 percent. They are not against aspirin, penicillin, or sulfa. They are not against pharmaceutical houses or druggists or the users of drugs produced and distributed by these groups. Yet there are "problems" directly related to these other, perhaps "good" drugs and there are groups trying to "do something" about them. Then there are those earnestly trying to discover new drugs and make them increasingly available. Further, methadone treatment provides an example of those attacking the "bad" drugs and their "bad" use by promoting the "good" use of "good" drugs. Finally, to round out the confusion, two historical episodes should not be forgotten. "Other drugs" were frequently used by professionals as well as victims in attempts to cope with alcoholism. The increase in the use of marijuana contributed to a decrease in the excessive use of alcohol among young adults in the late 1960s and early 1970s.

The phrase "war on drugs" presents a fair example of an abuse of metaphor. It is clear that various politicians found the phrase useful to indicate their sincerity and the high priority allotted to "doing something." It is also clear that this is not and is not intended to be a "war"; far from it. Indeed, any use of the military in this program is very questionable.[2] That any of the enemy (whoever they may be) are to be treated as other than full citizens is also questionable. Even if the outstanding "leaders" of the "enemy" are foreigners with their own armies, no attempt to shoot such an enemy will be allowed. Obviously, this is no war.

But in broad, general terms, what is the nature of the program directed toward "drugs"? Is it medical, educational, law enforcement, neighborhood public relations, or what? To some who are deeply concerned about this set of problems, it is none of the above. They perceive drug problems as manifestations of poverty and the growing schism in American society and culture between the "haves" and "have nots." Seeing the problems in this light makes enforcement, educational, and therapeutic endeavors (which make up 95 percent or more of the current program) rather futile. The "war," as they see it, is being directed against symptoms and must necessarily fail.

The confusion of "should be" with "is" appears frequently in alcohol problem attacks. An obvious example is the slogan (unfortunately, for many, a real belief) "drinking and driving don't mix." As has frequently been pointed out, the two, particularly in this country, are closely related. Many American drinkers rarely drink except when visiting others, having others visit them, at commercial places of eating or amusement, or when on "trips." Few Americans walk or ride on horseback to visit friends or patronize restaurants. They drive there and they drive back. The two customs are inextricably mixed, although drinking is much more closely associated with driving than is driving with drinking. Recognition of this readily observable and measurable reality might well suggest different approaches from those utilized by those adopting the fiction that "drinking and driving don't mix." Are there similar slogans and beliefs in the movement against "drugs"? If so, are they influencing policy and action?

Utilizing friends', supporters', and patrons' assertions as independent evidence to support their own programs has been a common practice in the alcohol programs of the past 100 years. The process takes on rather frightening overtones when government agencies control not only their own action policies but also the so-called independent research. Is this occurring in current movements to attack "drugs"?

Use of the "*post hoc ergo propter hoc*" fallacy is hardly limited to analyses of social problems. Some explanations of alcoholism, such as alcohol prices, heredity, or broken families, blatantly demonstrate this ancient form of faulty reasoning. If this weakness does not appear in reports and proposals in the field of programs to attack drugs, one could only be amazed.

Acceptance of the assertions of a famous astrologer, biochemist, or wonderful halfback about the effects of price change on purchase or about the efficacy of silent prayer in treatment of addictions is not the mark of intellectual maturity, acuity, or discrimination. Antialcohol groups over the past century have all too frequently manifested this sort of reasoning, even making it the basis for their communication. In the case of alcohol or drug problems, any such single cause or single-minded statement is almost certainly wrong and almost certainly misleading. One would hope that executives and legislators of the 1990s would be dubious of such narrow advice.

During the period 1945–1970, movements to "do something" about alcohol problems came to be dominated by two perceptions: first, that the overwhelming majority of alcohol problems can be described and defined as alcoholism: second, that alcoholism is a disease. No definition, disciplined observation, evidence, analysis, or objective testing was involved in reaching these assertions. The movement was humanitarian. It was opposed to earlier Dry attitudes and, perhaps more importantly, was motivated to generate interest or action toward its target issues within the American public. The movement's adoption of the disease label was based on political and mass media tactics, reinforced for some by the writings of Jellinek. The majority of supporters, indeed, claimed that alcoholism was a spiritual disease. It is doubtful that, however defined, the disease of alcoholism was the appropriate label to account for major auto accidents involving alcohol or to describe "too much" drinking and accompanying behavior by teenagers. By the early 1950s, the notion that most "common drunks" were addicts was being seriously questioned on the basis of direct observation, careful definition, and openly reasoned (and testable) analysis of hundreds of cases.

Whether the often casual, narrowly oriented, and frequently contentious modes of reasoning and consequent policy making often exhibited by the alcohol problem control movements are characteristic of the "other drug" programs poses a serious question for these new warriors.

Can any suggestions be offered by the old warriors to those now planning on a national level to attack the problems of the "other" drugs in the 1990s and the next century?

Suggestions about administration; about techniques of enforcement, education, and therapy; or about mass media development could all be produced. Experience of this sort should not be disregarded. And, granted that suggestions from the aged and aging warriors would be considered, it is also probable that this sort of information and advice would be the most likely to be accorded attention. But let us not be too optimistic. If only the problems of alcohol or of these "other" drugs were susceptible to significant resolution through such simple steps!

An initial suggestion would relate to a very basic attitude and evaluation: we do not have any very effective understanding of what this "problem" is all about. Unfortunately, some of those most motivated to "do something" are convinced not only that they know all about the "problem," but also that they know the "answer."

For the great majority of those perceiving the pains, costs, and frustrations of whatever the problems may be, the advent of these all-knowing and energetic leaders to resolve the problem would seem a happy occurrence. Yet the contention that any particularly new programmatic strategy has occurred in the alcohol problems field in the past 100 years (apart from unrelated technological changes) is difficult to maintain. After 10 or more generations of observation it is evident that, despite these great leaders, great knowledge, and great programs, the problems are even more extensive and more painful than they ever were. But it still seems that large segments of the society will applaud and in more concrete ways adopt the next proposed program, even though it is almost always similar to some earlier attempt.

For example, formally educating 7- to 17-year-olds about the dangers and evils of consuming alcoholic beverages has been a major policy for at least 150 years. This is a major thrust of the current "other drugs" movement. Arresting individuals who, after alcohol consumption, exhibit markedly disapproved behavior in public has been a major policy for alcohol movements for far more than 150 years. It is a major thrust of the current "War on Drugs." Whether either of these strategies has had any effect (desired or not) remains an unknown, but the lack of reduction in alcohol-related problems over many decades is unquestioned, despite the presence of these strategies.

One possible policy adoption could immediately eliminate both education and arrest programs. While such a reaction appears ridiculous, the lack of any evidence to support such programs after 150 years of experience is also rather ridiculous. Thus, one cannot deny the possible attraction of such an impulsive and slashing innovation. This author would entirely oppose such an idea.

It might be suggested that the continuing use of strategies that show no evidence of reducing, let alone preventing, the "problems" must be serving some positively evaluated functions for large segments of the population. Such strategies and programs are enormously expensive, are maintained in almost every segment of the country, and have persisted for generations. The suggestion that such programs are without a function is hardly credible. That they form a major segment of the drug attack policy (along with treatment and sales control) is quite clear. The resulting suggestion for policy is hardly amazing, but would represent an almost radical innovation in these fields—namely, that these programs, both past and current, be studied.

Haven't these programs or movements been studied in the past? Granted, one can use a very loose interpretation of the word *study.* Frequently enemies of a given program have described its pathetic failure, its horrible by-products, and its awful expense. The proponents have described the same program's magnificent progress, its super heroes, and its extraordinary efficiency. Sometimes either proponents or opponents have hired "independent scientists" to study their programs. To no one's amazement the consequent reports (if ever seen) show just how right the patrons of the study have been.

The proposal being suggested here refers to questioning, observation, data gathering, analysis, testing, and communication that is not conducted or controlled by action-policy groups. In other words, the proposal is not only for study, but is for professionally disciplined study.

But a further suggestion, one beyond developing study of programs and movements, would be study of the "problem." Haven't there been studies of "the problem"? Yes, but in this instance the proposal will ignore those studies made by various problem-attacking (or problem-denying) groups except as such statements and reports may be viewed themselves as data to be studied. Most of these so-called studies are of such poor caliber for the stated topic that their chief use would be to serve as negative guides.

Let it be clearly understood that the suggestion focuses upon "the problems." What are they? How do you know? What of it? Physiology, psychology, economics, social structure, customs, history, biochemistry, ethics and law, medicine, pedagogy, and religion will all be relevant. But none of these topics, research disciplines, or professional action disciplines is the key, the most important, or the basic determinant for gaining better understanding about these "problems." A matter worth study in and of itself is why there is such success in gaining policy dominance by a chemical, a fiscal, a therapeutic, a legal, or a religiously oriented approach. These rather simplistic, often self-serving, and almost always emotionally charged credos offer little to increased understanding. Such foci practically guarantee the persistence of narrow orientations, repeated failure in applications, and, all too often, confrontations.

A basic suggestion is to add a new thrust of key importance to the new "war," namely development of knowledge and understanding. At the same time, this suggestion lays down criteria for the nature of that development. The criteria discussed relate to the location of power for this development of knowledge and understanding. One is to the effect that the "action-policy" groups should not be in control. A second criterion is that a single branch of knowledge, whether "basic" or "applied," should not be in control.

A third criterion for this process of generating knowledge relates to its focus upon drugs, drug use, or users as such, not on just the awful, horrendous, illegal, sickly, or the unusual. If astronomers were to limit study to exploding stars or if biologists did research only on what were called diseases and freaks, there would presumably be violent critical reaction. The alcohol and drug fields of study need to mature out of their rather primitive orientation to the deviant and exceptional patterns of substance use.

More general criteria of research quality could include, for example:

- Recognition of cultural and social category diversity.
- Temporal orientation, including both the usual historical perception and the longitudinal approach.
- Adequacy of representativeness.
- Definitions of key terms.
- Reliability of data, including specification of the presence and absence of disciplined observation.

The weaknesses of research on alcohol-related problems in this and other societies are formidable. Certainly there has been repeated recognition of the research weaknesses by professional students of the societal problem phenomena in the area of alcohol. There is little reason to assume that the situation in research on "other drug" problems is particularly different.

A final criterion for the suggested knowledge generation is that it should not compete in any serious financial fashion with other major thrusts of the new "war" effort. If the educational program is to cost $700 million annually, if the therapeutic efforts are to require $1 billion, and if the enforcement development is to cost $2 billion, then for those programs the new "war" would require about $4 billion a year. If this should be the total, it is suggested that no more than 2 percent or $80 million should be added to the overall budget. The suggestion is directed at effort outside the current and traditional operations.

Because some readers may wonder how this new study would vary from the old, a few examples are offered. The first relates to the difference between older studies and new proposals for studies of "efficiency of program." Almost all older studies of alcohol education revolved around such topics as numbers of curriculum hours, numbers of students, numbers of hours of such study per student, number of grades or classes given such education, and/or volume of teaching materials. Considered by some to be of surprising sophistication, there have even been studies of what proportion of the material taught was remembered by the students three days or even three months after the teaching event. Examples of more recent efforts could be comparative studies of multiple schools or school districts on each of the above matters, indicating, for example, which administrative scheme was effective in which manner. The sort of study suggested might be labeled as study of the educational process as contrasted to studies of goal attainment. Indeed, the question of the relationship of this process to the problem or to the stated purpose of the program is never even raised by this sort of study. Study of "process" is, of course, much simpler than study of function in relation to purpose. But such more serious study can be pursued.

In the field of law enforcement and whatever the problem(s) of public drunkenness may be, traditional studies are based upon police department records, not upon whatever it is that might be defined as public drunkenness. Police departments can have any of a wide variety of policies on both drunkenness arrests and the recording of such arrests. For example, the policy may be to make arrests only if the drunkard or someone nearby is "seriously endangered" (as in the case of possible death by freezing in winter). Such policies may vary in severity and may vary from year to year. Moreover, whatever the official policy may be, the different squads and different individual policemen may or may not act in exact compliance with a given rule. Statistical manipulations of the resulting figures collected from 5 or 50 departments can even be silly. This writer studied a department with more than 400 policemen that made annual reports of drunkenness arrests through six precincts. For two years each of the six precincts had annual numbers of such arrests that were expressed in even tens (over half in even fifties and hundreds); it was quite a coincidence. By contrast, the sort of study suggested here would not ask police departments to conduct studies for the researchers; strangely enough, the researchers would make their own studies.

In the area of treatment it is sufficient to state that studies of therapy are largely in the realms of either cynicism or advertising until (1) there is an agreed upon working definition of alcoholism, (2) there is an agreed definition of "recovery" or "successful" treatment, and (3) there is agreement upon both the representativeness of samples and also of time lapses required for such labeling. As in other areas of "alcohol studies," much of the reporting, analyses, and discussion deals with therapeutic process and often has but minor concern with what in societal terms would seem to be a major "problem."

Some of the other traditional studies have dealt with "risk" or with the "costs" of the problem. These are also quite fanciful in nature and almost always seem to show close to complete support for whatever program the sponsoring group is favoring.

The new thrust for knowledge generation is pointed toward providing the new warriors with some developing usable knowledge about the "problems" they are trying to attack, about the nature of possible alternative goals they might adopt, and about means for measuring the impact of programs upon problems. The old warriors did not have and do not have these knowledges.

What about the old thrusts? What positive understanding can be gained by these new programmers from the long experience (almost 200 years) of their predecessors? As to the three major thrusts described as education, enforcement, and therapy, the evidence is poor and frequently irrelevant. Although it is clear that alcohol problems were not reduced, let alone prevented, during 150 to 175 years of their activation, it is equally unclear that these programs had no effect or that other programs of education, enforcement, and therapy might not have more or less desirable effects.

Two other major thrusts of the past that are relevant to the present consisted of international or diplomatic efforts and of governmentally controlled legalized usages, especially actions dealing with production and sale. Currently the governmentally controlled and legalized production and sale policy in regard to certain "major" drugs is not being adopted, perhaps not even seriously considered, by the new warriors. Records of the impact of Alcoholic Beverage Control boards upon alcohol-related problems may have contemporary relevance.

Just how one would measure the "success" of an international effort on whatever may be defined as "American drug problems" is quite difficult to imagine. Of course, studies of the diplomatic process and reports on imports or drug production from a particular country could become available, but important as these matters may seem to a particular administrator, they need not bear any evidence about the persistence of drug problems in this country. Inasmuch as these sorts of programs are important for the attack on current problems, it is clear that their impact on "problems" could only be measured if the problems were defined in measurable terms, and if the goals were measurably described. It may be taken for granted that all such programs are meant to "do good" and reduce "evils," but these expressions are not measurable goals. The proposed increase of knowledge and understanding about problems and programs would be as useful for these two types of programs as for the others.

Further questions of great significance are centered on the attitudes, motivations, associated energies, and the accompanying organization, communication, and financial and other support that both led to these programs and also persisted despite failure. It is upon these same forces that the new warriors will have to rely. These are their major resources.

Clearly, the attitudes and motivations underlying initiation and support of different programs in the alcohol problems field can be very different. Such differences obtain not only between different groups, for example, the controlled sales groups, Alcoholics Anonymous, and the Woman's Christian Temperance Union, but also show variations within the same group, for example, the flaming disagreements within the Methodist temperance organization in the 1940s. Without development of such groups, without the organized energy, communication, leadership, and persistence of such groups, it is difficult to see how any movements would have occurred at all. Yet little is known and there is little concern to gain such knowledge about the nature, structure, points of weakness, sequences of development, and sources of power in such movements.

These resources of attitude and motivation are extensive, long-lived, and profound in American society and culture. It is not only the protection and instigation of diversity of opinion and the protection and instigation of the freedom of expression and association; it is also a matter of a widespread and deep-seated belief in the achievable possibility that "things can be better." This includes the notion that "bad things can be reduced, even minimized." In some societies such attitudes and motivations would require violent revolution for their expression. In this society it is clear that individual and small group behaviors and expressions can manifest both diverse goals and also marked dissatisfactions with almost every aspect of the status quo, whether in terms of property, religion, governmental actions, structure, or power. This occurs without destroying the society or any of its major institutions, without allowing dominance by any one institution and without losing the expression of individual diversity and individual motivation. Very short-term goals do indeed determine some behaviors. Further, some individuals and groups exhibit extreme greed, arrogance, and insensitivity to both other individuals and groups and also to the total society. But the past 200 years of this country's history show that individual and group diversity and individual group motivation to "do something" about recognized societywide problems and to "achieve better ways of life" are powerful forces—forces that persist despite great counterpressures to maintain the status quo. These forces persist despite frustrating failure of attempted changes.

For the new warriors this enduring set of motivations represents a tremendous source of strength. However, just as there is a basic need for greater understanding and knowledge about the structures and processes of whatever it is that is labeled as a problem, so there is a basic need to gain greater knowledge and understanding of these social attitudes and motivations that not only allow, but even command the emergence of their programs as well as their maintenance and survival. Knowledges of biochemistry and physiology, essential and effective as they may be, can hardly explain a societal problem, let alone devise a program to attack that problem. Similarly, organizational tactics, important as they may be, can hardly explain a societal movement, let alone its alleged success or failure.

Perhaps the most significant message the old warriors can provide for their successors is that trite statement that it will not be an easy or short war. But there could be a marked difference between explanations of why and how it would be neither brief nor simple. The difficulties of this new war are not a matter of not enough money; of enough activity by teachers, police, and physicians; of sufficient legislation or mass media production; of enough study of the liver, genes, price fluctuations, and police arrests. All these are quite simple, and possibly useful for some purposes, and without question can achieve popular support sufficient for even large-scale activation.

This old warrior is suggesting that it will not be short or simple for a much more basic set of reasons. First, not enough is known about the problem or problems. Second, there is clear inadequacy of statement of goals, especially measurably achievable goals. Third, not enough is known about the programs utilized or about the structures and processes that lie behind those programs.

There is no suggestion that acquisition of the requisite knowledge, consequent understanding, and emerging development of testable applications will be a popular undertaking, offering a solution in the form of a "quickie" or a "cheapie." However, it can be rather confidently predicted that it will not require a century to produce meaningful and measurable results. Rather, it can be confidently predicted that it will not cost more, presumably far less, than the approaches attempted to date. And it can confidently be predicted that, no matter what degree of "success" or "failure" emerges in the following 10 or 20 years, there can be verifiable, relevant, and communicable knowledge about what occurred—a dramatic difference from past performance.

What are labeled the "other drug" problems do manifest some differences from the "alcohol problems," differences that seem to favor the new warriors. The old warriors never were equipped with meaningful, measurable, or practicable goals. Methods of gaining relevant knowledge about problems and programs have shown significant advances over the past 50 to 60 years. For these reasons there is optimism in estimating the chances of the new warriors. Their attack will require some innovative thinking, some time, and some leadership. It already has great public support. So, more power to you, the new warriors, from the battlers of the past.

Notes

1. Closely related to this manifest weakness is the petulant and sometimes vicious warfare existing between various theoretical approaches. The biologic, psychologic, psychiatric, sociologic, spiritual, and economic "partisans" illustrate a lack of maturity that is at times almost appalling. How there could be such a thing as alcoholism free of important dimensions of a chemical, biologic, physiologic, psychologic, or social nature is impossible to see. But the hostility between such groups is painfully obvious.
2. The congressionally proposed use of surplus U.S. military planes (after sale to another country) to spray lethal chemicals (also from the United States) on agricultural crops in other countries is very close to an exception to this denial of the propriety of use of the word "war." See the "International Narcotics Control Act of 1989," 1989 HR Bill 3611, pp. 24 ff.

References

Bacon, S.D. (1984). Alcohol issues and social science. *Journal of Drug Issues, 14*(1), 7–29.

Bacon, S.D. (1987). Alcohol problem prevention: A "common sense approach. *Journal of Drug Issues, 17*(4), 369–393.

CHAPTER 17

Selden D. Bacon: A Bibliography of His Work

CATHERINE WEGLARZ

Bacon, S.D. (1939). *The early development of American municipal police: A study of the evolution of formal controls in a changing society.* Thesis. Yale University, New Haven, CT.

Bacon, S.D., & Roth, F.L. (1943). *Drunkenness in wartime Connecticut.* Hartford, CT: Connecticut War Council.

Bacon, S.D. (1943). Sociology and the problems of alcohol: Foundations for a sociologic study of drinking behavior. *Quarterly Journal of Studies on Alcohol, 4,* 402–445. ALSO AS: Bacon, S.D. (1944). *Sociology and the problems of alcohol: Foundations for a sociologic study of drinking behavior.* New Haven, CT: Quarterly Journal of Studies on Alcohol. ALSO AS: Bacon, S.D. (1946). *Sociology and the problems of drinking: Foundations for a sociological study of drinking behavior (Yale University, memoirs of the Section of Studies on Alcohol, No. 1).* New Haven, CT: Hillhouse Press.

Bacon, S.D. (1944). Inebriety, social integration, and marriage. *Quarterly Journal of Studies on Alcohol, 5,* 86–125, 303–339.

Bacon, S.D. (1944). The sociological approach to the problems of alcohol. *Federal Probation, 8,* 20–23.

Bacon, S.D. (1945). *Inebriety, social integration and marriage.* New Haven, CT: Quarterly Journal of Studies on Alcohol.

Bacon, S.D. (1945). Excessive drinking and the institution of the family. In *Alcohol, science and society: Twenty-nine lectures with discussions as given at the Yale Summer School of Alcohol Studies* (pp. 223–238). New Haven, CT: Quarterly Journal of Studies on Alcohol. ALSO AS: Bacon, S.D. (1954). Excessive drinking and the institution of the family. In *Alcohol, science and society: Twenty-nine lectures with discussions as given at the Yale Summer School of Alcohol Studies* (pp. 223–228). New Haven, CT: Quarterly Journal of Studies on Alcohol. ALSO AS: Bacon, S.D. (1972). Excessive drinking and the institution of the family. In *Alcohol, science and society: Twenty-nine lectures with discussions as given at the Yale Summer School of Alcohol Studies* (pp. 223–238). Westport, CT: Greenwood Press.

Bacon, S.D. (1945). Alcohol and complex society. In *Alcohol, science and society: Twenty-nine lectures with discussions as given at the Yale Summer School of Alcohol Studies* (pp. 179–200). New Haven, CT: Quarterly Journal of Studies on Alcohol. ALSO AS: Bacon, S.D. (1954). Alcohol and complex society. In *Alcohol, science and society: Twenty-nine lectures with discussions as given at the Yale Summer School of Alcohol Studies* (pp. 179–200). New Haven, CT: Quarterly Journal of Studies on Alcohol. ALSO AS: Bacon, S.D. (1972). Alcohol and complex society. In *Alcohol, science and society: Twenty-nine lectures with discussions as given at the Yale Summer School of Alcohol Studies* (pp. 179–200). Westport, CT: Greenwood Press.

Bacon, S.D. (1945). Alcoholism and social isolation. In M. Bell (Ed.), *Cooperation in crime control* (pp. 209–234). New York: National Probation Association.

Bacon, S.D. (1945). The alcoholic and the jail. *Prison World, 7,* 3, 26.

Bacon, S.D. (1945). A student of the problems of alcohol and alcoholism views the motion picture, *The Lost Weekend. Quarterly Journal of Studies on Alcohol, 6,* 402–405.

Bacon, S.D. (1945). New legislation for the control of alcoholism: The Connecticut law of 1945. *Quarterly Journal of Studies on Alcohol, 6,* 188–204.

Maltbie, W.M., Banay, R.S., & Bacon, S.D. (1945). Penal handling of inebriates. In *Alcohol, science and society: Twenty-nine lectures with discussions as given at the Yale Summer School of Alcohol Studies* (pp. 373–385). New Haven, CT: Quarterly Journal of Studies on Alcohol. Also as: Maltbie, W.M., Banay, R.S., & Bacon, S.D. (1954). Penal handling of inebriates. In *Alcohol, science and society: Twenty-nine lectures with discussions as given at the Yale Summer School of Alcohol Studies* (pp. 373–385). New Haven, CT: Quarterly Journal of Studies on Alcohol. Also as: Maltbie, W.M., Banay, R.S., & Bacon, S.D. (1972). Penal handling of inebriates. In *Alcohol, science and society: Twenty-nine lectures with discussions as given at the Yale Summer School of Alcohol Studies* (pp. 373–385). Westport, CT: Greenwood Press.

Bacon, S.D. (1946). A judge's viewpoint on narcotic addicts before the court with comment on its applicability to alcoholism. *Quarterly Journal of Studies on Alcohol, 6,* 567–572.

Bacon, S.D. (1946). Alcoholism: A major social problem. *Public Welfare, 4,* 146–150.

Bacon, S.D. (1947). Sociological concepts useful for a program of action on problems of alcoholism. *Quarterly Journal of Studies on Alcohol, 8,* 334–339 (Section from *Medical, legal and social approaches to the problem of inebriety: Proceedings of a conference sponsored jointly by the Research Council on Problems of Alcohol and the New York Academy of Medicine, January 8, 1947).*

Bacon, S.D. (1947). Inebriety and social reform. In *You and Alcohol* (pp. 17–20). New York: Columbia Broadcasting System.

Bacon, S.D. (1947). Alcoholism: Its extent, therapy and prevention. *Federal Probation, 11,* 24–32.

Bacon, S.D. (1947). Alcoholism: Nature of the problem. *Federal Probation, 11,* 3–7.

Bacon, S.D. (1947). Connecticut Commission on Alcoholism. *Quarterly Journal of Studies on Alcohol, 7,* 619–623.

Bacon, S.D. (1947). The mobilization of community resources for the attack on alcoholism. *Quarterly Journal of Studies on Alcohol, 8,* 473–487.

Bacon, S.D., & Miller, D.P. (1947). The Connecticut Commission on Alcoholism. *Connecticut State Medical Journal, 11,* 742–747.

Bacon, S.D. (1948). Alcoholism in industry. *Industrial Medicine, 17,* 161–167.

Bacon, S.D. (1948). The social impact of alcoholism. *Connecticut State Medical Journal, 12,* 1105–1110.

Bacon, S.D., Jackson, A.H., et al. (1948). The Connecticut Commission on Alcoholism, report, 1947–1948, to the Governor. *Quarterly Journal of Studies on Alcohol, 9,* 480–494.

Bacon, S.D., Jackson, A.H., et al. (1949). The Connecticut Commission on Alcoholism, report, 1948–1949, to the Governor. *Quarterly Journal of Studies on Alcohol, 10,* 535–547.

Bacon, S.D. (1949). The administration of alcoholism rehabilitation programs. *Quarterly Journal of Studies on Alcohol, 10,* 1–47. Also as: Bacon, S.D. (1949). The administration of alcoholism rehabilitation programs. In H. Emerson (Ed.), *Administrative medicine* (pp. 157–182). New York: Nelson.

Bacon, S.D., Jackson, A.H. et al. (1950). The Connecticut Commission on Alcoholism, report, 1949–1950, to the Governor. *Quarterly Journal of Studies on Alcohol, 11,* 677–694.

Straus, R., & Bacon, S.D. (1951). Alcoholism and social stability: A study of occupational integration in 2,023 male alcoholism clinic patients. *Quarterly Journal of Studies on Alcohol, 12,* 231–260. Also as: Straus, R., & Bacon, S.D. (1951). *Alcoholism and social stability: A study of occupational integration in 2,023 male alcoholism clinic patients.* New Haven, CT: Hillhouse Press.

Bacon, S.D. (1951). Studies of drinking in Jewish culture. I. General introduction. *Quarterly Journal of Studies on Alcohol, 12,* 444–450.

Bacon, S.D. (1951). Alcoholism and industry. *Civitan Magazine, 31,* 3–8.

Bacon, S.D. (1952). Alcoholism, 1941–1951: A survey of activities in research, education, therapy. I. Introduction. *Quarterly Journal of Studies on Alcohol, 13,* 421–424.

Bacon, S.D. (1952). Alcoholism, 1941–1951: A survey of activities in research, education, therapy. IV. Social science research. *Quarterly Journal of Studies on Alcohol, 13,* 453–460.

Bacon, S.D., Bingham, C.T., Sherman, N., Conway, G.C., & Tiebout, H.M. (1952). The Connecticut Commission on Alcoholism, report, 1950–1951, to the Governor. *Quarterly Journal of Studies on Alcohol, 13,* 153–159.

Bacon, S.D., Tiebout, H.M., Sherman, N., Leary, G., & Griswold, D. (1952). The annual report of the Connecticut Commission on Alcoholism for the year 1951–1952. *Quarterly Journal of Studies on Alcohol, 13,* 663–670.

Bacon, S.D. (1953). Facts about alcoholism (interview). *U.S. News and World Report, 35,* 36–40.

Henderson, R.M., & Bacon, S.D. (1953). Problem drinking: The Yale Plan for business and industry. *Quarterly Journal of Studies on Alcohol, 14,* 247–262. Also as: Henderson, R.M., & Bacon, S.D. (1955). Problem drinking: The Yale Plan for business and industry. *Industrial Nurses Journal, 2,* 16–23.

Straus, R., & Bacon, S.D. (1953). *Drinking in college.* New Haven, CT: Yale University Press. Also as: Straus, R., & Bacon, S.D. (1971). *Drinking in college.* Westport, CT: Greenwood Press.

Bacon, S.D. (1954). *Inebriety, social integration and marriage (Yale University, memoirs of the Section on Alcohol Studies, No. 2).* New Haven, CT: Hillhouse Press.

Bacon, S.D. (1954, March). *Research and study.* Paper presented at the annual meeting of the National Committee on Alcoholism, New York, NY.

Bacon, S.D. (1954). Nature of the problem. In *Proceedings of the 1st annual Alberta Conference on Alcohol Studies* (pp. 41–44). Edmonton, Alberta, Canada: Alberta Alcoholism Foundation.

Bacon, S.D. (1954). Psychological effects of alcohol. In *Proceedings of the 1st annual Alberta Conference on Alcohol Studies* (pp. 29–33). Edmonton, Alberta, Canada: Alberta Alcoholism Foundation.

Bacon, S.D. (1954). Nature of social problems. In *Proceedings of the 1st annual Alberta Conference on Alcohol Studies* (pp. 13–18). Edmonton, Alberta, Canada: Alberta Alcoholism Foundation.

Bacon, S.D. (1954). A sociologist looks at A.A. In *Proceedings of the 1st annual Alberta Conference on Alcohol Studies* (pp. 69–72). Edmonton, Alberta, Canada: Alberta Alcoholism Foundation.

Bacon, S.D. (1954). Community action on the problem of alcoholism. In *Proceedings of Kansas Conference on Alcoholism* (pp. 69–77). Topeka, KS: University of Kansas.

Bacon, S.D. (1954). Alcoholism—Nature of the problem. In *Proceedings of Kansas conference on Alcoholism* (pp. 6–22). Topeka, KS: University of Kansas.

Bacon, S.D. (1954). Alcoholism—Nature of the problem. *Public Health News* (Trenton, NJ), *35,* 70–81.

Bacon, S.D., Tiebout, H.M., Leary, G., & Oaks, C.K. (1954). Annual report of the Connecticut Commission on Alcoholism, July 1, 1952–June 30, 1953. *Quarterly Journal of Studies on Alcohol, 15,* 169–176.

Bacon, S.D. (1955). Current research on problems of alcoholism. Reports of the section chairman of the Research Institute of Alcoholism, Madison, Wisconsin, 9–10 October 1954. V. Report of the Section on Sociological Research. *Quarterly Journal of Studies on Alcohol, 16,* 551–564.

Bacon, S.D. (1955). The definition of an intoxicating beverage: Editorial introduction. *Quarterly Journal of Studies on Alcohol, 16,* 313–315.

Bacon, S.D. (1955). The development of studies on alcohol and alcoholism at Yale. *Connecticut Review on Alcoholism, 6,* 17–20.

Bacon, S.D., & McCarthy, R.G. (1955). *Alcohol facts for college students.* Montgomery, AL: State of Alabama Department of Education.

Bacon, S.D. (1957). A sociologist looks at A.A. *Minnesota Welfare, 10,* 35–44.

Bacon, S.D. (1957). Evolutions dans la conception americaine de l'alcoolisme durant les vingt dernieres annees [Evolution of the American conception of alcoholism during the last 20 years]. *International Journal of Alcohol and Alcoholism, 2,* 17–32.

Bacon, S.D. (1957). Social settings conducive to alcoholism. *Journal of the American Medical Association, 164,* 177–181. Also as: Bacon, S.D. (1957). Social settings conducive to alcoholism. In American Medical Association, Committee on Alcoholism, Council on Mental Health, *Manual of alcoholism* (pp. 60–74). Chicago, American Medical Association. Also as: Bacon, S.D. (1958). Social settings conducive to alcoholism. *New York State Medical Journal, 58,* 3493–3499.

Bacon, S.D., ed. (1958). Understanding alcoholism. *Annals of the American Academy of Political and Social Science, 315.* Also as: Bacon, S.D., ed. (1961). Understanding alcoholism. (*Annals of the American Academy of Political and Social Science, 315.*) Philadelphia: American Academy of Political and Social Science.

Bacon, S.D. (1958). Alcoholics do not drink. *Annals of the American Academy of Political and Social Science, 315,* 55–64.

Bacon, S.D. (1958). New light on alcoholism. *New York Times Magazine,* February 9.

Bacon, S.D. (1959). Prevention can be more than a word. In *Realizing the potential in state alcoholism programs. Proceedings of the Northeast States Conference on Alcoholism* (pp. 5–18). Hartford, CT: Connecticut Commission on Alcoholism.

Bacon, S.D. (1959). Community self-evaluation in the attack upon alcoholism. In *Summary reports from the conference on Community Resources for Rehabilitation of the Alcoholic* (pp. 77–93). University, MS: University of Mississippi.

Bacon, S.D. (1959). The nature of alcoholism. In *Summary reports from the conference on Community Resources for Rehabilitation of the Alcoholic* (pp. 3–20). University, MS: University of Mississippi.

Bacon, S.D. (1959). The interrelatedness of alcoholism and marital conflict. *American Journal of Orthopsychiatry, 29,* 513–518.

Bacon, S.D., & Woodbury, C. (1961). What should you teach your child about drinking? *McCall's, 89,* 228.

Bacon, S.D. (1962). Alcohol and complex society. In D.J. Pittman & C.R. Snyder (Eds.), *Society, culture and drinking* (pp. 78–93). New York: Wiley.

Bacon, S.D. (1962). Alcoholism. In E. Josephson (Ed.), *Man alone: Alienation in modern society* (pp. 393–401). New York: Dell Publishing.

Bacon, S.D. (1962). Alcohol, alcoholism, and crime: An overview. In D.W. Haughey & N.A. Neiberg (Eds.), *Conference on alcohol, alcoholism and crime* (pp. 5–27). Boston, MA.

Bacon, S.D. (1962). The Rutgers Center of Alcohol Studies: A tentative conceptualization of purpose (notes and comment). *Quarterly Journal of Studies on Alcohol, 23,* 321–324.

Straus, R., & Bacon, S.D. (1962). The problems of drinking in college. In D.J. Pittman & C.R. Snyder (Eds.), *Society, culture and drinking patterns* (pp. 246–258). New York: Wiley.

Bacon, S.D. (1963). Introduction. In C. Jackson, *The lost weekend* (pp. xv-xix). New York: Time, Inc.

Bacon, S.D. (1963). State problems on alcoholism—A critical review. In *Selected papers delivered at the fourteenth annual meeting of the North American Association of Alcoholism Programs* (pp. 1–18). Washington, D.C.: North American Association of Alcoholism Programs.

Bacon, S.D. (1963). E.M. Jellinek, 1890–1963. *Quarterly Journal of Studies on Alcohol, 24,* 587–590.

Bacon, S.D. (1963). Alcohol, alcoholism and crime. *Crime and Delinquency, January,* 1–14.

Bacon, S.D. (1963). Changes in alcoholism rehabilitation, 1928–1963. *Academy of Medicine of New Jersey Bulletin, 9,* 166–178.

Bacon, S.D., & Greenberg, L.A. (1963). Drinking and driving: A new look at an old problem. *Brewers Digest, 39,* 28–30. ALSO AS: Bacon, S.D., & Greenberg, L.A. (1965). Drinking and driving: A new look at an old problem. In *Royal commission into the sale, supply, disposal or consumption of liquor in the state of Victoria. Appendix to Report part I* (pp. 80–83). Victoria, Australia.

Bacon, S.D., & Jones, R.W. (1963). *The relationship between the Alcoholic Beverage Control Law and the problems of alcohol.* New York: Moreland Commission on the ABC Law.

Bacon, S.D. (1964). Raymond G. McCarthy, 1901–1964. *Quarterly Journal of Studies on Alcohol, 25,* 413–416.

Warkov, S., Bacon, S.D., & Hawkins, S.C. (1965). Social correlates of industrial problem drinking. *Quarterly Journal of Studies on Alcohol, 26,* 58–71.

Bacon, S.D. (1966). Education on alcohol: A background statement. In *Alcohol education: Proceedings of a conference* (pp. 7–15). Washington, D.C.: U.S. Department of Health, Education and Welfare.

Bacon, S.D. (1966). American experiences in legislation and control dealing with the use of beverage alcohol. In *National conference on Legal Issues in Alcoholism and Alcohol Usage* (pp. 123–141). Boston: Boston University, Law-Medicine Institute.

Bacon, S.D. (1966). The social impact of alcoholism. *Transactions of the 54th National Safety Congress, Chicago, 24,* 73–78.
Bacon, S.D. (1966). State program on alcoholism. *Inventory, 15,* 25–27.
Bacon, S.D. (1966). Sociology of alcoholism. *Scientific American, 197,* 69–70.
Bacon, S.D. (1966). Our thinking about alcohol: The sociohistorical background to our present efforts to educate. *Journal of Alcohol Education, 12,* 2–15.
Bacon, S.D. (1966). Le mouvement antialcoolique classique aux Etats-Unis et son influence actuelle. *Alcool ou Sante, no. 79–80,* 32–49.
Bacon, S.D. (1967). Drinking driver control: Sociological aspects. In *Proceedings of the 1967 Eastern Region Military-Civilian Traffic Safety conference* (pp. 56–62). Harrisburg, PA.
Bacon, S.D. (1967). The classic Temperance Movement of the U.S.A.: Impact today on attitudes, action and research. *British Journal of Addiction, 62,* 5–18.
Bacon, S.D. (1967). Education on alcohol: A background statement. In U.S. Department of Health, Education, and Welfare, Secretary's Committee on Alcoholism, *Alcohol Education. Proceedings of a conference* (pp. 7–15). Washington, D.C.: U.S. Government Printing Office.
Bacon, S.D. (Ed.). (1968). Studies of drinking and driving. *Quarterly Journal of Studies on Alcohol, Suppl. No. 4.*
Bacon, S.D. (1968). Studies of drinking and driving: Introduction. *Quarterly Journal of Studies on Alcohol, Suppl. No. 4,* 1–10.
Bacon, S.D. (1968). Traffic accidents involving alcohol in the U.S.A.: Second-stage aspects of a social problem. *Quarterly Journal of Studies on Alcohol, Suppl. No. 4,* 11–22.
Bacon, S.D. (1968). The "temperance" tradition's historical impact on alcoholism programs. In U.S. Civil Service Commission. Bureau of Retirement and Insurance. *The first step: A report of a conference on drinking problems* (pp. 13–16). Washington, D.C.: Government Printing Office.
Bacon, S.D. (1968). Alcohol and problems of highway safety: The role of social science research. In *Driver behavior, cause and effect. Proceedings of the second annual Traffic Safety Research Symposium of the Automobile Insurance Industry* (pp. 145–164). Washington, D.C.: Insurance Institute for Highway Safety.
Bacon, S.D. (1968). The drinking driver. *Analogy, Spring,* 18–19.
Bacon, S.D. (1968). Alcoholism and the criminal justice system. *Law and Society Review, 2,* 489–495.
Zylman, R., & Bacon, S.D. (1968). Police records and accidents involving alcohol. *Quarterly Journal of Studies on Alcohol, Suppl. No. 4,* 178–211.
Bacon, S.D. (1969). Thoughts on the establishment of a school for criminal justice at Rutgers University. In *Report of the Committee on the School of Criminal Justice, section I* (pp. 1–39).
Bacon, S.D. (1969). Meeting the problems of alcohol in the U.S.A. In *International Congress on Alcohol and Alcoholism, proceedings, 28th, v 2* (pp. 29–36). Highland Park, NJ: Hillhouse Press. ALSO AS: Bacon, S.D. (1970). Meeting the problems of alcoholism in the United States. In E.D. Whitney (Ed.), *World Dialogue on alcohol and drug dependence* (pp. 134–145). Boston: Beacon Press.
Bacon, S.D. (1969). The 28th International Congress on Alcohol and Alcoholism: purposes and functions. In *International Congress on Alcohol and Alcoholism, proceedings, v 2* (pp. xii-xvii). Highland Park, NJ: Hillhouse Press.
Bacon, S.D. (1969). Relevance of the social problems of alcohol for coping with problems of drugs. In J.R. Wittenborn, H. Brill, J.P. Smith, & S.A. Wittenborn

(Eds.), *Drugs and youth: Proceedings of the Rutgers Symposium on Drug Abuse* (pp. 44–51). Springfield, IL: Thomas.

Bacon, S.D. (1969). Drug abuse and alcohol abuse—The social problem perspective. *Prosecutor, 5,* 32–36.

Bacon, S.D. (1969). Introduction. In D. Cahalan, I.H. Cisin, & H. Crossley, *American drinking practice* (pp. xv-xxvi). New Brunswick, NJ: Rutgers Center of Alcohol Studies.

Bacon, S.D. (1970). Ways of thinking about alcohol use and about alcohol problems. In *Proceedings of a conference on Public Health Teaching about Alcoholism* (pp. 12–23). New York: American Public Health Assocation.

Bacon, S.D. (1970). A summary of proceedings in matched-professions discussion groups. In *Proceedings of a conference on Community Response to Alcoholism and Highway Crashes* (p. 119). Ann Arbor, MI: University of Michigan.

Bacon, S.D. (1970). College drinking: So what and what next? In G.L. Maddox (Ed.), *The domesticated drug: Drinking among collegians* (pp. 457–474). New Haven, CT: College & University Press.

Bacon, S.D. (1970). Alcohol, alcoholism and crime. In B. Cohen. *Crime in America* (pp. 202–210). Itasca, IL: Peacock.

Bacon, S.D. (1970). Criticisms and two critical needs. *Behavioral Research in Highway Safety, 1,* 55–56.

Bacon, S.D. (1970). Fragmentation of alcohol problem research. In *International Congress on Alcoholism and Drug Dependence, 29th interdisciplinary seminars on medical and psychiatric aspects of alcoholism and drug dependence: Proceedings of Melbourne sessions* (pp. 19–28). Melbourne, Australia: University of Melbourne. ALSO AS: Bacon, S.D. (1971). Fragmentation of alcohol problem research. In *International Congress on Alcoholism and Drug Dependence, proceedings, 29th* (pp. 481–495). Melbourne, Australia: University of Melbourne.

Bacon, S.D. (1971). The McCarthy Memorial Collection in 1971. *Quarterly Journal of Studies on Alcohol, 32,* 472–477.

Bacon, S.D. (1971). A private consultation on addiction. In *Project Health, alcohol: Drug of choice* (pp. 30–35). Chicago: Searle Education Systems.

Bacon, S.D. (1971). The role of law in meeting problems of alcohol and drug use and abuse. In *International Congress on Alcoholism and Drug Dependence, proceedings, 29th* (pp. 162–172). Melbourne, Australia: University of Melbourne.

Bacon, S.D. (1971). Drug abuse and alcohol abuse—The social problem perspective. In *Drug dependence and abuse reference book of the National District Attorneys Association* (pp. 212–216). Chicago: National District Attorneys Association.

Bacon, S.D. (1973). Drug abuse and alcohol abuse—The social problems perspective. In B.Q. Hafen (Ed.), *Drug abuse: Psychology, sociology, pharmacology* (pp. 79–83). Provo, UT: Brigham Young University Press.

Bacon, S.D. (1973). One society's reaction to problems attached to a particular drug—Alcohol. In R.A. Bowen (Ed.), *Anglo-American Conference on Drug Abuse: Society's reaction—Medicine's responsibility. Proceedings of a conference sponsored by the Royal Society of Medicine and the Royal Society of Medicine Foundation* (pp. 36–43). London, Royal Society of Medicine.

Bacon, S.D. (1973). Highway crashes, alcohol problems, and programs for social controls. In P.G. Bourne & R. Fox (Eds.), *Alcoholism: Progress in research and treatment* (pp. 311–335). New York: Academic.

Bacon, S.D. (1973). The process of addiction to alcohol: Social aspects. *Quarterly Journal of Studies on Alcohol, 34,* 1–27. ALSO AS: Bacon, S.D. (1973). Social aspects of the process of addiction to alcohol (Abstract). In *Report on a conference on the Epidemiology of Drug Dependence* (pp. 84–88). Copenhagen: World Health Organization.

Bacon, S.D. (1973). The problems of alcoholism in American society. In D. Malikin, *Social disability: Alcoholism, drug addiction, crime and social disadvantage with contributions by S.D. Bacon, J.D. Case, K. Jackson, E.W. Gordon and the Committee on Relation of Addictions* (pp. 8–30). New York: New York University Press.

Bacon, S.D., & Fillmore, K. (1973). A follow-up study of drinking in college. In R. Straus (Ed.), *Alcohol and society (Psychiatric Annals, 3)* (pp. 15–23). New York: Insight Communications.

Bacon, S.D. (1974). Role of drinking in U.S.A. today. *Stateways, 3,* 19, 22.

Bacon, S.D. (1975) The study of alcohol: Rutgers Center seeks answers to many questions in its search for knowledge. *FAH Review, 8,* 22–23.

Bacon, S.D. (1975). One society's reaction to problems attached to a particular drug—Alcohol. *The Alcoholism Digest, 47,* Abstract No. 50456.

Bacon, S.D. (1976). Concepts. In W.J. Filstead, J.J. Rossie, & M. Keller (Eds.), *Alcohol and alcohol problems: New thinking and new directions* (pp. 57–134). Cambridge, MA: Balinger/Lippincott.

Bacon, S.D. (1976). Defining adolescent alcohol use: Implications for a definition of adolescent alcoholism (reports of meetings). *Journal of Studies on Alcohol, 37,* 1014–1019. ALSO AS: Bacon, S.D. (1980). Defining adolescent alcohol use: Implications for a definition of adolescent alcoholism. In J.E. Mayer & W.J. Filstead (Eds.), *Adolescence and alcohol* (pp. 45–50). Cambridge, MA: Balinger.

Bacon, S.D. (1976). Research and the Center of Alcohol Studies. In *Medical-scientific conference on Work in Progress in Alcoholism, 6th (Annals of the New York Academy of Sciences, 273,* 81–86).

Marden, P., Zylman, R., Fillmore, K.M., & Bacon, S.D. (1976). Comment on "A national study of adolescent drinking behavior, attitudes and correlates." *Journal of Studies on Alcohol, 37,* 1346–1358.

Bacon, S.D. (1977). About Mark Keller on his retirement. *Journal of Studies on Alcohol, 38,* v-x.

Bacon, S.D. (1977). Words about alcohol: Their meaning in study, public relations, and policy-making. In P.A. O'Gorman, S. Stringfield, & L. Smith (Eds.), *Defining adolescent alcohol use: Implications toward a definition of adolescent alcoholism. Proceedings of a congress* (pp. 208–220). New York: Council on Alcoholism.

Bacon, S.D. (1978). Commentary (on papers by Andrew J. Gordon, Joy Leland, and Richard Stivers). *Medical Anthropology, 2,* 137–146.

Bacon, S.D. (1978). On the prevention of alcohol problems and alcoholism. *Journal of Studies on Alcohol, 39,* 1125–1147.

Bacon, S.D. (1979). Alcohol research policy: The need for an independent, phenomenologically oriented field of study. *Journal of Studies on Alcohol, Suppl. No. 8,* 2–26.

Bacon, S.D., et al. (1979). Discussion—Session 4. *Journal of Studies on Alcohol, Suppl. No. 8,* 289–332.

Bacon, S.D., et al. (1979). Discussion—Session 1. *Journal of Studies on Alcohol, Suppl. No. 8,* 46–74.

Bacon, S.D., et al. (1979). Discussion. *Journal of Studies on Alcohol, Suppl. No. 8,* 333–334.

Fillmore, K.M., Bacon, S.D., & Hyman, M. (1979). *The 27-year longitudinal study of drinking by students in college, 1949–1976: Final report.* Springfield, VA: U.S. National Technical Information Service.

Bacon, S.D. (1984). Alcohol issues and social science. *Journal of Drug Issues, 14,* 7–29.

Bacon, S.D. (1985). Journal interview, 9: Conversation with Selden D. Bacon. *British Journal of Addiction, 80,* 115–120.

Bacon, S.D. (1987). Alcohol problem prevention: A "common sense" approach. *Journal of Drug Issues, 17,* 369–393.

Bacon, S.D. (1991). An old warrior looks at the new. In P.M. Roman (Ed.), *Alcohol: The development of sociological perspectives on use and abuse* (pp. 359–379). New Brunswick, NJ: Rutgers Center of Alcohol Studies.

Contributors

Florence Kellner Andrews, Ph.D., Department of Sociology and Anthropology, Carleton University, Ottawa, Canada.

Selden D. Bacon, Ph.D., Director Emeritus, Center of Alcohol Studies, Rutgers University, New Brunswick, New Jersey.

Kaye Middleton Fillmore, Ph.D., Institute for Health and Aging, University of California, San Francisco, California.

Dwight B. Heath, Ph.D., Department of Anthropology, Brown University, Providence, Rhode Island.

Harry G. Levine, Ph.D., Department of Sociology, Queens College of the City University of New York, Flushing, New York.

Armand L. Mauss, Ph.D., Department of Sociology, Washington State University, Pullman, Washington.

James D. Orcutt, Ph.D., Department of Sociology, Florida State University, Tallahassee, Florida.

David J. Pittman, Ph.D., Department of Psychology, Washington University, St. Louis, Missouri.

Paul M. Roman, Ph.D., Institute for Behavioral Research and Department of Sociology, University of Georgia, Athens, Georgia.

Robin Room, Ph.D., Alcohol Research Group, Berkeley, California.

Earl Rubington, Ph.D., Department of Sociology and Anthropology, Northeastern University, Boston, Massachusetts.

William J. Sonnenstuhl, Ph.D., New York State School of Industrial and Labor Relations, Cornell University, Ithaca, New York.

Robert Straus, Ph.D., Department of Behavioral Science, College of Medicine, University of Kentucky, Lexington, Kentucky.

Melvin H. Tremper, Ph.D., Office of Substance Abuse, State of Maine, Augusta, Maine.

Harrison M. Trice, Ph.D., New York State School of Industrial and Labor Relations, Cornell University, Ithaca, New York.

Catherine Weglarz, M.L.S., Librarian, Center of Alcohol Studies, Rutgers University, New Brunswick, New Jersey.